Algebra and Discrete Mathematics   Vol. 5

GEOMETRY OF CRYSTALLOGRAPHIC GROUPS

Second Edition

Algebra and Discrete Mathematics

ISSN: 1793-5873

The series ADM focuses on recent developments in all branches of algebra and topics closely connected. In particular, it emphasizes combinatorics, set theoretical methods, model theory and interplay between various fields, and their influence on algebra and more general discrete structures. The publications of this series are of special interest to researchers, post-doctorals and graduate students. It is the intention of the editors to support fascinating new activities of research and to spread the new developments to the entire mathematical community.

Algebra and Discrete Mathematics Vol. 5

GEOMETRY OF CRYSTALLOGRAPHIC GROUPS

Second Edition

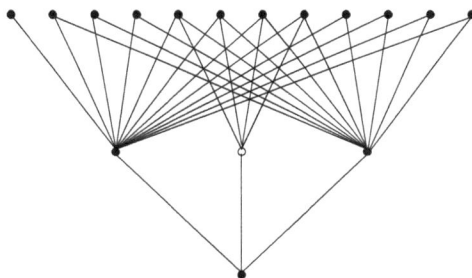

Andrzej Szczepański
University of Gdańsk, Poland

World Scientific

NEW JERSEY · LONDON · SINGAPORE · BEIJING · SHANGHAI · HONG KONG · TAIPEI · CHENNAI · TOKYO

Published by

World Scientific Publishing Co. Pte. Ltd.
5 Toh Tuck Link, Singapore 596224
USA office: 27 Warren Street, Suite 401-402, Hackensack, NJ 07601
UK office: 57 Shelton Street, Covent Garden, London WC2H 9HE

Library of Congress Control Number: 2024032816

British Library Cataloguing-in-Publication Data
A catalogue record for this book is available from the British Library.

Algebra and Discrete Mathematics— Vol. 5
GEOMETRY OF CRYSTALLOGRAPHIC GROUPS
Second Edition

ISBN 978-981-12-8659-9 (hardcover)
ISBN 978-981-12-8660-5 (ebook for institutions)
ISBN 978-981-12-8661-2 (ebook for individuals)

For any available supplementary material, please visit
https://www.worldscientific.com/worldscibooks/10.1142/13685#t=suppl

For Margaret

Preface

Crystallographic groups are groups which act in a nice way and via isometries on some n-dimensional Euclidean space. They got their name, because in three dimensions they occur as the symmetry groups of a crystal (which we imagine to extend to infinity in all directions).

To be precise, an n-dimensional crystallographic group Γ is a discrete subgroup of the group of isometries $\mathrm{Isom}(\mathbb{R}^n) = E(n) = O(n) \ltimes \mathbb{R}^n$ of Euclidean n-space, having a compact fundamental domain. This means that Γ acts properly discontinuous on $\mathbb{R}^n$ and there is a compact subset $D \subseteq \mathbb{R}^n$, such that $\mathbb{R}^n = \Gamma D$.

The algebraic structure of a crystallographic group is well understood, thanks to the three Bieberbach theorems. By the first theorem, we know that the subgroup of pure translations inside a crystallographic group Γ is normal and of finite index in Γ. Moreover, this translation subgroup contains n linearly independent translations leading to a short exact sequence of groups

$$1 \to \mathbb{Z}^n \to \Gamma \to G \to 1,$$

where $\mathbb{Z}^n$ corresponds to the translation subgroup of Γ and $G \cong \Gamma/\mathbb{Z}^n$ is finite. Such a short exact sequence gives rise to a representation $h_\Gamma \colon G \to GL(n, \mathbb{Z})$, called the holonomy representation and the group Γ can then be described using the second cohomology of G with coefficients in $\mathbb{Z}^n$. As a consequence, we will in this text describe properties of crystallographic groups using the language of representation theory and cohomology of finite groups.

The torsion free crystallographic groups Γ, called Bieberbach groups, have a special geometric importance, because in this case, the quotient space $M^n = \mathbb{R}^n/\Gamma$ is a manifold with fundamental group Γ. In fact, as Γ acts via isometries on $\mathbb{R}^n$, the manifold M^n inherits the Riemannian structures from Euclidean space, making it into a manifold with sectional curvature zero, i.e. a flat manifold. It turns out that the group G is the holonomy group of this manifold and hence we will call G the holonomy group of Γ. Moreover, any compact flat manifold is obtained in this way, so there is a one to one correspondence between the geometric world of flat manifolds and the algebraic world of Bieberbach groups. In fact, many parts of this book can be considered as a dictionary between the geometric properties of M^n and the algebraic properties of Γ.

In the first two chapters we introduce the basic definitions and prove the Bieberbach Theorems (Theorem 2.4). This part covers well known facts, definitions and compiles information mainly from [23] and [148, Part 3]. The third Bieberbach Theorem states that in any dimension there are only a finite number of isomorphism classes of crystallographic groups. It therefore makes sense to try to classify them and this is the theme of the third chapter. We consider three general classification methods. The first one, the most algebraic, is based on Zassenhaus' algorithm from 1948, [151], the second one is called *Calabi's method*, because it was proposed by E. Calabi in 1957 [22] and the last one was defined in 1970 by Auslander and Vasquez [145]. We present a full classification of all 2-dimensional crystallographic groups (subsection 3.2) and we give a list, using the elementary classification algorithm, of ten 3-dimensional flat manifolds i.e. 3-dimensional Bieberbach groups (subsection 3.3). We include it for two reasons. The 2-dimensional case is classical, moreover the list of all ten torsion free 3-dimensional crystallographic groups has been known since around 1933 [56], [107], but there are some recent astronomical observations which suggest that the physical universe might actually be a 3-dimensional flat manifold (platycosm) [29]. The next two chapters (4 and 5) are related to the *Calabi method* of classification. We present there one of the most interesting results (Theorem 4.4) of recent years (1986), namely the characterization of the so called *primitive* groups. A finite group is called primitive, when it can be realized as the holonomy group of a Bieberbach group with a finite abelianization. These groups, and the corresponding flat manifolds, play a crucial role in Calabi's classification method.

Another interesting result, which is useful in the context of Calabi's method is a theorem (Theorem 5.9) from 1996, which gives a necessary and sufficient condition for infiniteness of the outer automorphism group of a crystallographic group. It interconnects classical representation theory with dynamical properties of flat manifolds. In chapters 6 to 9, we present the most interesting results (in our opinion) from recent years. Again the idea

in all of these chapters is to exploit the possibility of translating some geometric condition on a flat manifold (existence of spin or Kähler structure)
into a pure algebraic, group theoretic language. In chapter 6 we consider the
existence question of spin structures on flat manifolds. A full answer is given
for flat manifolds with a cyclic holonomy group and in the 4-dimensional
case. Here, we want to emphasize that in contrast to the results of Chapters 4, 5 and 7, the existence of a spin structure does not only depend on
the holonomy representation or the holonomy group itself but also on other
properties of the Bieberbach group, see Example 6.24. As an application,
we mention some methods for calculating the Dirac operator in this case.
Here we assume some knowledge of differential geometry and especially the
Levi-Civita connection.

The next chapter (7) gives necessary and sufficient conditions for the
existence of a Kähler structure on an even-dimensional flat manifold. We
also present a complete list of such manifolds in complex dimension.

In chapter 8 relations between the flat (Euclidean) world and the hyperbolic one are considered. In particular, for any Bieberbach group Γ of
dimension n there is a discrete subgroup $\Gamma' \subseteq \mathrm{Isom}(\mathbb{H}^{n+1})$, such that Γ is a
subgroup of Γ', and the space $\mathbb{H}^{n+1}/\Gamma'$ has a finite volume. Here some final
steps of a proof, related to separability of subgroups, are omitted and we
refer the reader to the original papers.

Since a complete classification of crystallographic groups in all dimensions is impossible, it makes sense to consider some families. The generalized Hantzsche-Wendt groups are an example of such a family. The
names Hantzsche and Wendt, denote German mathematicians who are the
authors of a paper from the years $1934/35$ [56] where the classification of
3-dimensional flat manifolds was completed. The family of manifolds presented here generalizes the unique oriented 3-flat manifold with a non-cyclic
holonomy group. In chapter 9 we introduce this abundant class and try to
convince the reader of its unusual properties.

We make no claim to completeness. For example there are omitted topics about generalization of crystallographic groups to nilpotent or solvable
world. We do not touch relations with the theory of fix points and classical
crystallography. One of our main ideas was to present a text as short as
possible. We give a list of conjectures and problems in the last chapter. In
Appendix I we present an alternative proof of the first Bieberbach Theorem.
This proof is due M. Gromov and is completely different from the proof in
Chapter 2.

In Appendix II the Burnside transfer theorem (Theorem 11.31) is presented with a proof, it is used mainly in chapter 4 in the proof of Theorem 4.4.

The last Appendix III gives an example of the Bieberbach group with a trivial centre and a trivial outer automorphism group.

The text is an extended version of the notes of the lectures which were given by the author at the Gdańsk University for graduate students in the academic-year 2004/5. We assume that the reader knows elementary facts from algebra, cohomology of groups and topology. After each chapter some exercises are given. (They are not always easy!) Except for the first two basic chapters and some part of Chapter 5, this text differs from the book of L. Charlap [23]. We want to emphasize that most theorems and facts presented here are from the last two decades. This is after the book of L. Charlap was published. We should also add that an essential source for us was the chapter about "Euclidean forms" in the J. Wolf's book, [148].

Various portions of the text were read by Jonathan Hillman and he provided many insightful comments and suggestions. We also thank P. Buser, K. Dekimpe, G. Hiss, F. E. A. Johnson, R. Lutowski, J. P. Rossetti and V. Schroeder for many interesting discussions about crystallographic groups. Moreover we would like to thank A. Gąsior for preparing diagrams. Some part of the work was done at the IHES and the University of Zürich. The author would like to thank the staff and personnel of these institutions for their help and friendly atmosphere.

Finally, the author wishes to express his gratitude to Bogumiła Klemp-Dyczek for her numerous helpful comments concerning the use of the English language.

Preface to the second edition

It is eleven years since the first edition of *Geometry of crystallographic groups* appeared. The main changes in this edition are a new result about automorphism of crystallographic groups (chapter 5.4) and about Hantzsche-Wendt groups/manifolds (chapter 9.5.1-9.5.3). Moreover some slight obscurities and several misprints have been corrected. The author would like to express thanks to R. Lutowski who have contributed his comments to the chapter 9.

Contents

1. Definitions

In this introductory chapter we define a language of the isometry group of the Euclidean space. We also introduce an elementary world of covering spaces, group actions and discrete subgroups.

Let $\mathbb{R}^n$ be a n-dimensional Euclidean space with the standard scalar product

$$\langle x, y \rangle = \sum_{i=1}^{n} x_i y_i,$$

where $x = (x_1, x_2, \ldots, x_n) \in \mathbb{R}^n$ and $y = (y_1, y_2, \ldots, y_n) \in \mathbb{R}^n$. This gives us a norm of vectors

$$\|x\| = \sqrt{\langle x, x \rangle} = \sqrt{\sum_{i=1}^{n} x_i^2}$$

and metric

$$d(x, y) = \|x - y\| = \sqrt{\sum_{i=1}^{n} (x_i - y_i)^2}.$$

Definition 1.1. A map $f \colon \mathbb{R}^n \to \mathbb{R}^n$ is an isometry if for any $x, y \in \mathbb{R}^n$,

$$d(x, y) = d(f(x), f(y)).$$

The next proposition follows directly from the definition.

Proposition 1.2. *The set of all isometries of $\mathbb{R}^n$ is a group with respect to composition of maps.*

$\square$

By $E(n)$ we shall denote the group of all isometries of the Euclidean space $\mathbb{R}^n$. We distinguish two types of isometries: translations and linear orthogonal maps.

Definition 1.3. Let $a \in \mathbb{R}^n$. A map $t_a \colon \mathbb{R}^n \to \mathbb{R}^n$ defined by the formula

$$t_a(x) = a + x$$

is called a translation of the space $\mathbb{R}^n$.

Proposition 1.4. *The set of all translations of $\mathbb{R}^n$ is a normal subgroup of $E(n)$. The function $a \to t_a$ defines an isomorphism of the additive group $\mathbb{R}^n$ and the group of all translations of $\mathbb{R}^n$.*

$\square$

Therefore, we shall also denote by $\mathbb{R}^n$ the group of all translations of the Euclidean space $\mathbb{R}^n$.

Definition 1.5. A linear map $A \colon \mathbb{R}^n \to \mathbb{R}^n$ is orthogonal if for any $x, y \in \mathbb{R}^n$,

$$\langle x, y \rangle = \langle A(x), A(y) \rangle.$$

Proposition 1.6. *A linear map $A \colon \mathbb{R}^n \to \mathbb{R}^n$ is orthogonal if and only if for any $x \in \mathbb{R}^n$,*

$$\|x\| = \|A(x)\|.$$

Proof. The implication '$\Rightarrow$' follows from the definition of the orthogonal map. For the proof of the other implication let us assume that A is a linear map and $\|x\| = \|A(x)\|$, for any $x \in \mathbb{R}^n$. Then A preserves a quadratic form $Q(x) = \|x\|^2$. It means, that A is an invariant of the bilinear form

$$\langle x, y \rangle = \frac{1}{2}(Q(x+y) - Q(x) - Q(y)) = \langle A(x), A(y) \rangle.$$

$\square$

Proposition 1.7. *The set of all orthogonal linear maps $\mathbb{R}^n \to \mathbb{R}^n$ is a group.*
$\square$

The above group is denoted by $O(n)$ and called the orthogonal group of Euclidean space of dimension n. Elements of the orthogonal group are often considered as matrices or linear maps of the spaces $\mathbb{R}^n$ and $\mathbb{C}^n$.

Let $e_1, e_2, \ldots, e_n$ be an orthonormal basis of the euclidean space $\mathbb{R}^n$. By definition any matrix of the orthogonal map in the above basis is an orthogonal matrix i.e. $A^{-1} = A^T$, where A^T is the transpose of A. This gives an isomorphism of the group of all orthogonal linear maps onto the set of all orthogonal matrices. Hence, we shall identify the group of linear orthogonal maps $O(n)$ with the group of all orthogonal matrices.

Proposition 1.8. *Any isometry of the space $\mathbb{R}^n$ is a composition of an orthogonal linear map and a translation. In other words if $f \colon \mathbb{R}^n \to \mathbb{R}^n$ is an isometry of $\mathbb{R}^n$, then there exists t_a and an orthogonal map $A \colon \mathbb{R}^n \to \mathbb{R}^n$ such that $f = t_a \circ A$.*

Proof. Let $a = f(0)$ and $A = t_{-a} \circ f$. Hence A is an isometry and $A(0) = 0$. We claim that A is a linear and orthogonal. In fact, A also preserves inner products (as in Proposition 1.6). Hence it maps the standard orthonormal basis of $\mathbb{R}^n$ to an orthonormal basis. Since there is a linear orthogonal map g which has the same effect on the standard basis, the isometry $F = g^{-1}A$ leaves invariant the standard basis. But it is then easy to see that F must be the identity (and so $A = g$ is linear orthogonal), since the components (coordinates) of any vector with respect to the standard basis are given by inner products. This proves the claim and $f = t_a \circ A$. $\square$

Let us introduce an useful construction.

Definition 1.9. Let H and K denote groups with group multiplication '$\circ$' and '$\star$' respectively. Moreover, assume that H is a subgroup of the automorphism group of the abelian group K. The semi-direct product $H \ltimes K$ of the groups H and K is the set of all pairs (h, k), $h \in H$, $k \in K$, with the following multiplication

$$(h_1, k_1)(h_2, k_2) = (h_1 \circ h_2, k_1 \star h_1(k_2)).$$

Example 1.10.
$$E(n) = O(n) \ltimes \mathbb{R}^n.$$

In fact, it is the set of all pairs $(A, t_a) = (A, a) \in O(n) \times \mathbb{R}^n$ with multiplication

$$(A, a)(B, b) = (AB, Ab + a),$$

where $(B, b) \in O(n) \times \mathbb{R}^n$.

The elements of the group $O(n)$ are sometimes called rotations. To distinguish the direct product from the semi-direct product we shall denote the second one by $\ltimes$.

Example 1.11. The affine group $A(n) = GL(n, \mathbb{R}) \ltimes \mathbb{R}^n$ is the semi-direct product of the group $\mathbb{R}^n$ and $GL(n, \mathbb{R})$.

Proposition 1.12. *There is the following sequence of subgroups*

$$E(n) \subset A(n) \subset GL(n + 1, \mathbb{R}).$$

Proof. From the definition $E(n) \subset A(n)$. Let $(A, a) \in A(n)$. We have a $(n+1) \times (n+1)$ matrix $\begin{bmatrix} A & a \\ 0 & 1 \end{bmatrix}$, which obviously defines an inclusion $A(n) \subset GL(n+1, \mathbb{R})$. $\qquad\square$

From the above $E(n)$ can be considered as a topological space with topology induced from the Euclidean space $\mathbb{R}^{(n+1)^2}$.

Definition 1.13. A subset X of an Euclidean space is discrete if any $x \in X$ has an open neighbourhood U_x such that $U_x \cap X = \{x\}$.

Note that this coincides with the standard definition in which any subset of X is closed and open. We put the following statement as an easy exercise.

Lemma 1.14. *A metric space X is discrete if and only if every convergent sequence $\{x_n\}$ in X is eventually constant.* [1]

$\qquad\square$

Definition 1.15. Let Γ be a subgroup of the group $E(n)$. Then Γ is discrete if it is a discrete subset of the Euclidean space $\mathbb{R}^{(n+1)^2}$. We say that Γ acts properly discontinuously on $\mathbb{R}^n$ if for any $x \in \mathbb{R}^n$ there is an open neighbourhood U_x such that the set

$$\{\gamma \in \Gamma \mid \gamma U_x \cap U_x \neq \emptyset\}$$

is finite. Moreover, Γ acts freely, if for any $x \in \mathbb{R}^n \{\gamma \in \Gamma \mid \gamma x = x\} = \{(I, 0)\}$.

A set of points $\{1/n\}_{n \in \mathbb{N}} \subset \mathbb{R}$ is an example of a discrete set, which shall not considered here. In most cases a discrete subset of Euclidean space will have a group structure.

Lemma 1.16. *Any discrete subgroup of the group $E(n)$ is closed in $E(n)$.*

Proof. Let Γ be a discrete subgroup of $E(n)$ and suppose that $E(n) \setminus \Gamma$ is not open. Then there is a γ in $E(n) \setminus \Gamma$ and γ_n in $V_\gamma^{1/n} \cap \Gamma$, for each $n \in \mathbb{N}$. $V_\gamma^{1/n}$ is a ball with centre in γ of radius $1/n$. As $\gamma_n \to \gamma$ in $E(n)$, we have $\gamma_n \gamma_{n+1}^{-1} \to 1$ in Γ. But $\{\gamma_n \gamma_{n+1}^{-1}\}$ is not eventually constant, which contradicts Lemma 1.14. Therefore, the set $E(n) \setminus \Gamma$ must be open, and so Γ is closed in $E(n)$. $\qquad\square$

---

[1] The sequence $\{x_n\}_{n \in \mathbb{N}}$ is eventually constant if there exists $N \in \mathbb{N}$ such that $\forall n > N, x_n = x_{n+1}$.

Lemma 1.17. *If Γ is a discrete subgroup of the group $E(n)$ and $V_0 \subset \mathbb{R}^n$ is an open disk centred at the point 0 and of radius r, then*

$$\{\gamma \in \Gamma \mid \gamma V_0 \cap V_0 \neq \emptyset\} \subset \Gamma \cap (O(n) \times V_0'),$$

where V_0' is a disk centred at 0 and of radius $2r$.

Proof. Let $\gamma = (A, a) \in \Gamma$ and $\gamma V_0 \cap V_0 \neq \emptyset$. Then there exist $x, x' \in V_0$, such that $\gamma x = Ax + a = x'$. Hence, from the triangle inequality $\|a\| = \|x' - Ax\| \leq \|x'\| + \|Ax\| < 2r$ and $\gamma \in O(n) \times V_0'$.

$\square$

Proposition 1.18. *Let Γ be a subgroup of the group $E(n)$. The following conditions are equivalent:*

(i) Γ acts properly discontinuously on $\mathbb{R}^n$;

(ii) $\forall x \in \mathbb{R}^n, \Gamma x$ is a discrete subset of $\mathbb{R}^n$;

(iii) Γ is a discrete subgroup of $E(n)$.

Proof. Let Γ act properly discontinuously on $\mathbb{R}^n$. We claim that Γ is discrete. Let elements $\{\gamma_n\}_{n=1}^{n=\infty} \subset \Gamma$ converge to the identity. By assumption there is a neighbourhood U_0 of $0 \in \mathbb{R}^n$ such that the set $\{\gamma_i \mid U_0 \cap \gamma_i U_0 \neq \emptyset\}$ is finite. Hence $\gamma_i = (I, 0)$ for large i and the sequence $\{\gamma_n\}$ is eventually constant. In general case if $\gamma_n \to \gamma$ then $\gamma_n \gamma^{-1} \to (I, 0)$ and from previous consideration, together with Lemma 1.14 the implication $(i) \to (iii)$ is proved. We shall use Lemma 1.17 for the proof of the reverse implication. Let $x \in \mathbb{R}^n$ be any point and V_x be a disk of radius r centred at x. By definition of properly discountinuous action and Lemma 1.17 we have

$$\{\gamma \in \Gamma \mid \gamma V_x \cap V_x \neq \emptyset\} = \{\gamma \in \Gamma \mid t_{-x} \gamma t_x V_0 \cap V_0 \neq \emptyset\} \subset t_{-x} \Gamma t_x \cap (O(n) \times V_0').$$

Since Γ is discrete, and by Lemma 1.16 also closed, the above set is finite and the implication $(iii) \to (i)$ is proved.

Let us assume the condition (i) (or equivalently (iii)). We have to prove that for any $x \in \mathbb{R}^n$ the set Γx is discrete. Suppose it is not. Then there is $y \in \mathbb{R}^n$ and a sequence $\{\gamma_i x = A_i x + a_i\}_{i=1}^{i=\infty}$, which is not eventually constant and converges to y. Since the group $O(n)$ is compact, the sequence $\{A_i\}_{i=1}^{i=\infty}$ converges to some $A \in O(n)$. We claim that the sequence $\{a_i\}_{i=1}^{i=\infty}$ converges to $-Ax + y$. In fact, the value

$$\|a_i + Ax - y\| \leq \|a_i + A_i x - y\| + \|Ax - A_i x\|$$

can be arbitrarily small for large i. Summing up, we showed that the sequence $\{\gamma_i\}_{i=1}^{i=\infty}$ converges to $\gamma = (A, -Ax+y)$ in $E(n)$. Hence, $\{\gamma_i\gamma_{i+1}^{-1}\}_{i=1}^{i=\infty}$ converges to the identity, cf. Lemma 1.16. Since Γ is discrete, a sequence $\{\gamma_i x\}$ is eventually constant, cf. Lemma 1.14. This contradicts our assumptions and proves the implication $(i) \to (ii)$. Finally, we prove that (i) follows from (ii). We shall use Lemma 1.14. Let $\{\gamma_n\}$ be a convergent sequence in Γ. Then, for any $x \in \mathbb{R}^n, \{\gamma_n x\}$ is a convergent sequence in Γx. By definition it is eventually constant. Hence a sequence $\{\gamma_n\}$ is also eventually constant. This finishes the proof. $\square$

Proposition 1.19. *A discrete subgroup of $E(n)$ acts freely on $\mathbb{R}^n$ if and only if it is torsion free (has no elements of finite order).*

Proof. Assume that a group Γ has an element γ of order k. For any $x \in \mathbb{R}^n$ the element $x + \gamma x + \gamma^2 x + \cdots + \gamma^{k-1} x$ is invariant under the action of γ. Hence the action of Γ is not free. The reverse implication follows from the equality of the sets

$$\{\gamma \in \Gamma \mid \gamma a = a\} = \Gamma \cap t_a(O(n) \times 0)t_{-a}$$

where $a \in \mathbb{R}^n$. In fact, since the orthogonal group $O(n)$ is compact and Γ is discrete the above set is always finite. $\square$

Let us introduce an important definition.

Definition 1.20. The Hausdorff topological space $\mathcal{M}$ is a manifold *of given class*, when the following conditions are satisfied:
1. $\mathcal{M}$ is covered by the family of open sets $U_i, i \in I$, such that each of them is mapped by a homeomorphism φ_i on an open subset of $\mathbb{R}^n$;
2. if $U_i \cap U_j \neq \emptyset$, then

$$\varphi_j \circ \varphi_i^{-1} \colon \varphi_i(U_i \cap U_j) \to \varphi_j(U_i \cap U_j)$$

is a map *of given class*.

The pair (U_i, φ_i) is a *map* of the manifold $\mathcal{M}$, and the set of all maps is *an atlas*.

Example 1.21. The spaces $\mathbb{R}^n$, $\mathbb{C}^n$ and projective spaces $\mathbb{R}P^n, \mathbb{C}P^n$ are examples of manifolds. The n-dimensional sphere

$$S^n = \{x \in \mathbb{R}^{n+1} \mid \ \|x\| = 1\}$$

and the n-dimensional torus

$$T^n = S^1 \times S^1 \times \cdots \times S^1 \simeq \mathbb{R}^n/\mathbb{Z}^n$$

are examples of manifolds, [148].[2] The Cartesian product of two manifolds is a manifold.

Definition 1.22. Let G be a differential manifold and simultaneously a group, such that the group structures $G \times G \to G, G \xrightarrow{-1} G$ are differential maps of the manifolds. Then G is called a Lie group.

Example 1.23. The sphere $S^1 \subset \mathbb{C}$, with the multiplication of complex numbers, is a Lie group. The Cartesian product of S^1, i.e. n-dimensional torus $T^n = (S^1)^n$, is a Lie group. The matrix groups $U(n), O(n), SL(n, \mathbb{R}), SL(n, \mathbb{C})$ are Lie groups.

Definition 1.24. Let Γ be a subgroup of $E(n)$. The orbit space of the action of Γ on $\mathbb{R}^n$ is defined to be the set of Γ-orbits $\mathbb{R}^n/\Gamma = \{\Gamma x \mid x \in \mathbb{R}^n\}$ topologized with the quotient topology from $\mathbb{R}^n$. The quotient map will be denoted by $\pi \colon \mathbb{R}^n \to \mathbb{R}^n/\Gamma$.

Lemma 1.25. *If Γ is a subgroup of $E(n)$, then the natural projection map $p \colon E(n) \to E(n)/\Gamma$ and the projection on the orbit space $\pi \colon \mathbb{R}^n \to \mathbb{R}^n/\Gamma$ are open and closed.*

Proof. It is a consequence of the definition of topology on the quotient topological space. $\qquad \square$

Proposition 1.26. *Let Γ be a subgroup of $E(n)$. Then the orbit space $\mathbb{R}^n/\Gamma$ is compact if and only if the space of cosets $E(n)/\Gamma$ is compact.*

Proof. By definition $E(n)/O(n) = \mathbb{R}^n$ and a group Γ acts on the space of cosets $E(n)/O(n)$ by $gO(n) \mapsto (\gamma g)O(n)$, where $\gamma \in \Gamma, g \in E(n)$. It is easy to see that the above action agrees with a standard action of Γ on $\mathbb{R}^n$. Next, let us note that the map $E(n)/\Gamma \to (E(n)/O(n))/\Gamma = \mathbb{R}^n/\Gamma$, given by

$$g^{-1}\Gamma \to \Gamma(gO(n))$$

is a continuous open map whose fibres (inverse images of points) are compact. Hence it follows easily (cf. Exercise 1.7) that $E(n)/\Gamma$ is compact if and only if $\mathbb{R}^n/\Gamma$ is compact. $\qquad \square$

---

[2]$\mathbb{Z}^n = \{z_1 e_1 + z_2 e_2 + \cdots + z_n e_n \mid z_i \in \mathbb{Z}, e_i \text{ is a basis of } \mathbb{R}^n, i = 1, 2, \ldots, n\}$ is a torsion-free, finitely generated abelian group of rank n. See Exercise 2.5.

An illustration of the above Proposition 1.26 is the following Lemma.

Lemma 1.27. *A space $E(n)/\Gamma$ is compact if and only if there exists a compact subset $D \subset E(n)$, such that $E(n) = D\Gamma$.*

Proof. Since $E(n)$ is a subset of $\mathbb{R}^{(n+1)^2}$, there exists a family of open sets $U_k \cap E(n)$, such that the family of sets $p(U_k \cap E(n))$ covers the compact space $E(n)/\Gamma$. Here U_k is an open disk centred at origin and of radius k. Hence there exists k_0, such that $p(U_{k_0} \cap E(n)) = E(n)/\Gamma$. Let D be a closure of the set $(U_{k_0} \cap E(n))$, i.e. $D = \overline{(U_{k_0} \cap E(n))}$. Finally, we have

$$E(n) = D\Gamma.$$

The proof of the reverse implication follows from the equality $D/\Gamma = E(n)/\Gamma$.
□

Definition 1.28. A subgroup $\Gamma \subset E(n)$ is cocompact, if the space $E(n)/\Gamma$ is compact.

Roughly speaking a fundamental domain for a group Γ of isometries in a metric space X is a subset of X which contains exactly one point from each of these orbits.

Definition 1.29. Let X be a metric space and G a subgroup of a group of its isometries. An open, connected subset $F \subset X$ is a *fundamental domain* if

$$X = \bigcup_{g \in G} g\bar{F}$$

and $gF \cap g'F = \emptyset$, for $g \neq g' \in G$.

1.1   Exercises

Exercise 1.1. Find the set of all isometries of the space $\mathbb{R}^2$.

Exercise 1.2. Give proofs of Lemma 1.14 (see [122, Lemma 2]), Lemma 1.25 and Proposition 1.12.

Exercise 1.3. Prove that if a group Γ of isometries of a metric space X has a fundamental domain, then Γ is a discrete subgroup of a group all isometries of X.

Exercise 1.4. Let Γ be a subgroup of $E(n)$. Prove that the following conditions are equivalent:

- $\forall x \in \mathbb{R}^n, \Gamma x$ is a discrete subset of $\mathbb{R}^n$;
- $\exists x \in \mathbb{R}^n, \Gamma x$ is a discrete subset of $\mathbb{R}^n$.

Hint: Compare a proof of Lemma 2.11.

Exercise 1.5. Let $f \colon \mathbb{R}^n \to \mathbb{R}^n$ be a linear map and $g \in O(n)$. Prove that

$$\|gf\| = \|f\| = \|fg\|.$$

Exercise 1.6. Let G be a connected Lie group. Prove that G has a countable basis of open sets.

Exercise 1.7. Let $f \colon X \to Y$ be a continuous and closed map from a topological space X to a topological space Y. Assume that for any $y \in Y$, $f^{-1}(y)$ is compact. Prove that X is compact if and only if Y is compact.

Hint: Show that f is closed if and only if for any $y \in Y$ and an open neighbourhood $f^{-1}(y) \subset U$ there exists an open set V, with $y \in V$, such that $f^{-1}(V) \subset U$.

2. Bieberbach Theorems

The central results on crystallographic groups were proved by Bieberbach in the years 1910-1912. The proof of the first Bieberbach theorem is the most difficult part of the chapter. In addition we prove a theorem of Zassenhaus from 1948 which treats crystallographic groups purely as abstract groups. Here we may apply the theory of group extensions using the language of finite group cohomology. Together with the Jordan-Zassenhaus theorem this is enough to prove the third Bieberbach theorem. Finally we prove the second Bieberbach theorem on the rigidity of crystallographic groups.

2.1 The first Bieberbach Theorem

Definition 2.1. A crystallographic group of dimension n is a cocompact and discrete subgroup of $E(n)$.

Remark 2.2. The first part of the eighteenth Hilbert problem was about the description of discrete and cocompact groups of isometries of $\mathbb{R}^n$.

Example 2.3. If $(B, \binom{1/2}{0}), (I, \binom{0}{1}) \in E(2)$, where $B = \begin{pmatrix} 1 & 0 \\ 0 & -1 \end{pmatrix}$, then the group $\Gamma \subset E(2)$ generated by the above elements is a crystallographic group of dimension two and the orbit space $\mathbb{R}^2/\Gamma$ is the Klein bottle.

The answer for the above Hilbert problem was given by the German mathematician L. Bieberbach.

Theorem 2.4 (Bieberbach).
1. *If $\Gamma \subset E(n)$ is a crystallographic group then the set of translations $\Gamma \cap (I \times \mathbb{R}^n)$ is a torsion free and finitely generated abelian group of rank n, and is a maximal abelian and normal subgroup of finite index.*
2. *Two crystallographic groups of dimension n are isomorphic if and only if they are conjugate in the group $A(n)$.*
3. *For any natural number n, there are only a finite number of isomorphism classes of crystallographic groups of dimension n.*

Remark 2.5. The original proof of the first Bieberbach Theorem was given in 1910 (see [11] and [12]). Then in 1911 G. Frobenius proved that up affine equivalence there are only a finite number of crystallographic groups of dimension n [47]. Finally in 1912 L. Bieberbach proved the second and third theorem, [13] and [14]. Next in 1938 a new proof by H. Zassenhaus [149] was given. Let us mention some other proofs: L. Auslander [5], [6], [7] (1960, 1961), J. Wolf [148] (1970), M. Gromov [54] (1978) and R. K. Oliver [108]. Gromov's proof was given for virtually nilpotent groups. Some version of it for crystallographic groups was elaborated by P. Buser and H. Karcher; see [19] - [21] and the Appendix. In these papers, we can also find some comments and comparison of the above proofs.

Proof.[1] We shall first prove several Lemmas.

Lemma 2.6. *There is a neighbourhood of the identity $U \subset O(n)$ such that for any $h \in U$ if $g \in O(n)$ commutes with $[g, h] = ghg^{-1}h^{-1}$, then g commutes with h.*

Proof. Let $\lambda_1, \lambda_2, \ldots, \lambda_r$ be the eigenvalues of a map $g \colon \mathbb{C}^n \to \mathbb{C}^n$ induced by an orthogonal matrix g, and $\mathbb{C}^n = V_1 \oplus V_2 \oplus \cdots \oplus V_r$ its invariant subspaces. Since $g[g, h] = [g, h]g$, $ghg^{-1}h^{-1} = hg^{-1}h^{-1}g$. Moreover for $i = 1, 2, \ldots, r$ and $\forall x \in V_i$ we have $gx = \lambda_i x$. Hence

$$ghg^{-1}h^{-1}x = hg^{-1}h^{-1}gx = hg^{-1}h^{-1}\lambda_i x = \lambda_i hg^{-1}h^{-1}x$$

and $hg^{-1}h^{-1}V_i \subset V_i$. Since h and g are isomorphisms, $h^{-1}V_i = gh^{-1}V_i$. This shows that $h^{-1}V_i$ is g-invariant and so

$$h^{-1}V_i = (h^{-1}V_i \cap V_1) \oplus (h^{-1}V_i \cap V_2) \oplus \ldots \oplus (h^{-1}V_i \cap V_r),$$

where $h^{-1}V_i \cap V_j = \{x \in h^{-1}V_i \mid gx = \lambda_j x\}$.

Let $w, v \in \mathbb{C}^n$ be such that $\|w\| = \|v\| = 1$ and $w \perp v$ in a sense of the Hermitian inner product. Hence $\|w - v\| = \sqrt{2}$. Moreover, let $\|h^{-1} - I\| < \epsilon = \sqrt{2} - 1$, $i \neq j$ and $0 \neq x \in (h^{-1}V_i \cap V_j)$. We can assume, that $\|x\| = 1$. From the definition, there is $y \in V_i$, such that $h^{-1}y = x$. But $x \in V_j$, and $\langle x, y \rangle = 0$.[2] Since

$$\sqrt{2} = \|x - y\| = \|h^{-1}y - y\| = \|(h^{-1} - I)y\| < \epsilon = \sqrt{2} - 1,$$

---

[1]The proof is based on [5] and [148].

[2]We use the fact that eigenvectors of different eigenvalues are perpendicular. In fact $\lambda_i \langle x, y \rangle = \langle Ax, y \rangle = \langle x, A^T y \rangle = \lambda_j \langle x, y \rangle$. Hence $\lambda_i = \lambda_j$, what is a contradiction and $\langle x, y \rangle = 0$.

we obtain a contradiction. Hence $h^{-1}V_i = V_i$, for $i = 1, 2, \ldots, r$, and $gh = hg_{|V_i}$. In fact, the matrix of g is diagonal. Since any element of $\mathbb{C}^n$ is a sum of elements from V_i, it follows that g and h commute. Let

$$U = \{h \in O(n) \mid \|I - h^{-1}\| < \epsilon\}.$$

$\square$

Lemma 2.7. *For some neighbourhood of the identity $U \subset O(n)$ and for any $g, h \in U$ the sequence*

$$[g, h], [g, [g, h]], [g, [g, [g, h]]], \ldots$$

converges to the identity.

Proof. Let U be a neighbourhood of the identity of radius $\epsilon < 1/4$. By definition (cf. Exercise 1.5) we have

$$\|[g, h] - I\| = \|gh - hg\| = \|gh - g - h + I - hg + h + g - I\| =$$

$$\|(g - I)(h - I) - (h - I)(g - I)\| \leq 2\|g - I\|\,\|h - I\| < \frac{\|h - I\|}{2}$$

for $g, h \in U$. Hence $[g, h] \in U$ and, by induction,

$$\|\underbrace{[g, [g, [g, \ldots, [g, h]\ldots]]]}_{n} - I\| \leq \frac{\|h - I\|}{2^n}.$$

$\square$

Lemma 2.8. *Let $G \subset O(n)$ be a connected subgroup and U be a neighbourhood of the identity, then the group $\langle G \cap U \rangle$ generated by the set $G \cap U$ is equal to G.*

Proof. The inclusion $\langle G \cap U \rangle \subset G$ is obvious. For the proof of opposite inclusion let us show that the set $\langle G \cap U \rangle$ is open. In fact, let $x \in \langle G \cap U \rangle$ and let $B(x, \epsilon) \subset U$ be any open disk centred at x and of radius ϵ. Then, for any $y \in B(x, \epsilon) \cap G$ we have

$$\|yx^{-1} - I\| = \|yx^{-1} - xx^{-1}\| = \|y - x\| < \epsilon. \tag{2.1}$$

Hence $yx^{-1} \in B(I, \epsilon) \cap G \subset U \cap G$, for ϵ small enough and $y = yx^{-1}x \in \langle G \cap U \rangle$. By definition the set $S = G \setminus \langle G \cap U \rangle$ is closed. Moreover, we claim that if $y \in S$, then $B(y, \epsilon) \cap G \subset S$. Indeed, suppose by contradiction there is $x \in B(y, \epsilon) \cap G$ such that $x \in \langle G \cap U \rangle$. From (2.1) we see that $y \in \langle G \cap U \rangle$, which is impossible. Summing up, we obtain that sets S and $\langle G \cap U \rangle$ are simultaneously closed and open. Since U is nonempty and G is connected, we have $S = \emptyset$.   □

Lemma 2.9. *There is an arbitrary small neighbourhood V of $I \in O(n)$ such that $\forall g \in O(n)$, $gVg^{-1} = V$.*

Proof. Let ϵ be a positive number and $V = B(I, \epsilon)$ be an open disk. By definition (cf. Exercise 1.5) $\forall g \in O(n)$ and $\forall h \in V$ we have

$$\|ghg^{-1} - I\| = \|g(h - I)g^{-1}\| = \|h - I\| < \epsilon.$$

Since $\forall g \in O(n), gVg^{-1} \subset V$ and $g^{-1}Vg \subset V$, $V = g(g^{-1}Vg)g^{-1} \subset gVg^{-1}$.
□

Definition 2.10. We shall call a neighbourhood U satisfying Lemmas 1, 2, 4 a stable neighbourhood of identity.

Lemma 2.11. *Let Γ be a crystallographic group and $x \in \mathbb{R}^n$. Then the linear space generated by the set $\{\gamma(x)\}, \gamma \in \Gamma$ is equal to $\mathbb{R}^n$.*

Proof. (see [5, page 1233]) Assume the lemma is false and that $x_0 \in \mathbb{R}^n$ exists such that $\{\gamma(x_0)\}$ lies in W, a proper linear subspace of $\mathbb{R}^n$. By a new choice of origin in $\mathbb{R}^n$ we may assume $O(n)$ leaves x_0 fixed. In fact, $\Gamma(x_0) = \Gamma(I, x_0)(I, -x_0)(x_0) = \Gamma(I, x_0)(0)$. Hence sets $(I, -x_0)\Gamma(I, x_0)(0)$ and $\Gamma(x_0)$ differ by translation $(I, -x_0)$ and define linear subspace of the same dimension. It follows that for $\gamma \in \Gamma, \gamma = (A, a)$ must have $a \in W$.

Since Γ is a group, $A(W) = W$ for all $A \in p_1(\Gamma)$. Let $W^\perp$ be the orthogonal complement of W. Let $x \in W^\perp$ be an element at a distance d from the origin. It is easy to see that for any $\gamma = (A, a) \in \Gamma, \langle \gamma(x), \gamma(x) \rangle = \langle x, x \rangle + \langle a, a \rangle$. Hence $\|x\| \leq \|\gamma(x)\|$. Summing up, points in $W^\perp$ at a distance d from origin stay at least at a distance d from 0. It follows that Γ cannot have a compact fundamental domain. This proves our assertion.   □

Lemma 2.12. *Let Γ be an abelian crystallographic group; then Γ contains only pure translations.*

Proof. Let $(B, b) \in \Gamma$, where $B \neq I$. Then we can always choose an origin and a coordinate system in $\mathbb{R}^n$ such that, $B = \begin{bmatrix} I & 0 \\ 0 & B' \end{bmatrix}$, where I is the $r \times r$ identity matrix, $B' - I$ is a nonsingular $s \times s$ matrix, $r + s = n$, and r can be equal to zero. Moreover we can assume that $b = (b', 0, ..., 0)$, where $b' \in \mathbb{R}^r$. Then, by Lemma 2.11, there exists an element $(C, (t_1, t_2)) \in \Gamma$, where $t_1 \in \mathbb{R}^r, t_2 \in \mathbb{R}^s$ and $t_2 \neq 0$. Then, since Γ is abelian and $BCb = CBb = Cb$

$$(B, b)(C, (t_1, t_2)) = (BC, (b' + t_1, B'(t_2))) = (CB, (Cb + t_1, t_2))$$

$$= (C, (t_1, t_2))(B, b).$$

Hence $B'(t_2) = t_2$, which contradicts a nonsingularity of $B' - I$. $\square$

Lemma 2.13. *Let Γ be a crystallographic group. Let $p_1 \colon E(n) \to O(n)$ be a projection onto the first factor. Then $p_1(\Gamma)_0$ is an abelian group.*

Proof. Let $U = B(I, \epsilon),$[3] where $\epsilon < 1/4$, and $\gamma_1 = (A_1, a_1), \gamma_2 = (A_2, a_2) \in (p_1^{-1}(U) \cap \Gamma)$. By recurrence we define for $i \geq 2$

$$\gamma_{i+1} = [\gamma_1, \gamma_i].$$

We have

$$\gamma_{i+1} = ([A_1, A_i], (I - A_1 A_i A_1^{-1})a_1 + A_1(I - A_i A_1^{-1} A_i^{-1})a_i).$$

Hence $A_{i+1} = [A_1, A_i]$ and $\|a_{i+1}\| \leq \|I - A_i\| \|a_1\| + \frac{1}{4}\|a_i\|$. From Lemma 2.7 we have $\lim_{i \to \infty} A_i = I$. Hence $\lim_{i \to \infty} a_i = 0$. Since Γ is discrete, $\gamma_i = (I, 0)$ for sufficient large i. However, from Lemma 2.6 we have $A_1 A_2 = A_2 A_1$. Hence any elements of the set $p_1(\Gamma)_0 \cap U$ commute and using Lemma 2.8, we prove commutativity of the group $p_1(\Gamma)_0$. $\square$

Let us finish the proof of the first Bieberbach Theorem. Assume first that $\Gamma \cap (I \times \mathbb{R}^n)$ is trivial. Then p_1 is an isomorphism of Γ into $O(n)$. Since $O(n)$ is compact, the closure of $p_1(\Gamma)$ can have only a finite number of components. Hence, since by Lemma 2.13 $p_1(\Gamma)_0$ is abelian, Γ contains a subgroup Γ_1 of finite index which is abelian. But then Γ_1, being of finite index in Γ, also is a crystallographic group. Hence, by Lemma 2.12, Γ_1 consists of pure translations. Thus we see that $\Gamma \cap (I \times \mathbb{R}^n)$ is nonempty.

---

[3]By Lemma 2.7 U is a stable neighbourhood.

Let $W \subset \mathbb{R}^n$ be the subspace of $\mathbb{R}^n$ spanned by the pure translations of Γ, i.e., by $\Gamma \cap (I \times \mathbb{R}^n)$. Then, $p_1(\Gamma)$ leaves W invariant because $\Gamma \cap (I \times \mathbb{R}^n)$ is normal in Γ. Note further that $p_1(\Gamma)_{|}W$ is a finite group, for otherwise it would contain elements arbitrarily close to the identity which would, under inner automorphism with a basis of $\Gamma \cap (I \times \mathbb{R}^n)$, force Γ to be nondiscrete. In fact, let $(A_i, a_i) \in \Gamma, i \in \mathbb{N}$ be an infinite sequence of elements such that $A_i \overset{i \to \infty}{\to} I$. Put

$$(B_i, b_i) = (I, e_k)(A_i, a_i)(I, -e_k) = (A_i, (I - A_i)e_k + a_i),$$

where $e_k \in \Gamma \cap (I \times \mathbb{R}^n)$. Then a sequence $(B_i, b_i)(A_i^{-1}, -A_i^{-1}(a_i)), i \in \mathbb{N}$ defines a nondiscrete subset of Γ. Moreover, we see that Γ induces an action on $\mathbb{R}^n/W$ which is obviously cocompact. We claim that it is also properly discountinuous. To prove a claim we shall use Proposition 1.18, (ii) and Exercise 1.4. We have decomposition $\mathbb{R}^n = W \oplus W^\perp$, where $W^\perp \simeq \mathbb{R}^n/W$. Let $pr_1 \colon \mathbb{R}^n \to W, pr_2 \colon \mathbb{R}^n \to W^\perp$ are projections. Let X be any discrete subset of $\mathbb{R}^n$. It can happen that sets $pr_1(X)$ and $pr_2(X)$ are not discrete subsets of W and $W^\perp$. Since $p_1(\Gamma)\,|_W$ is finite, we can concentrate on elements $\gamma \in \Gamma$ such that $p_1(\gamma)$ acts as identity on W. From Proposition 1.18, (ii) the orbit $\Gamma(0)$ is discrete in $\mathbb{R}^n$. By contradiction let us assume that $pr_2(\Gamma(0))$ is not discrete at $W^\perp$ and $y \in W^\perp$ is an accumulation point of $pr_2(\Gamma(0))$. Let $pr_2(\gamma_i(0)) \to y$, where $\gamma_i \in \Gamma, i \in \mathbb{N}$. Using elements from $\Gamma \cap (I \times \mathbb{R}^n)$ we can define a sequence of elements of $\bar{\gamma}_i \in \Gamma, i \in \mathbb{N}$ such that $\forall i \in \mathbb{N}, pr_1(\gamma_i(0)) \subset C \subset W$, where C is a compact set. Here we use an obvious fact that $\Gamma \cap \mathbb{R}^n$ is a cocompact subgroup of W, cf. Exercise 2.5 (3). It is easy to see that a set $\{\bar{\gamma}_i(0)\}, i \in \mathbb{N}$ has an accumulation point at a discrete set $\Gamma(0)$. We get contradiction and our claim is proved.

Hence Γ is a crystallographic group on $\mathbb{R}^n/W$ with no pure translations. By the above, this implies the zero dimension of $\mathbb{R}^n/W$. $\square$

2.2   Proof of the second Bieberbach Theorem

Let us start to prove the second Bieberbach Theorem. Let $h \colon \Gamma \to \Gamma'$ be any isomorphism of crystallographic groups of dimension n.

The restriction $h\,|_{(\Gamma \cap \mathbb{R}^n)}$ to the subgroup of translations defines a linear map $A \in GL(n, \mathbb{R})$. Let $(\gamma, a) \in \Gamma$. Put $h(\gamma, a) = (r, c) \in \Gamma'$. We have

$$h((\gamma, a)(I, e_i)(\gamma^{-1}, -\gamma^{-1}a)) = (I, A\gamma(e_i)).$$

Hence $rA(e_i) = A\gamma(e_i)$, for $i = 1, 2, \ldots, n$, and so $r = A\gamma A^{-1}$. We can conjugate h by some matrix from the group $GL(n, \mathbb{R})$ such that the matrix A will be the identity. Let $h(\gamma, a) = (\gamma, a_\gamma) \in \Gamma'$. It is easy to observe that $\bar{h} \colon \Gamma \to A(n)$, given by the formula $\bar{h}(\gamma, a) = (\gamma, a - a_\gamma)$, is a homomorphism and $\ker \bar{h} = \Gamma \cap \mathbb{R}^n$, see [152, p. 123] and Exercise 2.10. From Proposition 1.19, there is a fixed point $x_0 \in \mathbb{R}^n$ of the action of the finite group $\bar{h}(\Gamma)$. We have $x_0 = (\gamma, a - a_\gamma)(x_0) = \gamma(x_0) + a - a_\gamma$. Hence, $a = -\gamma(x_0) + a_\gamma + x_0$, for all $\gamma \in \Gamma$. Finally, for any $x \in \mathbb{R}^n$, we get

$$(I, x_0)(\gamma, a_\gamma)(I, -x_0)x = (I, x_0)(\gamma(x - x_0) + a_\gamma) =$$

$$\gamma(x) - \gamma(x_0) + a_\gamma + x_0 = (\gamma, a)x.$$

Then $(I, x_0)\Gamma'(I, -x_0) = \Gamma$ (cf. [152, p. 123]) and the isomorphism h has the required properties.

□

Example 2.14. Let $\Gamma \subset E(n)$ be a torsion free crystallographic group. The orbit space $\mathbb{R}^n/\Gamma$ is a manifold. If Γ is not torsion free then the orbit space $\mathbb{R}^n/\Gamma$ is an orbifold.

The above manifolds (orbifolds) shall be called *flat*. The word *flat* comes from Riemannian geometry where the Euclidean space $\mathbb{R}^n$ at each point has sectional curvature equal to zero (see [148]).

In linear algebra an orientation means a choice of some basis, where two orientations are equivalent if the change of basis matrix has positive determinant.

Definition 2.15. A manifold is oriented if it has an atlas (U_i, φ_i) such that all maps $\varphi_i \varphi_j^{-1}$ preserve the orientation of $\mathbb{R}^n$.

Example 2.16. $\mathbb{R}^n, \mathbb{C}^n, \mathbb{C}P^n, S^n, T^n$ are orientable manifolds. $\mathbb{R}P^n$ (n even) and the Klein bottle are not orientable manifolds.

Let us introduce a crucial definition for this course.

Definition 2.17. A Bieberbach group is a torsion free crystallographic group.

2.3   Proof of the third Bieberbach Theorem

2.3.1   Cohomology group language

From the first Bieberbach theorem any crystallographic group $\Gamma \subset E(n)$ defines a short exact sequence of groups

$$0 \to \mathbb{Z}^n \to \Gamma \overset{p}{\to} G \to 0, \tag{2.2}$$

where $\mathbb{Z}^n = \Gamma \cap \mathbb{R}^n$ is subgroup of all translations and G is a finite group. We have to say that relations between the above group extension and the second cohomology group are crucial. Then, we partially introduce some definitions and facts from cohomology theory. However, we understand that sometimes it can be not enough. Then, we send the reader to the K. S. Brown book, [18].

Let G be a finite group, written multiplicatively and R be a commutative ring. Let RG or $(R[G])$ be the free R-module generated by the elements of G. Thus an element of RG is uniquely expressible in the form $\sum_{g \in G} \alpha(g)g$, where $\alpha(g) \in R$ and $\alpha(g) = 0$ for almost all g. The multiplication in G extends uniquely to a R-bilinear product $RG \times RG \to RG$; this makes RG a ring, called the R group ring of G. There is a well known correspondence between R-representations of G and modules over RG, see [31]. In our book we shall use the above correspondence mainly for $R = \mathbb{Z}, \mathbb{Q}, \mathbb{R}, \mathbb{C}, \mathbb{F}_2$, where $\mathbb{F}_2$ is the two-elements field.

Let M be any $\mathbb{Z}G$-module, (also called G-module). In this section we introduce the cohomology theory of the group G with coefficients in M.

Let $C^n(G, M), n \geq 1$, be the group of all functions $f \colon G^n \to M$ with the property that $f(g_1, g_2, ..., g_n) = 0$ if some g_i equals 1. Then f is called a n-dimensional cochain. The usual addition of functions makes $C^n(G, M))$ an abelian group. Let $C^0(G, M) = M$ and $C^n(G, M) = 0$ for $n < 0$. The coboundary operator

$$\delta^n \colon C^n(G, M) \to C^{n+1}(G, M)$$

is defined by

$$(\delta^n f)(g_0, g_1, ..., g_n) = g_0 \cdot f(g_1, ..., g_n)$$

$$+ \sum_{j=1}^{n} (-1)^j f(g_0, ..., g_{j-2}, g_{j-1}g_j, g_{j+1}, ..., g_n)$$

$$+(-1)^{n+1}f(g_0, ..., g_{n-1}) \text{ for } n \geq 1;$$

$(\delta^0 m)(g_1) = g_1 \cdot m - m; \delta^n = 0$ for $n < 0$. $(C^n(G, M), \delta^n)_{n \in \mathbb{Z}}$ is a cochain complex, i.e. $\delta^{n+1}\delta^n = 0$. The group

$$\ker \delta^n / \delta^{n-1}(C^{n-1}(G, M)) = H^n(G, M) \tag{2.3}$$

is called the n-th cohomology group of G with coefficients in M.[4] We shall use the following properties of the homology functor. For the proof of the second one we send the reader to [18, page 71].

Proposition 2.18. *Let M, G be as above. If $|G| = m$ is invertible in M (e.g., if M is a $\mathbb{Q}G$-module), then $H^n(G, M) = 0$ for all $n > 0$.*

Proof. It is enough to show that the homomorphisms

$$\nu \colon C^q(G, M) \to C^q(G, M), f \mapsto m \cdot f$$

induce the trivial homomorphism $\nu^q \colon H^q(G, M) \to H^q(G, M)$ for $q \geq 1$, i.e. $\nu^q(H^q(G, M)) = 0$. Let $g_1, ..., g_q \in G$ and $f \in \ker \delta^q$ be a cocycle. Then

$$0 = \sum_{h \in G} (\delta^q f)(g_0, g_1, ..., g_{q-1}, h) = \sum_{h \in G} g_0 f(g_1, ..., g_{q-1}, h)$$

$$+ \sum_{h \in G} \sum_{j=1}^{q-1} (-1)^j f(g_0, g_1, ..., g_{j-2}, g_{j-1}g_j, g_{j+1}, ..., g_{q-1}, h)$$

$$+ (-1)^q \sum_{h \in G} f(g_0, g_1, ..., g_{q-1}h) + (-1)^{q+1} \sum_{h \in G} f(g_0, g_1, ..., g_{q-1}).$$

This implies for $K(x_1, ..., x_{q-1}) = \sum_{h \in G} f(x_1, ..., x_{q-1}, h)$:

$$\pm m f(g_0, ..., g_{q-1}) = g_0 K(g_1, ..., g_{q-1})$$

$$+ \sum_{j=1}^{q-1} (-1)^j K(g_0, ..., g_{j-2}, g_{j-1}g_j, g_{j+1}, ..., g_{q-1}) + (-1)^q K(g_0, ..., g_{q-1}).$$

K is a cochain since it vanishes if some $g_i = 1$. From the above equation and definition of δ^{q-1} it follows that

$$\nu(f)(g_1, ..., g_q) = m \cdot f(g_1, ..., g_q) = \pm(\delta^{q-1}K)(g_1, ..., g_q).$$

$\square$

---

[4]See also Definition (3.7) in the next Chapter.

Proposition 2.19 ([18, Proposition 6.1(ii') on page 72]). *For any exact sequence* $0 \to M' \xrightarrow{i} M \xrightarrow{j} M'' \to 0$ *of G-modules and any integer n there is a natural map* $\sigma \colon H^n(G, M'') \to H^{n+1}(G, M')$ *such that the sequence*

$$0 \to H^0(G, M') \to H^0(G, M) \to H^0(G, M'') \xrightarrow{\sigma}$$
$$\to H^1(G, M') \to H^1(G, M) \to \cdots$$

is exact.

$\square$

Let $G \subset GL(n, \mathbb{Z})$ be a finite group. An obvious action defines G-modules $\mathbb{Z}^n, K^n, K^n/\mathbb{Z}^n$ and a short exact sequence

$$0 \to \mathbb{Z}^n \to K^n \to K^n/\mathbb{Z}^n \to 0,$$

where $K = \mathbb{Q}, \mathbb{R}$. From the above Propositions we have:

Corollary ([92, p. 117]). *The following groups are isomorphic*

$$H^2(G, \mathbb{Z}^n) \simeq H^1(G, \mathbb{Q}^n/\mathbb{Z}^n) \simeq H^1(G, \mathbb{R}^n/\mathbb{Z}^n). \tag{2.4}$$

$\square$

Let us define the group $\mathrm{Der}(G, M)$ of derivations (or crossed homomorphisms)

$$\{f \colon G \to M \mid \forall g_1, g_2 \in G, f(g_1 g_2) = g_1 f(g_2) + f(g_1)\},$$

with the action of addition. It has a subgroup $\mathrm{P}(G, M)$ of the principal derivations (or principal crossed homomorphisms)

$$\{f \colon G \to M \mid \exists m \in M, \forall g \in G, f(g) = gm - m\}.$$

It is not difficult to prove (see [18, Proposition 2.3 on page 89]) that

$$H^1(G, M) \simeq \mathrm{Der}(G, M)/\mathrm{P}(G, M). \tag{2.5}$$

Let Γ be a crystallographic group from the above short exact sequence of groups (2.2). We define the representation of holonomy

$$h_\Gamma \colon G \to GL(n, \mathbb{Z}), \tag{2.6}$$

where $\forall g \in G, h_\Gamma(g)(e_i) = \bar{g}e_i\bar{g}^{-1}$. Here $e_i \in \mathbb{Z}^n$ is the standard basis and $p(\bar{g}) = g$, for $\bar{g} \in \Gamma$.

There exists a correspondence between crystallographic groups of dimension n, with holonomy group G and elements of $H^1(G, \mathbb{Q}^n/\mathbb{Z}^n)$, where G acts on $\mathbb{Z}^n$ via the holonomy representation h_Γ, (cf. (2.6)).

In fact, let $\Gamma \subset E(n)$ be a crystallographic group with holonomy G. It is easy to see that $\Gamma = \{(A, s(A) + t)\} \subset E(n)$, where $A \in p(\Gamma)$ and $t \in \Gamma \cap \mathbb{R}^n \simeq \mathbb{Z}^n$. Moreover,

$$s(A) \in \{(a_1, a_2, \ldots, a_n) \in \mathbb{R}^n \mid \forall i \in \{1, 2, \ldots, n\}, 0 \leq a_i < 1\}.$$

We define an element

$$[s()] \in H^1(G, \mathbb{R}^n/\mathbb{Z}^n) \simeq H^1(G, \mathbb{Q}^n/\mathbb{Z}^n) \tag{2.7}$$

which corresponds to Γ. Conversely, assume there is given a faithful representation $h_\Pi \colon G \to GL(n, \mathbb{Z})$ which defines G-module $\mathbb{Z}^n$. Then each element

$$[s()] \in H^1(G, \mathbb{Q}^n/\mathbb{Z}^n) \overset{(2.4)}{\simeq} H^1(G, \mathbb{R}^n/\mathbb{Z}^n))$$

defines the set $\Pi = \{(A, s(A) + z)\} \subset A(n)$, where $z \in \mathbb{Z}^n$. It is an exercise to check that Π is a crystallographic group with a holonomy representation h_Π. When for the above 1-cocycle $[s()] = 0$ holds[5] we have an isomorphism of groups

$$\Pi \simeq G \ltimes \mathbb{Z}^n, \tag{2.8}$$

where semidirect product is defined by h_Π. In fact, if $s = 0$ then it is clear. In other case an isomorphism between Π and semidirect product is a conjugation by some translation. For this topics see monographs [92, Theorem 4.1, p. 112] and [18, Theorem 3.12, p. 93].

Before we prove the third Bieberbach Theorems let us present a theorem of H. Zassenhaus from 1948, [151].

Theorem 2.20 (Zassenhaus). *A group Γ is isomorphic to a crystallographic group of dimension n if and only if Γ has a normal, free abelian subgroup $\mathbb{Z}^n$ of finite index which is a maximal abelian subgroup of Γ.*

---

[5]This happens if $H^1(G, \mathbb{Q}^n/\mathbb{Z}^n) = 0$.

Proof. If Γ is a crystallographic group of dimension n, then a subgroup of all translations T is a normal, abelian subgroup of Γ. Since $T \subset \{I\} \times \mathbb{R}^n \simeq \mathbb{R}^n$ is a discrete and cocompact subgroup of $\mathbb{R}^n$, $T \simeq \mathbb{Z}^n$. For the proof that T is a maximal abelian subgroup, take any element $(A, a) \in \Gamma \subset E(n)$ which commutes with any translation of Γ. It is clear that $A = I$, (see Lemma 2.12).

Let Γ be any abstract group which defines the following short exact sequence of groups

$$0 \to \mathbb{Z}^n \to \Gamma \xrightarrow{p} G \to 0, \tag{2.9}$$

with a finite group G. Since $\mathbb{Z}^n$ is a maximal abelian subgroup, the holonomy representation h_Γ is an injection. The free abelian group $\mathbb{Z}^n$ can be considered as a subgroup of $\mathbb{R}^n$. Hence we have Diagram 2.1, where all the vertical arrows are monomorphic maps.

To define the middle, vertical sequence of groups

$$\Gamma \xrightarrow{j} \overline{\Gamma} \to A(n) = GL(n, \mathbb{R}) \ltimes \mathbb{R}^n,$$

we use the representation of holonomy (2.6). Moreover, $\overline{\Gamma}$ is defined as a pushout of monomorphisms $i \colon \mathbb{Z}^n \to \Gamma$ and $i' \colon \mathbb{Z}^n \to \mathbb{R}^n$. [6]

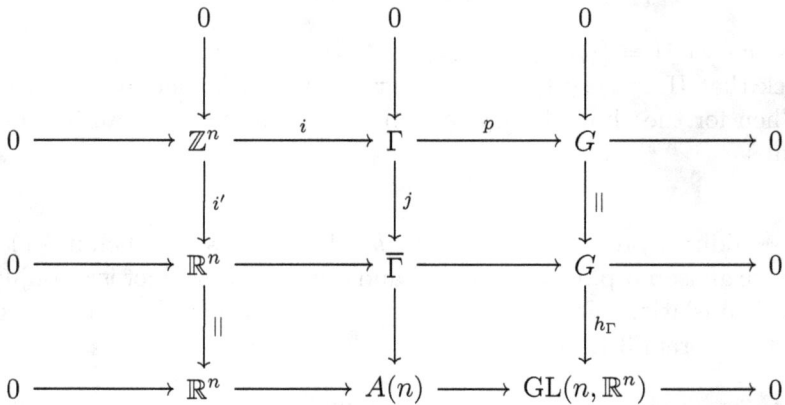

Diagram 2.1

We claim that the middle, horizontal sequence of groups splits. This is saying that $\overline{\Gamma} \simeq G \ltimes \mathbb{R}^n$, where an action of G on $\mathbb{R}^n$ is defined by h_Γ. In fact, from

---

[6] $\overline{\Gamma}$ is the largest quotient of $\Gamma \ltimes \mathbb{R}^n$ such that the left-hand-up square of Diagram 2.1 commutes, (see [18, page 94]).

Proposition 2.18 $H^2(G, \mathbb{R}^n) = 0$. Now, from (2.8) it follows that our claim is true. To finish the proof, we shall use a well known fact, that any finite subgroup of $GL(n, \mathbb{R})$ is conjugate to a finite subgroup of $O(n)$, see [31, page 256] or Exercise 2.15. □

Remark 2.21 ([96], [145, Theorem 3.1]). If, in the above Theorem of Zassenhaus, Γ is torsion free we can only assume that Γ has an abelian finitely generated subgroup A of finite index.

Proof of Remark. Let

$$\Delta(\Gamma) = \{x \in \Gamma \mid |\Gamma : C_\Gamma(x)| < \infty\}, \tag{2.10}$$

where $C_\Gamma(x) = \{\gamma \in \Gamma \mid \gamma x = x\gamma\}$. In other words, it is a subgroup of Γ consisting of all elements whose conjugacy classes are finite sets. Sometimes it is called FC-centre of Γ. Since $A \subset \Delta(\Gamma)$, we see that $\Delta(\Gamma)$ is finitely generated. We claim that $\Delta(\Gamma)$ has a central subgroup $Z(\Delta(\Gamma))$ of finite index. In fact, let $x_1, \ldots, x_n$ be a set of generators of $\Delta(\Gamma)$. We have (cf. [111, Lemma 1.5, page 116])

$$C_\Gamma(\Delta(\Gamma)) = \bigcap_{i=1}^{n} C_\Gamma(x_i),$$

and $|\Gamma : C_\Gamma(\Delta(\Gamma))| < \infty$. Since $Z(\Delta(\Gamma)) = \Delta(\Gamma) \cap C_\Gamma(\Delta(\Gamma))$, we see that our claim is proved. Finally, we observe that $[\Delta(\Gamma), \Delta(\Gamma)]$, the commutator subgroup of $\Delta(\Gamma)$ is finite. In fact, let $y_1, \ldots, y_k$ be coset representatives for $Z(\Delta(\Gamma))$ in $\Delta(\Gamma)$ and set $c_{ij} = y_i y_j y_i^{-1} y_j^{-1} = [y_i, y_j]$. From the above claim, it is easy to see that these are, in fact, all elements of the commutator of $\Delta(\Gamma)$. Let us go back to the proof of remark. Let $\tilde{A}$ be a maximal abelian subgroup of Γ which contains A. Clearly $\tilde{A} \subset \Delta(\Gamma)$. From the above and from our assumption, since $\Delta(\Gamma)$ is an abelian group, we have $\tilde{A} = \Delta(\Gamma) = \mathbb{Z}^n$ for some n. This finishes the proof. □

Definition 2.22. The group G from the above theorem will be called the holonomy group of the given crystallographic group Γ.

Let us start to prove the third Bieberbach Theorem. The central role is played here by the Theorem of Jordan-Zassenhaus (see [31]), which says that for any finite group G, there exists a finite number of non isomorphic G-modules $\mathbb{Z}^n$. We want to mention, (see [31, page 559]) that in some sense it is a generalization of the theorem on the finiteness of the number of ideal classes in an algebraic number field. [7] We send the reader to the paper of W. Gaschütz [51] for more information about the Jordan-Zassenhaus theorem. [8] We start with observation about the properties of finite groups of integer matrices. It can be considered as a weak form of the Jordan-Zassenhaus Theorem.

Proposition 2.23. *For any n, the number of isomorphism classes of finite subgroups of $GL(n, \mathbb{Z})$ is finite.*

Proof ([43]). Let p be an odd prime number. We claim that the kernel of the natural homomorphism

$$\phi \colon GL(n, \mathbb{Z}) \to GL(n, \mathbb{Z}/p\mathbb{Z})$$

is torsion free. Let the matrix $A \neq I$ belong to $\ker \phi$. We have $A = I + pB$, where $B \in M(n, \mathbb{Z})$. [9] Assume, that there exists a prime number q such that

$$A^q = (I + pB)^q = I + pqB + \binom{q}{2} p^2 B^2 + \cdots + p^q B^q = I.$$

After reduction, we get

$$qB + \binom{q}{2} pB^2 + \binom{q}{3} p^2 B^3 + \cdots + p^{q-1} B^q = 0.$$

Let α be a maximal number such that p^α divides all elements of B. Since any element of pB^2 is divisible by $p^{2\alpha+1}$, it follows that elements of qB are divisible by $p^{2\alpha+1}$. Hence, if $p \neq q$, then from the maximality of α we get $2\alpha + 1 \leq \alpha$, what is absurd. Moreover, if $p = q$, then $2\alpha \leq \alpha$ and $\alpha = 0$. Finally, we have

$$B + \binom{p}{2} B^2 + \binom{p}{3} pB^3 + \cdots + p^{p-2} B^p = 0.$$

Since the prime number p is odd, we conclude that $p \mid \binom{p}{2}$ and p divides other elements of the above sum. Hence p divides elements of B and $\alpha \geq 1$.

---

[7]In [31, page 559] the authors wrote "...the Jordan-Zassenhaus theorem is a sweeping generalization of the theorem on the finiteness of the number of ideal class in an algebraic number field".

[8]We thank G. Nebe for this remark.

[9]$M(n, \mathbb{Z})$ is the set of square matrices of dimension n with integer coefficients.

This proves our claim. Summing up, we can say that any finite subgroup of $GL(n, \mathbb{Z})$ has trivial intersection with the kernel of ϕ and is isomorphic to some finite subgroup of $GL(n, \mathbb{Z}/p\mathbb{Z})$. □

Unfortunately, the above Proposition is not enough for the proof of the third Bieberbach Theorem. We have to prove the following.

Theorem 2.24 (Jordan-Zassenhaus). *For any n, the number of conjugacy classes of finite subgroups of $GL(n, \mathbb{Z})$ is finite.*

The above Theorem 2.24 is equivalent to

Theorem 2.25 ([31, Theorem 79.1, p. 559]). *Let G be a finite group and N a finite dimensional $\mathbb{Q}G$ module, then there exists only a finite number of isomorphism classes of $\mathbb{Z}G$ submodules of N.*

We shall only present main steps of a proof.

1. N is an irreducible $\mathbb{Q}G$ module;

   In this case there is very original a short proof proposed by W. Gaschütz from 2005 [51]. There is used the theorem of Minkowski [105] and I. Schur lemma.

2. N is a finite dimensional. There is used the above first point and binding function [31, p. 560] or theory of $\mathrm{Ext}_{\mathbb{Z}G}(A, B)$ groups, where A, B are $\mathbb{Z}G$ modules, see [51, p. 154].

To finish the proof of the third Bieberbach Theorem let us observe, that for any given G-module $\mathbb{Z}^n$ the number of short exact sequences

$$0 \to \mathbb{Z}^n \to \Gamma \to G \to 0$$

is bounded by (see [92, Proposition 5.3, p. 117]) the number of elements of the finite group $H^2(G, \mathbb{Z}^n)$, see section 2.3.1. This finishes the proof of the third Bieberbach Theorem. □

2.4 Exercises

Exercise 2.1. Show, that if a continuous map $f \colon X \to Y$ is closed then the topological space X is compact if and only if Y is compact.

Exercise 2.2. Prove that if Γ is a torsion free crystallographic group of dimension n, then projection onto the orbit space

$$\pi \colon \mathbb{R}^n \to \mathbb{R}^n/\Gamma$$

is a covering map and the space $\mathbb{R}^n/\Gamma$ is a manifold.

Exercise 2.3. Let Γ be a crystallographic group of dimension n. Prove that $p_1(\Gamma) \subset O(n)$ is a discrete subgroup of the orthogonal group.
Hint: Apply the fact that Γ is a closed subset of $E(n)$, see Lemma 1.16.

Exercise 2.4. 1. Prove that if $G \subset O(n)$ is a subgroup, then its closure $\overline{G}$ also is a subgroup of $O(n)$.

2. Describe discrete subgroups of $\mathbb{R}^n$. Which ones are cocompact?

   Hint: Compare Exercise 2.5 (5).

Exercise 2.5. 1. Prove that a maximal, connected and abelian subgroup of $O(n)$ is a toral subgroup, i.e. $(S^1)^k$, for some $k \in \mathbb{N}$.

2. Prove, that if G is a closed subgroup of $O(n)$, then G_0 is a normal subgroup of finite index in G (compare [1, Theorem 2.26-7]).

3. Let $\Gamma \subset E(n)$ be a discrete subgroup of $E(n)$ and W be a linear space generated by translations subgroup $\Gamma \cap (I \times \mathbb{R}^n) \subset \Gamma$. Prove, that $\Gamma \cap (I \times \mathbb{R}^n)$ is a cocompact subgroup of W.

4. Prove, that for any $(A, a) \in E(n)$ there is $b \in \mathbb{R}^n$, such that $t_b(A, a)t_{-b} = (A, c)$ and $A(c) = c$.

5. Prove, that a discrete and cocompact subgroup of $\mathbb{R}^n$ is isomorphic to $\mathbb{Z}^n$.

Exercise 2.6. 1. Let $G_1 \subset G \subset E(n)$ be a sequence of subgroups. Prove, that if G is cocompact and $G_1 \trianglelefteq G$ has finite index then G_1 is cocompact.

2. Let Γ be a subgroup of $E(n)$. Prove, that the function $F \colon \mathbb{R}^n/\Gamma \to \mathbb{R}$, defined by the formula $F([x]) = inf_{\gamma \in \Gamma}\{\|\gamma x\|\}$, is well defined and continuous.

Exercise 2.7. Prove that in the Zassenhaus Theorem (Theorem 2.20) the maximality of $\mathbb{Z}^n$ in Γ is equivalent to the faithfulness of the holonomy representation h_Γ.

Exercise 2.8. Let $H_1, \ldots, H_n$ be subgroups of Γ of finite index. Prove that $H = H_1 \cap H_2 \cap \cdots \cap H_n$ has finite index in Γ.

Exercise 2.9 (Fact due to Schur, [111, Lemma 1.4, p. 115]). Let Γ be a group with a central subgroup $Z(\Gamma)$ of finite index. Prove that the commutator subgroup $[\Gamma, \Gamma]$ of Γ is finite.

Exercise 2.10. Prove that the map $\overline{h}\colon \Gamma \to A(n)$, defined in the proof of the Second Bieberbach Theorem is a homomorphism and $\ker \overline{h} = \Gamma \cap \mathbb{R}^n$.
Hint: See [152, p. 123].

Exercise 2.11. Prove that any finite abelian subgroup of $O(n)$ is diagonalizable over the complex numbers.

Exercise 2.12. Complete the proof of the Jordan-Zassenhaus Theorem.

Exercise 2.13. Let $A\colon \mathbb{R}^n \to \mathbb{R}^n$ be a selfadjoint operator which positive in the sense that $\langle x, A(x)\rangle \geq 0$ for all $x \in \mathbb{R}^n$. Prove that there exists a unique positive selfadjoint operator $B\colon \mathbb{R}^n \to \mathbb{R}^n$ with $B^2 = A$.
Hint: Compare with [73, Square root lemma 2.3, p. 5].

Exercise 2.14. Let F_1 and F_2 denote two Euclidean scalar products on $\mathbb{R}^n$. Let G be a finite subgroup of $O(F_1, n)$ and $O(F_2, n)$ and assume that the underlying real representations of G are isomorphic. Prove that there exists a G-equivariant isometry $A\colon (\mathbb{R}^n, F_1) \to (\mathbb{R}^n, F_2)$.
Hint: Let $g\colon R^n \to R^n$ be a G-equivariant isomorphism. Introduce an self adjoint linear operator $A\colon \mathbb{R}^n \to \mathbb{R}^n$ by the formula $\langle x, A(y)\rangle = \langle g^{-1}(x), g^{-1}(y)\rangle; x, y \in \mathbb{R}^n$. Then show that $B \circ g$ is an G-equivariant isometry, where B is such that $B^2 = A$, see Exercise 2.13.

Exercise 2.15. Prove that any finite subgroup of $GL(n, \mathbb{R})$ is conjugate to a finite subgroup of $O(n)$.
Hint: Use averaging over a finite group. It means consider a new Euclidean scalar product

$$\langle x, y\rangle_G = \frac{1}{|G|} \sum_{g \in G} \langle gx, gy\rangle,$$

and apply Exercises 2.13 and 2.14.

3. Classification Methods

Since for each dimension there is only a finite number of isomorphism classes of crystallographic groups, we start this part with a review of the classification methods. We have to say that relations between the group extension and the second cohomology group will be very useful. We specially describe Auslander-Vasquez method, which seems to be completely forgotten. We finish the chapter with an elementary classification in dimensions 2 and 3. In dimension 3 we present a complete list of ten Bieberbach groups. First, we sketch an elementary, geometric algorithm which gives all ten groups. Next, we describe, step by step each of them as a subgroup of $A(3)$.

For any crystallographic group Γ we have the following short exact sequence (see Theorem 2.2 and (2.9))

$$0 \to \mathbb{Z}^n \to \Gamma \xrightarrow{p} G \to 0, \tag{3.1}$$

where $\mathbb{Z}^n$ is the maximal abelian subgroup and G is finite.

Now, we propose a simple criterion for recognizing when a crystallographic group is torsion free.

Theorem 3.1. *Let Γ be an n-dimensional crystallographic group, which corresponds to a cocycle $\alpha \in H^2(G, \mathbb{Z}^n)$. Then Γ is torsion-free if and only if, for any cyclic group of prime order p, $\mathbb{Z}_p \subset G$, the image of the restriction homomorphism $res_{\mathbb{Z}_p}^G(\alpha) \in H^2(\mathbb{Z}_p, \mathbb{Z}^n)$ is not zero.*

Proof. Let us define a restriction homomorphism $res_{\mathbb{Z}_p}^G \colon H^1(G, \mathbb{R}^n/\mathbb{Z}^n) \to H^1(\mathbb{Z}_p, \mathbb{R}^n/\mathbb{Z}^n)$. If $[f] \in H^1(G, \mathbb{R}^n/\mathbb{Z}^n)$, then $res_{\mathbb{Z}_p}^G([f]) = [f_|]$, where $f_|$ is the restriction of f to a subgroup $\mathbb{Z}_p \subset G$. We use an isomorphism $H^1(G, \mathbb{R}^n/\mathbb{Z}^n) \simeq H^2(G, \mathbb{Z}^n)$. Let Γ be torsion free. Assume there is an element $g \in G$ of prime order p, such that $res_{\langle g \rangle}^G(\alpha) = 0$. It follows that the short exact sequence of groups

$$0 \to \mathbb{Z}^n \to p^{-1}(\langle g \rangle) \to \langle g \rangle \simeq \mathbb{Z}_p \to 0$$

splits. By definition, there exists a homomorphism

$$h \colon \langle g \rangle \simeq \mathbb{Z}_p \to p^{-1}(\langle g \rangle),$$

such that $ph = id_{(\langle g \rangle \simeq \mathbb{Z}_p)}$, where $\langle g \rangle \subset G$ is the cyclic subgroup of order p. Hence, Γ has a torsion, what is impossible. The proof of the reverse implication is similar. □

3.1   Three methods of classification

Let n be a natural number. From the third Bieberbach Theorem we know that there is only a finite number of isomorphism classes of n-dimensional crystallographic groups. Hence the classification problem makes sense. We want to present three methods of the classification.

The first way of classification is called *the Zassenhaus algorithm* [148], [151]. It has the following steps.

1. Describe all finite subgroups of $GL(n, \mathbb{Z})$.

2. Describe all faithful G-modules $\mathbb{Z}^n$. [1]

3. Calculate $H^2(G, \mathbb{Z}^n) = H^1(G, \mathbb{R}^n/\mathbb{Z}^n)$ for all finite groups G from 1. and G-modules $\mathbb{Z}^n$ from 2.

4. Recognize which crystallographic groups from 3 are isomorphic.

The second way of the classification is called *the induction method of Calabi* [148] and is used only in the torsion free case. It has the following steps.

1. Classify all torsion free crystallographic groups of dimension less than n.

2. Describe all torsion free crystallographic groups Γ of dimension n, whose abelianization $\Gamma/[\Gamma, \Gamma]$ is finite, (see next chapter).

3. Describe all torsion free crystallographic groups Γ of dimension n, defined by the short exact sequence of groups

$$0 \to \Gamma_{n-1} \to \Gamma \to \mathbb{Z} \to 0,$$

   where Γ_{n-1} is any torsion free crystallographic group of dimension $(n-1)$.

The third way of the classification is called *the Auslander-Vasquez method* [132] and is used only for torsion free groups with given holonomy group G also. See Diagram 3.2.

---

[1] A G-module is faithful, if the corresponding representation of group G, $G \to GL(n, \mathbb{Z})$ is faithful (is one to one).

1. Calculate the Vasquez invariant $n(G)$ and describe the Bieberbach group(s) Γ_G of dimension $n(G)$.

2. Describe all torsion free crystallographic groups Γ of dimension $n \geq n(G)$, defined by the short exact sequence of groups, (see middle vertical row of Diagram 3.2)

$$0 \to \mathbb{Z}^{(n-n(G))} \to \Gamma \to \Gamma_G. \tag{3.2}$$

3. Classify all torsion free crystallographic groups $\leq n(G)$ with holonomy group G.

The first method is the most widely known and effective. The classification of all crystallographic groups up to dimension 5 has been done by using the *Zassenhaus algorithm*, (see [24], [151]). There is also given a list (see [24]) of all torsion free crystallographic groups of dimension 6.

However, let us mention that the order of the cohomology group $H^2(G, \mathbb{Z}^6)$ can be equal to $2^{30} = 1073741824$, for some $G \subset GL(6, \mathbb{Z})$, (see [110, p. 16]). Hence, already in dimension six the classification is not possible without computers. In fact, the first method provides an algorithm. As we know, the step 1 of this method has been performed up to dimension smaller than 30. Moreover, there is an algorithm for calculation of the second cohomology group. In the computer packet GAP [49] there is a subpacket CARAT (see [109]), based on *the Zassenhaus algorithm*, which gives a list of low dimensional crystallographic groups with given holonomy. For more information we refer to the paper by W. Plesken [113], (see also [16] and [24]).

3.1.1 The methods of Calabi and Auslander-Vasquez

The second method of the classification follows from the next proposition. By means of it E. Calabi classifies all four dimensional flat manifolds (see [22] and [65]).

Proposition 3.2 (Calabi). *Let Γ be a torsion free crystallographic group of dimension n with an epimorphism $f : \Gamma \to \mathbb{Z}$. Then $\ker f = \Gamma'$ is a torsion free crystallographic group of dimension $n - 1$.*

Proof. An exactness of the middle vertical sequence of Diagram 3.1 follows from the definition of Γ. Since $\Gamma' \cap \mathbb{Z}^n \subset \mathbb{Z}^n$, it follows that $\Gamma' \cap \mathbb{Z}^n$ is a free abelian group and index $|\Gamma' : \Gamma' \cap \mathbb{Z}^n|$ is finite. Hence, and from Remark 2.21

Γ' is a torsion free crystallographic group. We claim, that the dimension of Γ' is equal to $n-1$. In fact, assume that the dimension of Γ' is smaller than $(n-1)$. From Diagram 3.1 it follows that the rank of $(\mathbb{Z}^n/(\Gamma' \cap \mathbb{Z}^n)$ is greater than 2. But this is impossible, because the homomorphism g is one to one. Analogously, if $\dim \Gamma' = n$, then the group $\mathbb{Z}^n/(\Gamma' \cap \mathbb{Z}^n)$ should be finite or trivial. In the first case we have a contradiction because g is an injection. In the second case we have a contradiction because f is a surjection and $\mathbb{Z}^n$ is a maximal subgroup of Γ. □

$$
\begin{array}{ccccccccc}
 & & 0 & & 0 & & 0 & & \\
 & & \downarrow & & \downarrow & & \downarrow & & \\
0 & \longrightarrow & \Gamma' \cap \mathbb{Z}^n & \xrightarrow{i_|} & \mathbb{Z}^n & \xrightarrow{f_|} & \mathbb{Z}^n/(\Gamma' \cap \mathbb{Z}^n) & \longrightarrow & 0 \\
 & & \downarrow & & \downarrow & & \downarrow g & & \\
0 & \longrightarrow & \Gamma' & \xrightarrow{i} & \Gamma & \xrightarrow{f} & \mathbb{Z} & \longrightarrow & 0 \\
 & & \downarrow & & \downarrow p & & \downarrow & & \\
0 & \longrightarrow & G' & \longrightarrow & G & \longrightarrow & G/G' & \longrightarrow & 0 \\
 & & \downarrow & & \downarrow & & \downarrow & & \\
 & & 0 & & 0 & & 0 & &
\end{array}
$$

Diagram 3.1

An extension

$$0 \to \Gamma' \to \Gamma \xrightarrow{f} \mathbb{Z} \to 0$$

always is a semi-direct product and it is completely determined by the way in which $\mathbb{Z}$ acts (by inner automorphisms) on Γ', i.e. by a homomorphism $\Phi\colon \mathbb{Z} \to \mathrm{Aut}(\Gamma')$. This map in turn is determined by a single element $\Phi(1) = \alpha \in \mathrm{Aut}(\Gamma')$. For $\alpha \in \mathrm{Aut}(\Gamma')$ let us denote by E_α the semi-direct product $\Gamma' \rtimes_\alpha \mathbb{Z}$ where the generator $1 \in \mathbb{Z}$ acts on Γ' via α.

Let H be any group and let $\Pi\colon \mathrm{Aut}(H) \to \mathrm{Out}(H) = \mathrm{Aut}(H)/\mathrm{Inn}(H)$ be the natural projection. We have:

Theorem 3.3 ([96]). *Let Γ be a torsion free crystallographic group and let $\alpha \in \mathrm{Aut}(\Gamma)$. The group $E_\alpha = \Gamma \rtimes_\alpha \mathbb{Z}$ is a crystallographic group if and only if $\Pi(\alpha) \in \mathrm{Out}(\Gamma)$ is of finite order.*

Proof. Our proof starts with the obvious observation that E_α is torsion free. Suppose that $\Pi(\alpha)$ is of finite order. We show that the group E_α has an abelian subgroup of finite index. Let $k \in \mathbb{N}$ be such that $\alpha^k \in \mathrm{Inn}(\Gamma)$.

Consider the diagram:

$$
\begin{array}{ccccccccc}
0 & \longrightarrow & \Gamma & \longrightarrow & \nu^{-1}(k\mathbb{Z}) & \xrightarrow{\ \nu\ } & \mathbb{Z} & \longrightarrow & 0 \\
& & \Big\downarrow{\scriptstyle =} & & \Big\downarrow & & \Big\downarrow{\scriptstyle k\cdot} & & \\
0 & \longrightarrow & \Gamma & \longrightarrow & E_\alpha & \xrightarrow{\ \nu\ } & \mathbb{Z} & \longrightarrow & 0.
\end{array}
$$

Since $\alpha^k \in \mathrm{Inn}(\Gamma)$, $\nu^{-1}(k\mathbb{Z}) = \Gamma \rtimes_{\alpha^k} \mathbb{Z} = \Gamma \times \mathbb{Z}$. But Γ has an abelian subgroup A of finite index. Hence $A \times \mathbb{Z} \subset \Gamma \times \mathbb{Z} = \nu^{-1}(k\mathbb{Z}) \subset E_\alpha$ is an abelian subgroup of finite index in E_α, and we can apply the Remark 2.21.

Conversely, suppose that E_α has an abelian subgroup A of finite index. We have the diagram:

$$
\begin{array}{ccccccccc}
0 & \longrightarrow & \Gamma \cap A & \longrightarrow & A & \xrightarrow{\ \nu\ } & \nu(A) & \longrightarrow & 0 \\
& & \Big\downarrow & & \Big\downarrow & & \Big\downarrow & & \\
0 & \longrightarrow & \Gamma & \longrightarrow & E_\alpha & \xrightarrow{\ \nu\ } & \mathbb{Z} & \longrightarrow & 0.
\end{array}
$$

Now $|E_\alpha : A| < \infty$ implies $|\mathbb{Z} : \nu(A)| = k < \infty$, i.e. $\nu(A) \simeq \mathbb{Z}$. Moreover, $|\Delta(\Gamma) : \Gamma \cap A| < \infty$, where $\Delta(\Gamma)$ is an FC-centre of Γ, see (2.10) and a proof of a Proposition 3.2. Set $\beta = \alpha^k \in \mathrm{Aut}(\Gamma)$. Then $A = (\Gamma \cap A) \rtimes_\beta \mathbb{Z}$ and since A is abelian we have $\beta\mid_{\Gamma \cap A} = id_{\Gamma \cap A}$.

We show that $\beta\mid_{\Delta(\Gamma)} = id_{\Delta(\Gamma)}$. To see this, take any $x \in \Delta(\Gamma)$ and write $\beta(x) = y$. For some $m \in \mathbb{N}$ it holds $mx, my \in \Gamma \cap A$ as $|\Delta(\Gamma) : \Gamma \cap A| < \infty$. Thus $mx = \beta(mx) = m\beta(x) = my$ with $mx, my \in \Gamma \cap A$. Hence $mx = my$ and $x = y$ as $\Delta(\Gamma)$ is torsion free abelian.

Now, we have $\beta \in \mathrm{Aut}(\Gamma)$ with $\beta\mid_{\Delta(\Gamma)} = id_{\Delta(\Gamma)}$. It remains to show that some power of β is an inner automorphism. By taking a suitable power of β we can assume that β induces an identity on $G = \Gamma/\Delta(\Gamma)$. Hence

$$
\begin{array}{ccccccccc}
0 & \longrightarrow & \Delta(\Gamma) & \longrightarrow & \Gamma & \longrightarrow & G & \longrightarrow & 0 \\
& & \Big\downarrow{\scriptstyle =} & & \Big\downarrow{\scriptstyle \beta} & & \Big\downarrow{\scriptstyle =} & & \\
0 & \longrightarrow & \Delta(\Gamma) & \longrightarrow & \Gamma & \longrightarrow & G & \longrightarrow & 0.
\end{array}
$$

By definition (see a proof of a Proposition 3.2) $\Delta(\Gamma)$ is the maximal abelian subgroup of Γ and we can apply Theorem 5.4 from the Chapter 5. In fact, $\beta \in \mathrm{Aut}^0(\Gamma)$ and since the group $H^1(G, \Delta(\Gamma))$ is finite, it follows that some power of β is an inner automorphism, as desired. $\square$

We send the reader to the next chapter for more information about the Calabi method.

To say a little more about the third method we have to introduce some new definitions and objects. Our main reference are monographs [92] and [18].

Definition 3.4. Let P, M, N be R-modules, where R is a ring. In most cases it will be a group ring. A module P is projective, if for any epimorphism $\pi \colon M \to N$ and homomorphism $\phi \colon P \to N$, there exists $\psi \colon P \to M$, such that $\phi = \pi\psi$.

Example 3.5. Free modules are projective.

Let us give a general definition of (co)homology groups of a finite group G. By a projective resolution F of the trivial G-module $\mathbb{Z}$, we understand any exact sequence of projective $\mathbb{Z}[G]$-modules F_i,

$$
\ldots \xrightarrow{h_i} F_i \xrightarrow{h_{i-1}} F_{i-1} \xrightarrow{h_{i-2}} \ldots \xrightarrow{h_1} F_1 \xrightarrow{h_0} F_0 \xrightarrow{\epsilon} \mathbb{Z} \to 0,
$$

where $i \in \mathbb{N}$, see [18, pages 10-35]. An exact sequence of G-modules means, that for any $i \in \mathbb{N}$, $\ker h_{i-1} = \mathrm{Im}\, h_i$ and $\ker \epsilon = \mathrm{Im}\, h_0$. Moreover, a homomorphism ϵ is an epimorphism.

Example 3.6. For any $n \geq 0$ let $C_n(G)$ denote a free abelian group with a set of free generators equal to G^{n+1}. Let $(g_0, g_1, \ldots, g_n) \in G^{n+1}$ be a generator. Put $g(g_0, g_1, \ldots, g_n) = (gg_0, gg_1, \ldots, gg_n)$. This defines a G module structure on $C_n(G)$. For $n \geq 1$, we define G-homomorphisms $d_n \colon C_n(G) \to C_{n-1}(G)$, by a formula

$$d_n(g_0, g_1, ..., g_n) = \sum_{i=0}^{n}(-1)^i(g_0, g_1, ..., \hat{g}_i, ..., g_n),$$

where $(g_0, g_1, ..., \hat{g}_i, ..., g_n) = (g_0, g_1, ...g_{i-1}, g_{i+1}, ..., g_n)$. Moreover, let $\epsilon \colon C_0 \to \mathbb{Z}$ be a epimorphism which maps all generators to 1. It is not difficult to prove that $C_* \to \mathbb{Z}$ is a free resolution of the trivial G-module $\mathbb{Z}$.

Definition 3.7.
$$H_i(G, M) = H_i(F \otimes_G M),$$
$$H^i(G, M) = H^i(\mathrm{Hom}_G(F, M)).$$

Remark 3.8. It can be proved that the above definition does not depend on the resolution, [18]. If we use the resolution from example 3.6, then we get the definition of cohomology groups from the previous chapter, see (2.3.1).

Example 3.9 ([18, page 58]). Let $C = \langle \sigma \rangle$ be a cyclic group of order n, and let M be any C-module. Then

$$H^2(C, M) = M^C / IM,$$

where $IM = \{(1 + \sigma + \sigma^2 + \cdots + \sigma^{n-1})m \mid m \in M\}$, and $M^C = \{m \in M \mid \sigma m = m\}$.

Let us formulate the Shapiro Lemma. First, we have to introduce the definition of an induced module.

Definition 3.10. ([18]) Let H be a subgroup of index n of a group G. Let M be any H-module. Then

$$\mathrm{Ind}_H^G M = \oplus_{g \in G/H} gM,$$

where for $\bar{g} \in G/H$ and $\bar{g}_i \in G/H, i = 1, 2, 3, \ldots, n$ we have

$$\bar{g}(\bar{g}_1 m_1, \bar{g}_2 m_2, \ldots, \bar{g}_n m_n) = (g\bar{g}_1 m_1, g\bar{g}_2 m_2, \ldots, g\bar{g}_n m_n) \in \oplus_{g \in G/H} gM.$$

Moreover, the action on factors is as follows: Let $gg_i = g_j h$, then $g(g_i m_i) = g_j h m_i$. Here a factor defines an element $g_i, i = 1, 2, \ldots, n$.

Remark 3.11. If G is finite, then the G-module $\mathrm{Ind}_H^G M$ is also denoted by $\mathrm{Coind}_H^G M$.

Theorem 3.12 (Shapiro's Lemma [18]). *Let M, G, H be as above. Then*

$$H_*(H, M) \simeq H_*(G, \mathrm{Ind}_H^G M)$$

and

$$H^*(H, M) \simeq H^*(G, \mathrm{Coind}_H^G M).$$

Proof. Let F be a projective resolution of the trivial G-module $\mathbb{Z}$. Then F can also be regarded as a projective resolution over H, so

$$H_*(H, M) \simeq H_*(F \otimes_H M).$$

But $F \otimes_H M \simeq F \otimes_G (\mathbb{Z}G \otimes_H M) \simeq F \otimes_G (\mathrm{Ind}_H^G M)$, whence the first isomorphism. The second isomorphism follows from the universal property of co-induction, see [18, Proposition 6.2]. $\qquad\square$

Let G be a finite group, and let $\mathcal{X}$ be the set of conjugacy classes of the cyclic subgroup of prime order p.

Definition 3.13. Let $n \geq 1$ be a natural number. A G-lattice is any G-module isomorphic to a free abelian group $\mathbb{Z}^n$.

Example 3.14. A G-module

$$S = \oplus_{C \in \mathcal{X}} \mathrm{ind}_C^G \mathbb{Z} \tag{3.3}$$

is a G-lattice.

The next definition is a consequence of Theorem 3.1.

Definition 3.15. An element $\alpha \in H^2(G, \mathbb{Z}^n)$ is *special*, if it defines a torsion free crystallographic group.

Let $h \colon M \to N$ be any G-homomorphism ($\mathbb{Z}G$-homomorphism) of G-modules M, N. Let $h_* \colon H^2(G, M) \to H^2(G, N)$ be the induced homomorphism on cohomology groups.

Definition 3.16 ([26]). Let M be a G-lattice. Say that a G-lattice L has the property $\mathcal{S}$ when for each *special* element $\alpha \in H^2(G, M)$ there exists a G-homomorphism $h \colon M \to L$, such that $h_*(\alpha) \in H^2(G, L)$ is *special*.

Proposition 3.17 ([26]). *The G-lattice $\oplus_{C\in\mathcal{X}} \operatorname{ind}_C^G \mathbb{Z}$ has the property $\mathcal{S}$.*

Proof. Let $C \in \mathcal{X}$ and $\alpha \in H^2(G, M)$ be *special* elements. First, we prove the existence of a C-homomorphism $\phi_C \colon M \to \mathbb{Z}$ such that $(\phi_C)_*(res_C^G(\alpha)) \neq 0$. Let $h \colon M \to M$ be a homomorphism such that

$$h(m) = \sum_{g\in C} gm.$$

We define a short exact sequence of G-modules

$$0 \to \ker h \to M \to h(M) \to 0. \tag{3.4}$$

It is easy to prove that $(h(M))^C = h(M)$ and $(\ker h)^C = 0$. Hence the map $h_* \colon H^2(C, M) \to H^2(C, h(M))$ is an injection. In fact, it follows from the long exact sequence of cohomologies (see Proposition (2.19) for a short exact sequence of modules (3.4)). Hence, we can assume that a group C acts trivially on M. Next, choose a basis in the lattice M, such that, there exists a C-isomorphism $M = \oplus_{i\in I}\mathbb{Z}$. Hence, we have an isomorphism $H^2(C, M) = \oplus_{i\in I}H^2(C, \mathbb{Z})$, where I is a finite set. Then, we can define a C-projection $\phi_C \colon M \to \mathbb{Z}$, such that $(\phi_C)_*(res_C^G(\alpha)) \in H^2(C, \mathbb{Z})$ has order p. From the Frobenius reciprocity theorem (see [126, Theorem 13, p. 73]) we have an isomorphism

$$\pi_* \colon Hom_{\mathbb{Z}G}(M, \operatorname{ind}_C^G \mathbb{Z}) \to Hom_{\mathbb{Z}C}(res_C^G M, \mathbb{Z}).$$

Let $\psi_C \in Hom_{\mathbb{Z}G}(M, \operatorname{ind}_C^G \mathbb{Z})$ be a map such that $\pi_*(\psi_C) = \phi_C$. Then

$$\pi_*(\psi_C)_*(res_C^G\alpha) = (\phi_C)_*(res_C^G\alpha) \neq 0$$

so $(\psi_C)_*(res_C^G\alpha) \neq 0$. We define

$$h = \oplus_{C\in\mathcal{X}}\psi_C \colon M \to S.$$

Since $res_C^G h_*(\alpha) = h_*(res_C^G\alpha)$, we have to prove that $h_*(res_C^G\alpha) \neq 0$ in $H^2(C, S)$, for all $C \in \mathcal{X}$. In fact, it is enough to observe that

$$(\psi_C)_*(res_C^G\alpha) \neq 0 \ in \ H^2(C, \operatorname{ind}_C^G \mathbb{Z}). \tag{3.5}$$

$\square$

Let us start to consider the third method of the classification of crystallographic groups. It comes from a topological construction. Denote by $\mathcal{M}$ a flat manifold with the fundamental group $\pi_1(\mathcal{M}) = \Gamma$. Let $T^k = \mathbb{R}^k/\mathbb{Z}^k$ be a flat torus on which a group Γ acts by isometries. Then Γ also acts by isometries on the space $\tilde{\mathcal{M}} \times T^k$, where $\tilde{\mathcal{M}}$ is the universal covering of $\mathcal{M}$. It is easy to show that the space $(\tilde{\mathcal{M}} \times T^k)/\Gamma$ is a flat manifold; we shall call it *a flat toral extension* of the manifold $\mathcal{M}$. We shall make the convention, that a point is the 0-dimensional torus, and hence any flat manifold can be a *flat toral extension* of itself. The following was proved by A. T. Vasquez in the year 1970. The idea was given by L. Auslander.

Theorem 3.18 ([145, Theorem 4.1]). *For any finite group G there exists a natural number $n(G)$ with the following property: if $\mathcal{M}$ is any flat manifold with holonomy group G, then $\mathcal{M}$ is a flat toral extension of some flat manifold of dimension $\leq n(G)$.*

We also have an algebraic version of the above theorem, whose proof follows from the definition and properties of the lattice with property $\mathcal{S}$.

Theorem 3.18A ([145, Theorem 3.6]). *For any finite group G there exists a natural number $n(G)$ with the following property: if Γ is a torsion free crystallographic group with holonomy group isomorphic to G, then the maximal abelian subgroup $A \subset \Gamma$ contains a normal subgroup A' such that Γ/A' is a Bieberbach group of dimension $\leq n(G)$.*

$\square$

Cliff and Weiss [26] gave a value of $n(G)$ for any finite p-group. It happens [132, Theorem 3], that the Vasquez invariant can be defined in a purely algebraic way.

Definition 3.19. The Vasquez invariant of a finite group G is

$$n(G) = \min\{rank_{\mathbb{Z}}(L) \mid L \text{ is } G\text{-lattice with property } \mathcal{S}\}.$$

Let L be a G-lattice with property $\mathcal{S}$ of rank $n(G)$ and let Γ be any torsion free crystallographic group of dimension $n \geq n(G)$ with holonomy group G and maximal abelian subgroup $\mathbb{Z}^n \subset \Gamma$. Then we have Diagram 3.2. The existence of the G-homomorphism $f\colon \mathbb{Z}^n \to L$ follows from the definition of the Vasquez invariant, see Definitions 3.16 and 3.19. A Bieberbach group Γ_G corresponds to a cocycle $f_*(\alpha) \in H^2(G, L)$, where $\alpha \in H^2(G, \mathbb{Z}^n)$ is defined by Γ, see (2.7). Definitions of all other groups and homomorphisms easily follows from properties of Γ, Γ_G and f. Hence, we have the third method of the classification of torsion free crystallographic groups with given holonomy, cf. (3.2). Unfortunately the value of the Vasquez invariant is known only for finite p-groups [26]. Moreover, a finite group G has $n(G) = 1$ if and only if its order is a product of distinct prime numbers [132].

$$
\begin{array}{ccccccccc}
& & 0 & & 0 & & 0 & & \\
& & \downarrow & & \downarrow & & \downarrow & & \\
0 & \longrightarrow & \ker f = \mathbb{Z}^{n-n(G)} & \stackrel{=}{\longrightarrow} & \mathbb{Z}^{n-n(G)} & \longrightarrow & 0 & \longrightarrow & 0 \\
& & \downarrow & & \downarrow & & \downarrow & & \\
0 & \longrightarrow & \mathbb{Z}^n & \longrightarrow & \Gamma & \longrightarrow & G & \longrightarrow & 0 \\
& & \downarrow f & & \downarrow & & \downarrow \| & & \\
0 & \longrightarrow & L & \longrightarrow & \Gamma_G & \longrightarrow & G & \longrightarrow & 0
\end{array}
$$

Diagram 3.2

3.2 Classification in dimension two

As an application of the above, we shall present an elementary classification of two dimensional crystallographic groups. We start from the description of all finite subgroups of $GL(2, \mathbb{Z})$.

Proposition 3.20. *Let $C_n \subset O(2)$ be a cyclic group of rotations of order n and let $D_n \subset O(2)$ be the dihedral group of order $2n$, generated by a reflection and a rotation of order n. If $G \subset GL(2, \mathbb{Z})$ is a finite subgroup, then G is conjugate to $C_2, C_3, C_4, C_6, D_1, D_2, D_3, D_4, D_6$. Moreover each of the groups D_1, D_2, D_3 has two conjugacy classes.*

Proof. [73]: Let G be a two-element subgroup of $GL(2, \mathbb{Z})$ and $M \in G \setminus \{I\}$. Let us consider a basis of $\mathbb{R}^2$, which defines a matrix M. Let $L \subset \mathbb{R}^2$ be the lattice (a discrete subgroup), whose elements are linear combinations of the above basis with integer coefficients. It is clear that $ML = L$. By the crystallographic restriction (see Exercise 3.10 and Lemma 3.22) we get $G = C_2$ or M is a reflection. Let $A = \{v \in \mathbb{R}^2 \mid Mv = v\}$, and $B = \{v \in \mathbb{R}^2 \mid Mv = -v\}$. Observe that if M is not reflection then A and B include non-zero elements of L. Let $0 \neq c \in L \backslash \{A \cup B\}$. Then an element $c + Mc \in A$ and since $c \notin B$, then it is non-zero. By analogy, since $c \notin B$, then element $c - Mc \in B$ is also non-zero. Let $a \in A$ and $b \in B$ be generators of lines A and B. Denote by L_M the lattice generated by a and b. We want to see if there exists another G-invariant lattice. We have $L_M \subset L$. Let $c \in L \backslash L_M$. We have

$$(c + ma + nb) + M(c + ma + nb) = c + Mc + 2ma,$$

$$(c + ma + nb) - M(c + ma + nb) = c - Mc + 2nb,$$

where $m, n \in \mathbb{Z}$. Hence $c + Mc + 2ma = 2la + a$ or $c + Mc + 2ma = 2la$ for some $l \in \mathbb{Z}$. In the first case, for $n' = m - l$, we have

$$a = c + Mc + 2n' = (c + n'a + nb) + M(c + n'a + nb),$$

and change c for $c' = c + n'a + nb$, we get $c' + Mc' = a$. Moreover $c' \equiv c$ (mod L_M). In the second case $c + Mc + 2n'a = 0 = c' + Mc'$. Hence $c' \in B$, which is an absurd. We can apply an analogous consideration to the element $c - Mc + 2nb$. Summing up, we can assume that $c + Mc = a, c - Mc = b$. As a consequence we have $c = 1/2(a + b)$ and $L = gen\{c, Mc\}$. Hence, the matrix M in the basis $\{a, b\}$ is equal to $D_1 = \begin{bmatrix} 1 & 0 \\ 0 & -1 \end{bmatrix}$, when in the basis $\{c, Mc\}$ is equal to $D_{1*} = \begin{bmatrix} 0 & 1 \\ 1 & 0 \end{bmatrix}$.

It is an easy exercise to show that D_1 and D_{1*} are not conjugate in $GL(2, \mathbb{Z})$. The lattice $L = gen\{a, b\}$ is called rectangular.[2] When the lattice $L = gen\{c, Mc\}$ is called rhombic.[3] It helps us to consider the next groups.

Let D_2 be a group generated by the reflections with axes a and b. Let D_{2*} be a group generated by the reflections with axes $c + Mc$ and $c - Mc$.

C_4 is generated by a rotation ρ of order four. From Lagrange's Lemma (Exercise 3.11) the lattice is generated by $a \in L \backslash \{0\}$ and $\rho(a)$. It is called the square lattice.

---

[2] a and b are orthogonal and D_1 is reflection in the line through a.
[3] c and Mc are equal length and D_{1*} is reflection in the line generated by $c + Mc$.

D_4 is the automorphism group of the square lattice. Let a, b be sides of any square of the lattice. D_4 is generated by the rotation of order four, which maps a to b and b to $-a$, and reflection with the axis a.

C_3 is generated by a rotation ρ of order three. From Lagrange's Lemma (Exercise 3.11) it follows that the lattice is generated by $a, \rho(a)$. The vectors $a, \rho(a), \rho^2(a)$ determine an equilateral triangle. When we add the vectors

$$-a, -\rho(a), -\rho^2(a),$$

then we obtain a hexagon. The above lattice is called hexagonal.

D_3 is generated by the above rotation of order three and three reflections with axes on three diagonals of a hexagon.

D_{3^*} is generated by the rotation of order three and three reflections with axes on edges connecting the middles of opposite edges of a hexagon.

We define C_6 as a group of rotations of order six. We can obtain the lattice from Lagrange's Lemma (Exercise 3.11). This is the hexagonal lattice L.

D_6 is equal to $\mathrm{Aut}(L)$, the automorphism group of the hexagonal lattice L. $\qquad \square$

We can start to describe and to classify the crystallographic groups of dimension two. It is well known (see for example [73]), that there are 17 crystallographic groups of dimension two, up to isomorphism. A list $\mathcal{A}$ of all possible holonomy groups (or a list of all finite subgroups of $GL(2, \mathbb{Z})$) is determined above. Then, we shall use the relation between extensions of groups and the second cohomology group $H^2(G, \mathbb{Z}^2)$, for $G \in \mathcal{A}$ and some lattice $\mathbb{Z}^2$, see [92, Theorem 4.1, p. 112], [18, Theorem 3.12, p. 93]. We have

$$H^2(C_i, \mathbb{Z}^2) = (\mathbb{Z}^2)^{C_i}/I\mathbb{Z}^2 = 0,$$

for $i = 2, 3, 4, 6$. Moreover, one can calculate that

$$H^2(D_1, \mathbb{Z}^2) = \mathbb{Z}_2, H^2(D_{1^*}, \mathbb{Z}^2) = 0.$$

It gives us eight crystallographic groups, of which one is the torus $\mathbb{R}^2/\mathbb{Z}^2$ and the second is related to the rotation about π.

We have

$$D_2 = \{A = \begin{bmatrix} 1 & 0 \\ 0 & -1 \end{bmatrix}, B = \begin{bmatrix} -1 & 0 \\ 0 & 1 \end{bmatrix}, AB = \begin{bmatrix} -1 & 0 \\ 0 & -1 \end{bmatrix}, \begin{bmatrix} 1 & 0 \\ 0 & 1 \end{bmatrix}\}.$$

From the theory of group cohomology, $H^2(D_2, \mathbb{Z}^2) = \mathbb{Z}_2 \oplus \mathbb{Z}_2$. Let us give an explicit description of 4 elements of this group:

$$s_1(A) = [0, 0], s_1(B) = [0, 0]; s_2(A) = [1/2, 0], s_2(B) = [0, 1/2];$$

$$s_3(A) = [1/2, 0], s_3(B) = [0, 0]; s_4(A) = [0, 0], s_4(B) = [0, 1/2].$$

It defines four crystallographic groups. It turns out that the two defined by the cocycles s_3 and s_4 are isomorphic. Moreover, we have

$$D_{2^*} = \left\{ A = \begin{bmatrix} 0 & 1 \\ 1 & 0 \end{bmatrix}, B = \begin{bmatrix} 0 & -1 \\ -1 & 0 \end{bmatrix}, AB = \begin{bmatrix} -1 & 0 \\ 0 & -1 \end{bmatrix}, \begin{bmatrix} 1 & 0 \\ 0 & 1 \end{bmatrix} \right\}.$$

Since $H^2(C, \mathbb{Z}^2) = 0$, for any subgroup $C \subset D_{2^*}$ of order two, $H^2(D_{2^*}, \mathbb{Z}^2) = 0$. It gives one crystallographic group.

Let us consider the group D_4, generated by the matrix $A = \begin{bmatrix} 0 & -1 \\ 1 & 0 \end{bmatrix}$ (a rotation of order four) and the matrix $B = \begin{bmatrix} 1 & 0 \\ 0 & -1 \end{bmatrix}$ (reflection with axis a). We have $H^2(D_4, \mathbb{Z}^2) = \mathbb{Z}_2$. It defines two crystallographic groups. The last three groups have a holonomy D_3, D_{3^*}, D_6. We can see, that for each of them, the second cohomology group is trivial. In fact, it is enough to use the property that restriction to the Sylow subgroup induces a monomorphism on cohomology groups (see [18, Theorem 10.3, p. 84]). Hence, the corresponding crystallographic groups are the semi-direct products of the given group with $\mathbb{Z}^2$.

With similar methods we can classify crystallographic groups of dimension three. In the torsion free case it is presented, for example in [29, p. 39-42]. For dimension four, five and six it is possible with support of a computer (see the commentary at the end of section 3.2, page 35).

3.3 Platycosms

We shall follow [29, Appendix I], see also [56] and [107]. By J. Conway and J. P. Rossetti [29], our cosmos is built of three dimensional flat manifolds so called *platycosms*, see [29]. Below we shall present (following J. Conway and J. P. Rossetti) an elementary method of the recognition of three dimensional, flat manifolds. An algorithm is outlined in Figure 1 and the accompanying explanations (i)-(vi). In the whole proof, we are assuming that we know a list of ten three dimensional flat manifolds (see [148]), and we are just proving that there are no more than these ten. There are also given names for any platycosm.

Let $\Gamma \subset E(3)$ be a torsion free crystallographic group.

(i) When there are screw motions[4] in Γ we have:

Lemma 3.21 (Splitting Lemma). *If* v *is the smallest screw vector in a given direction, say vertical, then the translation lattice* $\mathcal{T}$ *decomposes as* $\langle Nv \rangle \oplus \mathcal{T}'$, *where* N *is the period of the corresponding screw motion* σ.

Proof of Lemma 3.21. It suffices to prove that there is no translation τ whose vertical part cv is strictly between 0 and Nv. But then $\sigma\tau^{-1}$ if $0 < c < 1$, or $\tau\sigma^{-m}$ if $m < c < m+1, (m = 1, \ldots, N-1)$ would be a vertical screw motion with a vector shorter than v. □

(ii) If all screw vectors are parallel we have a cyclic point group whose order N is the least common multiple of their periods and which acts on the 2-dimensional lattice $\mathcal{T}'$. Obviously, our group is generated by the shortest screw motion of this period together with the translations of $\mathcal{T}'$. The identification with c2 (dicosm, screw motion of order 2), c3 (tricosm, screw motion of order 3), c4 (tetracosm, screw motion of order 4) or c6 (hexacosm, screw motion of order 6) follows from the well known lemma, (see Proposition (3.20) and Exercise 3.10):

Lemma 3.22 (Barlow's Lemma). *If a rotation of order* N *fixes a 2 dimensional lattice* $\mathcal{T}'$, *then* $N = 1, 2, 3, 4, 6$.

Proof of Lemma 3.22. Apply the rotation to a minimal non-zero vector of the lattice. We obtain a *"star"* of N vectors, the difference of adjacent members of which will be a shorter vector if $N \geq 7$. A rotation of order 5 would combine with reflection to produce one of order 10. □

(iii) Two non parallel screw motions σ_1 and σ_2 would make $\mathcal{T}$ decompose in two ways, say $\langle v_1 \rangle \oplus \mathcal{T}_1 \simeq \langle v_2 \rangle \oplus \mathcal{T}_2$, showing immediately that v_1 and v_2 are orthogonal. If σ_i had a period other than 2, then σ_j and $\sigma_i^{-1}\sigma_j\sigma_i$ would

---

[4]A Bieberbach group of rank 3 has a subgroup generated by a suitable 2-dimensional lattice of translations together with an orthogonal screw motion. We say a screw motion has period N if the lowest power of it that is a translation is N.

not be orthogonal. Finally, if σ_1 and σ_2 are orthogonal and of period 2, then $\sigma_1\sigma_2$ is a screw motion in the direction of v_3 orthogonal to v_1 and v_2. These together with the translations must generate our group, and can be identified with the translation subgroup of c22 (didicosm).

(iv) Any glide reflection maps to a reflection through the origin when one ignores translations.[5] The product of two of them is therefore a translation or a screw motion since the product of two reflection through o is the identity or a non-trivial rotation according to whether their planes are parallel or not.

(v) If all glide reflections are in parallel planes - call them basal - the holonomy group has order 2, so a group is generated by a single glide reflection g, together with its translations $L(v), v \in \mathcal{T}$. We have equation $g^{-1}L(v)g = L(r(v))$, where r is the reflection part of g, which reflects with respect to the basal plane. Let us resolve any translation vector into basal and perpendicular parts $v_1 + v_2$, than we see that $r(v) = v_1 - v_2$. Hence $v_1 - v_2, 2v_2 = (v_1 + v_2) - (v_1 - v_2)$ also are translation vectors. If v_2 itself (and so v_1) is translation vector, then the translation lattice[6] $\mathcal{T}$ decomposes into its basal and perpendicular parts, leading to $+a1$, (first amphicosm). Otherwise, adjoining the translation through v_2 embeds our platycosm in a copy of $+a1$, and easily identifies the original platycosm with $-1a$, (second amphicosm).

(vi) If two glide reflections g, g' have non-parallel planes intersecting in a line which will be called vertical, then their product will be a screw motion s whose screw vector v will be vertical. We choose things so that v is as short as possible. Then we can multiply by a power of s to reduce g until the vertical component of its glide vector λv has $|\lambda| \leq \frac{1}{2}$. But then λ must be 0 since g^2 is a translation and by Splitting Lemma 3.21 its vertical component must be a multiple of Nv, where $N \geq 2$ is the period of s. The glide vector of the new g' (defined by $gg' = s$) will have the same vertical component v as s. But now the translation vector $(g')^2$ has vertical component $2v$, showing (by Splitting Lemma 3.21) that s must have period 2, from which it follows that the planes of g and g' are orthogonal. We can now see that the translation lattice is generated by 3 orthogonal translations, say $(x, y, z) \mapsto (x + a, y, z), (x, y +$

---

[5]By glide reflection we understand an isometry of the euclidean space $\mathbb{R}^3$ obtained by the composition of a reflection in a plane π with a translation by a vector in π.

[6]A subgroup of all translations.

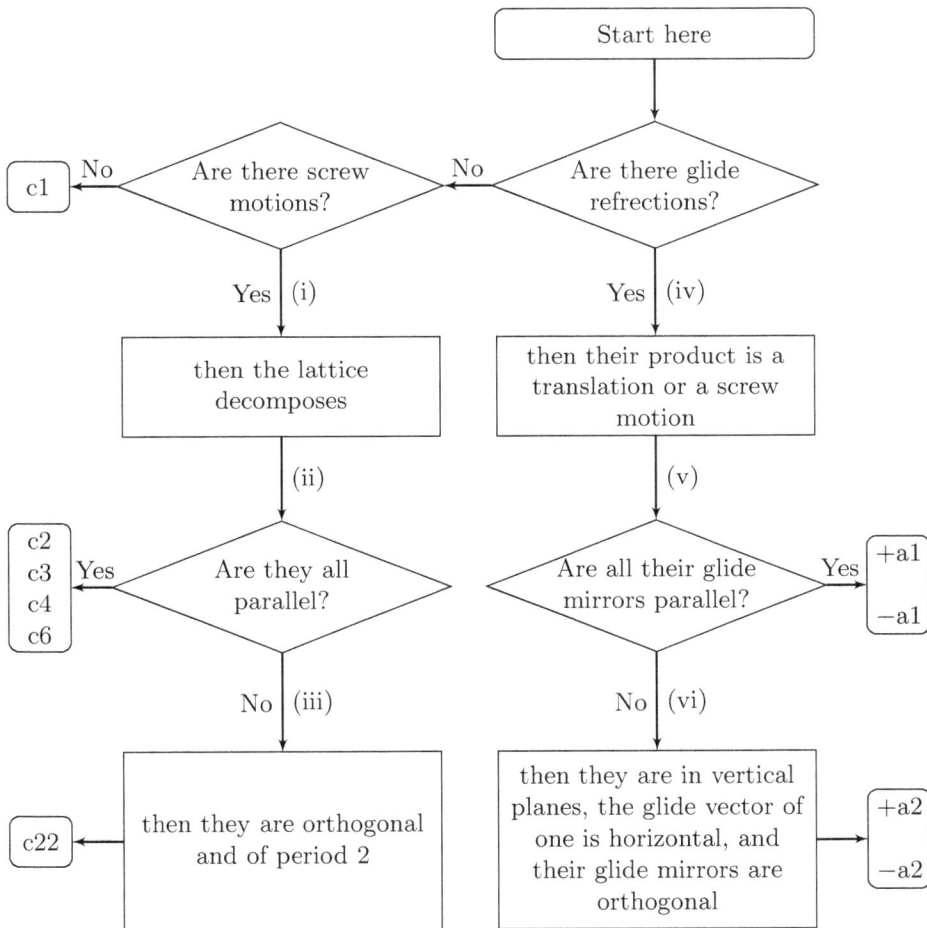

Figure 3.1: Guide to the proof

$b, z), (x, y, z + c)$. We can also see that the holonomy group has order 4, i.e. that g, g', together with translations, must generate our group. For, if not, there would be a glide mirror not parallel to either of those of g, g', and so perpendicular to both, by the previous paragraph. But then the product of the corresponding glide reflection with s would have the form

$$(x, y, z) \mapsto (\text{constant} - x, \text{constant} - y, \text{constant} - z),$$

which fixes a point. In these coordinates

$$s\colon (x,y,z) \to (-x,-y,z+\frac{1}{2}c)$$

$$g\colon (x,y,z) \to (x+\frac{1}{2}a, \lambda b - y, z)$$

where we may take $0 \leq \lambda < 1$ by compounding g with translations. From these we find

$$g' = g^{-1}s\colon (x,y,z) \to (\frac{1}{2}a - x, y - \lambda b, z + \frac{1}{2}c)$$

$$(g')^2\colon (x,y,z) \to (x, y - 2\lambda b, z + c)$$

showing that 2λ must be an integer, so either $\lambda = 0$, which gives $+a2$ (first amphidicosm); or $\lambda = \frac{1}{2}$, which gives $-a2$ (second amphidicosm). In the former case, our screw vector $(0,0,\frac{1}{2}c)$ is also a glide vector (of g') – in the latter case, no screw vector is a glide vector.

Let us give the list of all torsion free crystallographic groups (flat manifolds) of dimension 2 and 3. We have only two flat manifolds of dimension two – the torus and Klein bottle (see Example 2.3).

It is known since 1933 ([56], [107], [142], [148]) that there are ten flat manifolds in dimension three. The above (see Figure 3.1) is presented a geometric classification.

Now, we want to describe all of them, case by case in language which is used at the first part of book. We have observed above that, all but 4 of the 10 Bieberbach groups of dimension three have a cyclic holonomy group (see also [148, p. 125]) three $(c2, +1, -1)$ with the holonomy group $\mathbb{Z}_2$ and three $(c3, c4, c6)$ with holonomy groups $\mathbb{Z}_3, \mathbb{Z}_4, \mathbb{Z}_6$, correspondingly. Moreover, we have the orientable torus $(c1)$ and three flat manifolds with holonomy group $\mathbb{Z}_2 \times \mathbb{Z}_2$ $(c22, +a2, -a2)$. The manifolds $c1, c2, c3, c4, c6, c22$ are orientable, cf. Figure 3.1.

Let $A_1 = \begin{bmatrix} 1 & 0 & 0 \\ 0 & -1 & 0 \\ 0 & 0 & -1 \end{bmatrix}$. Let us define

$$\Gamma_1 = \langle (A_1, \begin{pmatrix} 1/2 \\ 0 \\ 0 \end{pmatrix}), (I, \begin{pmatrix} 0 \\ 1 \\ 0 \end{pmatrix}), (I, \begin{pmatrix} 0 \\ 0 \\ 1 \end{pmatrix}) \rangle \subset E(3).$$

The orientable platycosm $\mathbb{R}^3/\Gamma_1$, with holonomy $\mathbb{Z}_2$, is called the *dicosm* and denoted by $c2$; $\mathcal{G}_2$ in [148].

Let $A_2 = \begin{bmatrix} 1 & 0 & 0 \\ 0 & 1 & 0 \\ 0 & 0 & -1 \end{bmatrix}$. Let us define

$$\Gamma_2 = <(A_2, \begin{pmatrix} 1/2 \\ 0 \\ 0 \end{pmatrix}), (I, \begin{pmatrix} 0 \\ 1 \\ 0 \end{pmatrix}), (I, \begin{pmatrix} 0 \\ 0 \\ 1 \end{pmatrix})> \subset E(3).$$

The non-orientable platycosm $\mathbb{R}^3/\Gamma_2$, with holonomy $\mathbb{Z}_2$, is called the *first amphicosm* and denoted by $+1a$; $\mathcal{B}_1$ in [148].

Let $A_3 = \begin{bmatrix} 1 & 0 & 0 \\ 0 & 0 & 1 \\ 0 & 1 & 0 \end{bmatrix}$. Let us define

$$\Gamma_3 = \langle(A_3, \begin{pmatrix} 1/2 \\ 0 \\ 0 \end{pmatrix}), (I, \begin{pmatrix} 0 \\ 1 \\ 0 \end{pmatrix}), (I, \begin{pmatrix} 0 \\ 0 \\ 1 \end{pmatrix})\rangle \subset E(3).$$

The non-orientable platycosm $\mathbb{R}^3/\Gamma_3$, with holonomy $\mathbb{Z}_2$, is called the *second amphicosm* and denoted $-a$; $\mathcal{B}_2$ in [148].

Let $C_1 = \begin{bmatrix} 1 & 0 & 0 \\ 0 & 1 & 0 \\ 0 & 0 & -1 \end{bmatrix}$, $C_2 = \begin{bmatrix} 1 & 0 & 0 \\ 0 & -1 & 0 \\ 0 & 0 & 1 \end{bmatrix}$. Let us define

$$\Gamma_4 = \langle(C_1, \begin{pmatrix} 1/2 \\ 0 \\ 0 \end{pmatrix}), (C_2, \begin{pmatrix} 0 \\ 0 \\ 1/2 \end{pmatrix}), (I, \begin{pmatrix} 0 \\ 1 \\ 0 \end{pmatrix})\rangle \subset E(3).$$

The non-orientable platycosm $\mathbb{R}^3/\Gamma_4$, with holonomy $\mathbb{Z}_2 \times \mathbb{Z}_2$, is called the *first amphidicosm* and denoted $+a2$; $\mathcal{B}_3$ in [148].

Let $D_1 = \begin{bmatrix} 1 & 0 & 0 \\ 0 & 1 & 0 \\ 0 & 0 & -1 \end{bmatrix}$, $D_2 = \begin{bmatrix} 1 & 0 & 0 \\ 0 & -1 & 0 \\ 0 & 0 & 1 \end{bmatrix}$. Let us define

$$\Gamma_5 = \langle (D_1, \begin{pmatrix} 1/2 \\ 1/2 \\ 0 \end{pmatrix}), (D_2, \begin{pmatrix} 0 \\ 0 \\ 1/2 \end{pmatrix}), (I, \begin{pmatrix} 0 \\ 1 \\ 0 \end{pmatrix}) \rangle \subset E(3).$$

The non-orientable platycosm $\mathbb{R}^3/\Gamma_5$, with holonomy $\mathbb{Z}_2 \times \mathbb{Z}_2$, is called the *second amphidicosm* and denoted $-a2$; $\mathcal{B}_4$ in [148].

Let $A_6 = \begin{bmatrix} 1 & 0 & 0 \\ 0 & -1 & 1 \\ 0 & -1 & 0 \end{bmatrix}$. Let us define

$$\Gamma_6 =< (A_6, \begin{pmatrix} 1/3 \\ 0 \\ 0 \end{pmatrix}), (I, \begin{pmatrix} 0 \\ 1 \\ 0 \end{pmatrix}), (I, \begin{pmatrix} 0 \\ 0 \\ 1 \end{pmatrix}) >\subset A(3).$$

The orientable platycosm $\mathbb{R}^3/\Gamma_6$, with holonomy $\mathbb{Z}_3$, is called the *tricosm* and denoted $c3$; $\mathcal{G}_3$ in [148].

Let $A_7 = \begin{bmatrix} 1 & 0 & 0 \\ 0 & 0 & 1 \\ 0 & -1 & 0 \end{bmatrix}$. Let us define

$$\Gamma_7 = \langle (A_7, \begin{pmatrix} 1/4 \\ 0 \\ 0 \end{pmatrix}), (I, \begin{pmatrix} 0 \\ 1 \\ 0 \end{pmatrix}), (I, \begin{pmatrix} 0 \\ 0 \\ 1 \end{pmatrix}) \rangle \subset E(3).$$

The orientable platycosm $\mathbb{R}^3/\Gamma_7$, with holonomy $\mathbb{Z}_4$, is called the *tetracosm* and denoted $c4$; $\mathcal{G}_4$ in [148].

Let $A_8 = \begin{bmatrix} 1 & 0 & 0 \\ 0 & 1 & -1 \\ 0 & 1 & 0 \end{bmatrix}$. Let us define

$$\Gamma_8 = \langle (A_8, \begin{pmatrix} 1/6 \\ 0 \\ 0 \end{pmatrix}), (I, \begin{pmatrix} 0 \\ 1 \\ 0 \end{pmatrix}), (I, \begin{pmatrix} 0 \\ 0 \\ 1 \end{pmatrix}) \rangle \subset A(3).$$

The orientable platycosm $\mathbb{R}^3/\Gamma_8$, with holonomy $\mathbb{Z}_6$, is called the *hexacosm* and denoted $c6$; $\mathcal{G}_5$ in [148].

$$\text{Let } B_1 = \begin{bmatrix} 1 & 0 & 0 \\ 0 & -1 & 0 \\ 0 & 0 & -1 \end{bmatrix}, B_2 = \begin{bmatrix} -1 & 0 & 0 \\ 0 & 1 & 0 \\ 0 & 0 & -1 \end{bmatrix}. \text{ Let us define}$$

$$\Gamma_9 = \langle\langle (B_1, \begin{pmatrix} 1/2 \\ 1/2 \\ 0 \end{pmatrix}), (B_2, \begin{pmatrix} 0 \\ 1/2 \\ 1/2 \end{pmatrix}) \rangle\rangle \subset E(3).$$

The orientable platycosm $\mathbb{R}^3/\Gamma_9$, with holonomy $\mathbb{Z}_2 \times \mathbb{Z}_2$, is called the *didicosm* and denoted $c22$; $\mathcal{G}_6$ in [148].

$$\Gamma_{10} = \mathbb{Z}^3.$$

$\mathbb{R}^3/\Gamma_{10}$ is the torus and is called the *cubical torocosm* and denoted $c1$; $\mathcal{G}_1$ in [148]. All the above names and notations are from [29].

3.4 Exercises

Exercise 3.1. Show that if the fundamental group of a flat manifold is contained in $SO(n) \ltimes \mathbb{R}^n$, then the manifold is orientable.

Exercise 3.2. Complete the proof of Theorem 3.1 (see [140, p. 17]).

Exercise 3.3. Prove that in Proposition 3.2 the assumption of Γ being torsion free is necessary.

Exercise 3.4. Let H be any group and let $\Pi\colon \text{Aut}(H) \to \text{Out}(H)$ be the natural projection. Prove that if $\alpha, \beta \in \text{Aut}(H)$ and $\Pi(\alpha) = \Pi(\beta)$ then the groups $H \rtimes_\alpha \mathbb{Z}, H \rtimes_\beta \mathbb{Z}$ are isomorphic.

Exercise 3.5. Let Γ be a torsion free crystallographic group of dimension n with a trivial center and let $\alpha \in \text{Aut}(\Gamma)$. Assume that $\Pi(\alpha) \in \text{Out}(\Gamma)$ is of infinite order. Prove that the maximal abelian subgroup of $E_\alpha = \Gamma \rtimes_\alpha \mathbb{Z}$ has rank n.

Exercise 3.6. Prove that for a finite group G and its subgroup H we have in isomorphisms of G-modules

$$\text{Ind}_H^G M \simeq \mathbb{Z}G \otimes_H M \simeq \text{Hom}_H(\mathbb{Z}G, M) \simeq \text{Coind}_H^G M,$$

where M is any G-module.

Exercise 3.7. Prove that the homomorphism (Frobenius reciprocity, see [126, Theorem 13, p. 73])

$$\pi_*\colon \mathrm{Hom}_{\mathbb{Z}G}(M, \mathrm{ind}_C^G \mathbb{Z}) \to \mathrm{Hom}_{\mathbb{Z}C}(res_C^G M, \mathbb{Z})$$

is an isomorphism, where $\pi\colon \mathrm{ind}_C^G \mathbb{Z} \to \mathbb{Z}$ is the projection (3.6) described below. Let $\mathcal{X}$ be the set of conjugacy classes of a cyclic subgroup of prime order p, of a finite group G. Let $C \in \mathcal{X}$ and let $\sum_{g \in G/C} n_g gC \in \mathrm{ind}_C^G \mathbb{Z}$. We define

$$\pi\left(\sum_{g \in G/C} n_g gC\right) = n_1. \tag{3.6}$$

Exercise 3.8. Prove Theorem 3.18A.

Exercise 3.9. Prove that any finite group is the holonomy group of some torsion free crystallographic group.

Exercise 3.10 (Crystallographic Restriction). Describe finite subgroups of $GL(2, \mathbb{Z})$. Hint: Any such group is conjugate to a subgroup of $O(2)$.

Exercise 3.11 (Lagrange's Lemma). Let $L \subset \mathbb{R}^2$ be a discrete and cocompact subgroup (lattice). Choose $a \in L \setminus \{0\}$ with minimal length and choose $b \in L \setminus \mathbb{R}a$ with minimal length. Prove that the set $\{a, b\}$ generates L.

Exercise 3.12. Prove that $H^2(D_2, \mathbb{Z}^2) = \mathbb{Z}_2 \oplus \mathbb{Z}_2$ and $H^2(D_4, \mathbb{Z}^2) = \mathbb{Z}_2$.

Exercise 3.13. Prove that for $i = 6, 8$ there exist elements $b_i \in A(3)$, such that $b_i \Gamma_i b_i^{-1} \subset E(3)$.

4. Flat Manifolds with $b_1 = 0$

In this chapter we concentrate on the second step of the Calabi method of classification. We shall give a necessary and sufficient condition for a finite group to be a holonomy group of a Bieberbach group with finite abelianization (primitive groups). Here we shall use the Burnside transfer Theorem (Theorem 11.31), which is formulated and proved in Appendix II. At the end we give a list of primitive groups.

From Proposition 3.2, the first and third step of the Calabi method of the classification, are clear. Let us concentrate on the second step of this method. It considers the class of torsion free crystallographic groups with no epimorphism onto $\mathbb{Z}$. We have the following characterization of this class.

Proposition 4.1 ([63]). *Let Γ be a torsion free crystallographic group with holonomy group G and maximal abelian subgroup $\mathbb{Z}^n$. Then the following conditions are equivalent:*

(i) *the abelianization $\Gamma/[\Gamma,\Gamma]$ is finite;*

(ii) *the center of Γ is trivial;*

(iii) *$(\mathbb{Z}^n)^G = 0$, where G acts on $\mathbb{Z}^n$ by the holonomy representation.*

Proof. It is clear that some power of any non-trivial element of the centre defines a non-zero element of $(\mathbb{Z}^n)^G$. The reverse implication is obvious. This proves the equivalence of (ii) and (iii). In [63, Corollary 1.3] it is proved that

$$\dim_{\mathbb{Q}}(\Gamma/[\Gamma,\Gamma] \otimes \mathbb{Q}) = \text{rank}(\mathbb{Z}^n)^G.{}^1$$

Hence we have the equivalence of (i) and (iii). □

Let M be a flat manifold with the fundamental group Γ.

Definition 4.2. The rank of the abelian group $H_1(M, \mathbb{Z}) = \Gamma/[\Gamma,\Gamma]$ is the first Betti number M (group Γ). It is denoted by $b_1(M)(b_1(\Gamma))$.

Definition 4.3 ([63]). If G is a finite group we say G is primitive if G is the holonomy group of a flat manifold M with $b_1(M) = 0$.

---

[1] Let M be a finitely generated abelian group. Then rank $M = \dim_{\mathbb{Q}}(M \otimes \mathbb{Q})$.

There exists a theorem on complete characterization of primitive groups. It was proved in 1987 by H. Hiller and C. H. Sah [63]. The next proofs were given in 1989 by W. Plesken [114] and separately by G. Cliff and A. Weiss [26].

Theorem 4.4 ([26], [63]). *Let G be a finite group, and let $\mathcal{X}$ be the set of conjugacy classes of the cyclic subgroup of prime order p. The following conditions are equivalent:*

(i) G is primitive;

(ii) the transfer from G to C is trivial for every subgroup C in $\mathcal{X}$;

(iii) no cyclic Sylow p-subgroup of G has a normal complement.

Proof ([26]). We shall need the following lemma.

Lemma 4.5 ([26, Lemma 5]). *Let C be a cyclic subgroup of G. Let $a_C \colon \operatorname{ind}_C^G \mathbb{Z} \to \mathbb{Z}$ be given by*

$$a_C \left(\sum_{t \in T_C} n_t tC \right) = \sum_{t \in T_C} n_t.$$

Then $a_{C*} \colon H^2(G, \operatorname{ind}_C^G \mathbb{Z}) \to H^2(G, \mathbb{Z})$ is 0 if and only if the transfer from G to C is trivial.

Proof of Lemma 4.5. The exact sequence

$$0 \to \operatorname{ind}_C^G \mathbb{Z} \to \operatorname{ind}_C^G \mathbb{Q} \to \operatorname{ind}_C^G \mathbb{Q}/\mathbb{Z} \to 0$$

gives us an isomorphism

$$\delta \colon H^1(G, \operatorname{ind}_C^G \mathbb{Q}/\mathbb{Z}) \to H^2(G, \operatorname{ind}_C^G \mathbb{Z}),$$

since $H^i(G, \operatorname{ind}_C^G \mathbb{Q}) = 0, i \geq 1$. Similarly, we have an isomorphism between $H^1(G, \mathbb{Q}/\mathbb{Z})$ and $H^2(G, \mathbb{Z})$. So we only need to prove the same assertion about the map $a_{C*} \colon H^1(G, \operatorname{ind}_C^G \mathbb{Q}/\mathbb{Z}) \to H^1(G, \mathbb{Q}/\mathbb{Z})$. We exhibit elements of $H^1(G, \operatorname{ind}_C^G \mathbb{Q}/\mathbb{Z})$ via the isomorphism

$$H^1(G, \operatorname{ind}_C^G \mathbb{Q}/\mathbb{Z}) \overset{res_C^G}{\to} H^1(C, res_C^G \operatorname{ind}_C^G \mathbb{Q}/\mathbb{Z}) \overset{\pi_{C*}}{\to} H^1(C, \mathbb{Q}/\mathbb{Z})$$

of Shapiro's Lemma, where π_C is the projection map (3.6) defined in the previous Chapter. For $\chi \in H^1(C, \mathbb{Q}/\mathbb{Z}) = \operatorname{Hom}(C, \mathbb{Q}/\mathbb{Z})$, we construct a 1-cocycle $\hat{\chi}$ on G with values in $\operatorname{ind}_C^G \mathbb{Q}/\mathbb{Z}$; to define $\hat{\chi}(g)$ for $g \in G$, for each $t \in T_C$ write

$$gt = t_g c_{g,t} \text{ for some } t_g \in T_C \text{ and some } c_{g,t} \in C.$$

Then define

$$\hat{\chi}(g) = \sum_{t \in T_C} \chi(c_{g,t}) t_g C.$$

It is easy to check that $\hat{\chi}$ is a cocycle, so we get a class $[\hat{\chi}] \in H^1(G, \mathrm{ind}_C^G \mathbb{Q}/\mathbb{Z})$, and this class maps to χ under the isomorphism of Shapiro's Lemma. Thus $a_{C*} = 0$ on H^1 if and only if $a_{C*}(\hat{\chi}) = 0$, for all $\chi \in \mathrm{Hom}(C, \mathbb{Q}/\mathbb{Z})$. But

$$a_{C*}(\hat{\chi})(g) = a_C(\hat{\chi}(g)) = a_C\left(\sum_{t \in T_C} \chi(c_{g,t}) t_g C\right) = \sum_t \chi(c_{g,t}) = \chi(V(g)),$$

where $V: G \to C$ is the transfer map. Thus $a_{C*}(\hat{\chi}) = 0$ for all χ if and only if $V(g) = 1$, for all $g \in G$, and the proof is complete. □

Let us start the proof of the implication $(i) \Rightarrow (ii)$ of Theorem 4.4. Suppose that G is the holonomy group of a torsion free crystallographic group Γ with trivial centre and maximal abelian subgroup $\mathbb{Z}^n$. Let $\alpha \in H^2(G, \mathbb{Z}^n)$ be the special element, which corresponds to Γ. From Proposition 3.17 we have a map $h: \mathbb{Z}^n \to S$ with $h_*(\alpha)$ special. Then $h_*(\alpha) = \oplus_{C \in \mathcal{X}} \epsilon_C$, with ϵ_C a non-zero element of $H^2(G, \mathrm{ind}_C^G \mathbb{Z})$. For each $C \in \mathcal{X}$ we have $a_C: \mathrm{ind}_C^G \mathbb{Z} \to \mathbb{Z}$ defined above, and we set $\oplus_{C \in \mathcal{X}} a_C = a: S \to \mathbb{Z}^r$, giving the exact sequence

$$0 \to \ker a \xrightarrow{i} S \xrightarrow{a} \mathbb{Z}^r \to 0$$

where G acts trivially on $\mathbb{Z}^r$ and $(\ker a)^G = 0$. From the Proposition 4.1, $(\mathbb{Z}^n)^G = 0$, so the composite

$$\mathbb{Z}^n \xrightarrow{h} S \xrightarrow{a} \mathbb{Z}^r$$

is 0. It follows that $a_{C*}(\epsilon_C) = 0$, so we deduce from Lemma 4.5 that the transfer is trivial.

Conversely, suppose that the transfer is trivial for every $C \in \mathcal{X}$. Let $\beta = \oplus \beta_C$ be our special element in $H^2(G, S)$. By Lemma 4.5, $a_{C*}(\beta_C) = 0$ for each C, so $a_{C*}(\beta) = 0$ in $H^2(G, \mathbb{Z}^r)$. We have the exact sequence

$$H^2(G, \ker a) \xrightarrow{i_*} H^2(G, S) \xrightarrow{a_*} H^2(G, \mathbb{Z}^r)$$

so $\beta = i_*(\alpha)$ for some $\alpha \in H^2(G, \ker a)$, and α must be special, since β is. Moreover, $(\ker a)^G = 0$, so α gives us the required crystallographic group, except that G might not act faithfully on $\ker a$ so the holonomy group of our extension of G by $\ker a$ may not be G. Take a $\mathbb{Z}G$-lattice L on which G acts faithfully, and for which $L^G = 0$; let $A = \ker a \oplus L$, and let $\gamma \in H^2(G, A)$ be given by $\alpha \oplus 0$. Then the extension G by A defined by γ gives us the desired crystallographic group. This finishes the proof of the equivalence (i) and (ii). For the proof of the equivalence $(ii) \Leftrightarrow (iii)$ we need the following lemma.

Lemma 4.6 (Transfer Theorem). *Let G be a finite group, and let C be a subgroup of G of prime order p. Then the following are equivalent:*

(a) a Sylow p-subgroup of G is cyclic and is contained in the centre of its normalizer;

(b) a Sylow p-subgroup of G is cyclic and has a normal complement;

(c) the transfer from G to C is onto.

Proof of Lemma 4.6. (a) implies (b) is a consequence of Burnside's transfer theorem, (see [38, §18, p. 92] and Theorem 11.31 in Appendix II). Assume (b). Let P be a cyclic Sylow p-subgroup of G containing C and let H be its normal complement. Then the transfer from G to C is the composite of the transfer from G to P with the transfer from P to C; this latter transfer is onto, since P is cyclic. The transfer from G to P is onto, since one may compute it using H as the set of coset representatives of P in G. This proves (c).

Now assume (c). Let P be a p-subgroup of G maximal with respect to having C in its centre. The transfer from G to C is the composite of the transfer from G to K and from K to C, for any subgroup K of G containing C, and so the transfer from K to C cannot be trivial for any such K. We shall use the following claim which we leave as an exercise.

Claim. *Let C be a normal subgroup of a group K with C having prime order p. Then the transfer from K to C is trivial if one of the following holds:*

(i) K is non-cyclic of order p^2;

(ii) K is the quaternion group of order 8;

(iii) K/C is cyclic of order prime to p and C is not central in K.

Returning to the proof of **Lemma 4.6**, if P is not cyclic, then either it contains a non-cyclic subgroup H of order p^2, or else, it is the generalized quaternion 2-group. In the first instance, let K be generated by C and one of the subgroups of H which is not C, and apply the claim. In the second instance, P has a unique central subgroup of order 2, namely C, and a quaternion subgroup K of order 8 containing C, and we apply the claim again.

Thus P is cyclic. If P is not a Sylow p-subgroup of G, then it is not a Sylow p-subgroup of $N = N_G(P)$. Since p does not divide the order of the automorphism group of C, every p-subgroup of N centralizes C, hence by maximality is contained in P. So P is a cyclic Sylow p-subgroup of G.

Finally, if $g \in N$ and $g \neq C_G(P)$, then conjugation by g induces a non-trivial automorphism of C, of order prime to p. Replace g by a suitable p-power so that it has order prime to p, and apply part (iii) of the claim to $K = \langle C, g \rangle$. This completes the proof of Lemma 4.6. □

Since the equivalence $(ii) \Leftrightarrow (iii)$ of Theorem 4.4 is the same as the equivalence (not (c)) $\Leftrightarrow$ (not (b)) of Lemma 4.6 then the proof of Theorem 4.4 is also finished. □

4.1 Examples of (non)primitive groups

Let us illustrate the above by examples.

Example 4.7. Let Γ be a torsion free crystallographic group with maximal abelian subgroup $\mathbb{Z}^n$ and cyclic holonomy group C. It defines an element in

$$\alpha \in H^2(C, \mathbb{Z}^n) = (\mathbb{Z}^n)^C / I\mathbb{Z}^n,$$

where $I\mathbb{Z}^n$ is a C-submodule of $\mathbb{Z}^n$, see Example 3.9. Since the rank of the centre of Γ is equal to the rank of $(\mathbb{Z}^n)^C$ *(see [63])*, we have that for Γ, it is impossible to be a torsion free and to have a trivial center. Hence $\Gamma/[\Gamma, \Gamma]$ is always infinite. Then cyclic groups are not primitive.

As an easy consequence of the above observation we have:

Proposition 4.8 ([63, section 2]). *The following groups are not primitive: metacyclic groups $\mathbb{Z}_n \rtimes \mathbb{Z}_m, (m, n) = 1$, dihedral groups D_n, n odd, alternating group A_4, Borel subgroup of upper triangular matrices in $SL(2, \mathbb{F}_q)$ and $SL(2, \mathbb{F}_3)$.* □

On the other hand we have:

Proposition 4.9 ([63, Proposition 3.3]). *For every prime p, $\mathbb{Z}_p \times \mathbb{Z}_p$ is a primitive group.*

Proof. We follow [63]. Let a, b denote generators of $\mathbb{Z}_p \times \mathbb{Z}_p$ and define an integral representation N of $\mathbb{Z}_p \times \mathbb{Z}_p$ by:

$$a \mapsto \begin{bmatrix} I & & & & \\ & \zeta & & & \\ & & \zeta & & \\ & & & \cdots & \\ & & & & \zeta \end{bmatrix} \quad b \mapsto \begin{bmatrix} \zeta & & & & \\ & I & & & \\ & & \zeta & & \\ & & & \cdots & \\ & & & & \zeta^{p-1} \end{bmatrix}$$

where ζ denotes the $(p-1) \times (p-1)$ integral matrix of the action of $e^{\frac{2\pi i}{p}}$ on $\mathbb{Z}[e^{\frac{2\pi i}{p}}]$, i.e. ζ is conjugate to the companion matrix of the cyclotomic polynomial $1 + X + X^2 + \cdots + X^{p-1}$. We can view N as a direct sum $\bigoplus_{k=0}^{p} N_k$ according to the blocks along the diagonal. As each N_k is a faithful representation of $\mathbb{Z}_p \times \mathbb{Z}_p$ with no non-zero invariants, so is N. Consider the cyclic subgroup C of $\mathbb{Z}_p \times \mathbb{Z}_p$ generated by $a^i b^j$, where $i + j > 0, 0 \leq i, j < p$. If i (resp. j) is zero, we let $N_C = N_0$ (resp. $N_C = M_p$). If $i, j > 0$, there is a k such that $(\zeta^k)^i = \zeta^{p-i}$ and we let $N_C = N_k$. We claim, that for any $C \subset \mathbb{Z}_p \times \mathbb{Z}_p$, the restriction map

$$H^2(\mathbb{Z}_p \times \mathbb{Z}_p, N_C) \to H^2(C, N_C)$$

is non-zero. We suppose $i, j > 0$ as the other two cases are similar. Let Q denote the quotient $\mathbb{Z}_p \times \mathbb{Z}_p / C$ and consider the Hochschild-Serre spectral sequence of the short exact sequence:

$$0 \to C \to \mathbb{Z}_p \times \mathbb{Z}_p \to Q \to 0$$

with coefficients in N_C, (cf. [18, page 171]). Since $a^i b^j = \zeta^i (\zeta^k)^j = \zeta^p = 1$ under this representation, the restriction of N_C to C is trivial so $H^2(C, N_C) = 0$ and the inflation-restriction sequence gives

$$0 \to H^2(Q, N_C) \to H^2(\mathbb{Z}_p \times \mathbb{Z}_p, N_C) \to H^2(C, N_C).$$

But Q acts on N_C via ζ, so $H^2(Q, N_C) = 0$ and thus it suffices to show $H^2(\mathbb{Z}_p \times \mathbb{Z}_p, N_C) \neq 0$. This can be done by using the isomorphism:

$$H^2(\mathbb{Z}_p \times \mathbb{Z}_p, N_C) = H^1(\mathbb{Z}_p \times \mathbb{Z}_p, (\mathbb{R}/\mathbb{Z}) \otimes_{\mathbb{Z}} N_C)$$

and writing down an explicit non-zero cocycle. $\square$

Corollary. *Any elementary abelian group* $(\mathbb{Z}_p)^k, k \geq 2$ *is primitive.*

□

4.2   Minimal dimension

Let G be a finite primitive group. By $\delta(G)$ we shall understand the minimal dimension of a flat manifold M with $b_1(M) = 0$, and holonomy group G. From the previous sections it is clear (cf. [63], [64]) that $\delta(G)$ admits a purely algebraic description. We recall that description here.

Proposition 4.10. *The number* $\delta(G)$ *is the minimal dimension of an integral representation* N *of* G *satisfying:*

   (i) N is a faithful representation;

  (ii) N carries a special class, i.e. there exists an $\alpha \in H^2(G, N)$ *whose restriction to each cyclic subgroup of* G *is non-zero;*

(iii) N has no non-zero G-invariant, i.e. $N^G = 0$.

□

Our first result is as follows:

Theorem 4.11 ([64, Theorem 1.1]). *If* p *is a prime,* $k > 1$, *then* $\delta(\mathbb{Z}_p^k) = (p-1)(k+p-1)$.

Proof. If $\pi\colon \mathbb{Z}_p^k \twoheadrightarrow \mathbb{Z}_p$ is a surjection, then the composition $\mathbb{Z}_p^k \twoheadrightarrow \mathbb{Z}_p \hookrightarrow \mathbb{Q}(\zeta)^*$, where ζ is a primitive p-th root of unity, determines an irreducible rational representation $L(\pi)_{\mathbb{Q}}$ [2] with underlying $\mathbb{Q}$-vector space $\mathbb{Q}(\zeta)$ of dimension $p-1$. All irreducible rational representations arise in this fashion, see [39, Theorem 1.3]. We denote by $L(\pi)$ an integral representation satisfying $L(\pi)_{\mathbb{Q}} \simeq L(\pi) \otimes_{\mathbb{Z}} \mathbb{Q}$. [3] The kernel of the action of $\mathbb{Z}_p^k$ on $L(\pi)$ is $\ker(\pi)$ of order p^{k-1}. We shall need:

Lemma 4.12. *The cohomology group* $H^2(\mathbb{Z}_p^k, L(\pi))$ *can be identified with* $\mathbb{F}_p$-*dual of* $H = \ker(\pi)$. *Furthermore, if* $f \in H^* \simeq H^2(\mathbb{Z}_p^k, L(\pi))$, *then under this identification one has:*

$$H \setminus \ker(f) = \{x \in \mathbb{Z}_p^k \setminus \{0\} \mid Res_{\langle x \rangle}^{\mathbb{Z}_p^k}(f) \neq 0\}.$$

---

   [2]A representation is irreducible, when the only proper submodule is the 0 module, (see Lemma 5.1).

   [3]An example of such an integral representation given in the last section was the one denoted by the matrix ζ.

Proof of Lemma 4.12. We write $\mathbb{Z}_p^k = H \times C$, where C is a subgroup of order p acting faithfully and irreducibly on $L(\pi)$. It follows that each $\alpha \in H^2(\mathbb{Z}_p^k, L(\pi))$ splits on C. The Hochschild-Serre spectral sequence ([18]) associated to the short exact sequence

$$1 \to C \to \mathbb{Z}_p^k \to H \to 1$$

gives an isomorphism $H^2(\mathbb{Z}_p^k, L(\pi)) \simeq H^1(H, H^1(C, L(\pi)))$. Since C is cyclic we have: $H^1(C, L(\pi)) \simeq L/(x-1)L$, where $C = \langle x \rangle$. As L is an ideal class in the cyclotomic field of p-th roots of unity and $x - 1$ generates the prime ideal of degree one, $H^1(C, L(\pi)) \simeq \mathbb{F}_p$. The group H acts trivially on C and $L(\pi)$, hence also trivially on $H^1(C, L(\pi))$, so

$$H^2(\mathbb{Z}_p^k, L(\pi)) \simeq H^1(H, \mathbb{F}_p) \simeq \mathrm{Hom}(H, \mathbb{F}_p).$$

The proof of the second assertion we leave as an exercise. $\qquad\qquad\square$

A subset S of $\mathbb{Z}_p^k$ is called a *hyperplane difference* if $S = H \setminus K$, where H is a hyperplane in $\mathbb{Z}_p^k$ and K is a hyperplane in H. For example, the set $H \setminus \ker(f)$ in the above lemma is a hyperplane difference. The basic fact about hyperplane difference that we require is:

Lemma 4.13. *If V is a vector space of dimension k over the finite field $\mathbb{F}_p$, then $V \setminus \{0\}$ can be covered by $k + p - 1$ hyperplane differences.*

Proof of Lemma 4.13. Let $f_1, f_2, \ldots, f_k$ be a dual basis for V. For each $i, 1 \le i \le k$, let $H_i = \ker(f_i)$ and $D_i = H_i \setminus (H_{i+1} \cap H_i)$, $i \bmod k$. For each $x \in \mathbb{F}_p^*$, let $D_x = \ker(f_1 + xf_2) \setminus H_1$. It is easy to check that these hyperplane differences satisfy the required property. $\qquad\qquad\square$

Let us go back to the proof of Theorem 4.11. To construct the $\mathbb{Z}_p^k$-lattice required by Proposition 4.10 we invoke Lemma 4.13 to cover $\mathbb{Z}_p^k$ by hyperplane differences $H_i \setminus K_i, 1 \le i \le k + p - 1$. Then we construct representations $\pi_i : \mathbb{Z}_p^k \to \mathbb{Z}_p^k/H_i$ and functionals $f_i : H_i \to H_i/K_i \simeq \mathbb{F}_p$. According to Lemma 4.12 this data yields classes in $H^2(\mathbb{Z}_p^k, L(\pi_i))$ supported on cyclic subgroups generated by elements in $H_i \setminus K_i$. Since the hyperplane differences cover $\mathbb{Z}_p^k$, the sum of these cohomology classes is special. Computing the degree of the resulting representation $\bigoplus_{i=1}^{k+p-1} L(\pi_i)$ we get, $\delta(\mathbb{Z}_p^k) \le (p-1)(k+p-1)$.

Now we show that the given number is a lower bound. Suppose N is a $\mathbb{Z}_p^k$-lattice satisfying conditions of Proposition 4.10. Let $N_{\mathbb{Q}}$ denote the corresponding rational representation $N \otimes_{\mathbb{Z}} \mathbb{Q}$. From condition (iii) of Proposition 4.10, we get the decomposition $N_{\mathbb{Q}} = L(\pi_1)_{\mathbb{Q}} \oplus L(\pi_2)_{\mathbb{Q}} \oplus \cdots \oplus L(\pi_n)_{\mathbb{Q}}$. According to condition (ii), $H^2(C, N) \neq 0$ for any cyclic subgroup $C \subset \mathbb{Z}_p^k$. Hence $(N_{\mathbb{Q}})^C \neq 0$. According to Lemma 4.12, $L(\pi)^C \neq 0$ if and only if $C \subset \ker(\pi)$. Hence we have to find a minimal number of surjections $\pi_i \colon \mathbb{Z}_p^k \twoheadrightarrow \mathbb{Z}_p$ such that:

(a) for any cyclic subgroup $C \subset \mathbb{Z}_p^k$, there is some $i, 1 \leq i \leq n$, such that $C \subset \ker(\pi_i)$ and

(b) the intersection $\bigcap_{i=1}^n \ker(\pi_i)$ is trivial.

The second condition (b) insures that N is faithful. Hence it remains to check that a version of Lemma 4.12 is the best possible.

Lemma 4.14. *If V is a k-dimensional vector space over the finite field $\mathbb{F}_p$ and $\{H_1, H_2, \ldots, H_n\}$ is a family of hyperplanes in V satisfying:*

$$\bigcup_{i=1}^n H_i = V \text{ and } \bigcap_{i=1}^n H_i = 0$$

then $n \geq k + p - 1$.

Proof of Lemma 4.14. Without loss of generality we can assume that the first k hyperplanes are given by the coordinates $x_i = 0$. Call a vector $v \in V$ positive if $v \notin \bigcup_{i=1}^k H_i$. These vectors have to be contained in $H_{k+1}, \ldots, H_n$. If H is any hyperplane in V, we claim H contains at most $(p-1)^{k-1}$ positive vectors. In fact, there is a projection p of V onto one of H_i's, $1 \leq i \leq k$, for which $p|H$ is injective. Furthermore, p maps positive vectors to positive vectors in $H_i = \mathbb{F}_p^{k-1}$. In H_i, there are exactly $(p-1)^{k-1}$ positive vectors. Hence to cover the $(p-1)^k$ positive vectors in V we need at least $(p-1)^k/(p-1)^{k-1} = p-1$ additional hyperplanes. This verifies that the given number is the lower bound. $\square$

With the above lemma we finish the proof of Theorem 4.11. $\square$

We turn to several interesting families of 2-groups: the generalized quaternion groups $Q_{2^n}, n \geq 3$, the dihedral groups $D_{2^n}, n \geq 2$, and the semi-dihedral groups $SD_{2^n}, n \geq 4$. These groups are defined by generators and relations as follows:

$$Q_{2^n} = \langle x, y \mid x^{2^{n-2}} = y^2, y^{-1}xy = x^{-1} \rangle,$$

$$D_{2^n} = \langle x, y \mid x^{2^{n-1}} = y^2 = 1, y^{-1}xy = x^{-1} \rangle,$$
$$SD_{2^n} = \langle x, y \mid x^{2^{n-1}} = y^2 = 1, y^{-1}xy = x^{-1+2^{n-2}} \rangle.$$

Good references are [38] and [39]. We recall the following basic fact.

Fact ([39, p. 46]). *If H is a non-cyclic finite p-group whose normal abelian subgroups are all cyclic, then $p = 2$ and H is isomorphic to $Q_{2^n}, n \geq 3$, $D_{2^n}, n \geq 2$, or $SD_{2^n}, n \geq 4$.*

$\square$

Each of these groups has four one-dimensional lattices: the trivial lattice $\mathbb{Z}$, and the lattices S_1, S_2, S_3 determined by being trivial on x, y, xy, respectively.   At first we turn to the rational representation theory of Q_{2^n}. The group Q_{2^n} admits a unique faithful irreducible representation $\mathcal{T}_{\mathbb{Q}}$ of degree 2^{n-1} (see [39, p. 47]). We let $\mathcal{T}$ denote an integral representation that satisfies $\mathcal{T} \otimes_{\mathbb{Z}} \mathbb{Q} \simeq \mathcal{T}_{\mathbb{Q}}$. Our first result is:

Proposition 4.15. $\delta(Q_{2^n}) = 2^{n-1} + 3$. *In particular, $\delta(Q_8) = 7$.*

Proof. We begin by constructing a lattice of the correct dimension. It will be a direct sum of the faithful integral representation $\mathcal{T}$ and a three-dimensional representation $\mathcal{V}$. Let Q_{2^n} act on $\mathcal{V} = \mathbb{Z}^3$ by

$$x \mapsto \begin{pmatrix} -1 & 1 & 1 \\ 0 & 1 & 2 \\ 0 & 0 & -1 \end{pmatrix}, y \mapsto \begin{pmatrix} 1 & -1 & 0 \\ 0 & -1 & 0 \\ 0 & 0 & -1 \end{pmatrix}.$$

We exhibit an element $s \in H^2(Q_{2^n}, \mathcal{V})$ which restricts non-trivially to the subgroup $C = \langle y^2 \rangle \subset Q_{2^n}$. We describe s as a 1-cocycle using the identification $H^2(Q_{2^n}, \mathcal{V}) \simeq H^1(Q_{2^n}, \mathbb{R}^3/\mathbb{Z}^3)$. The cocycle $s \colon Q_{2^n} \to \mathbb{R}^3/\mathbb{Z}^3$ is defined on the given generators by:

$$s(x) = [0, 2^{-n+2}, 0], \quad s(y) = [0, 1/2, 1/2].$$

It is not difficult to check that s saves the relations, that define Q_{2^n} and also

$$s(y^2) = ys(y) + s(y) = [-1/2, 0, 0] \neq 0.$$

Since y^2 is contained in every non-trivial subgroup of Q_{2^n}, s defines a special class in $H^2(Q_{2^n}, \mathcal{V})$. Hence $\mathcal{W} = \mathcal{T} \oplus \mathcal{V}$ is a proper Q_{2^n}-lattice and we conclude that:

$$\delta(Q_{2^n}) \leq \deg(\mathcal{W}) = \deg(\mathcal{T}) + \deg(\mathcal{V}) = 2^{n-1} + 3.$$

To prove that this upper bound is exact, suppose $\mathcal{M}$ is another Q_{2^n}-lattice with the above properties and consider $\mathcal{M}_{\mathbb{Q}} = \mathcal{M} \otimes_{\mathbb{Z}} \mathbb{Q}$. Since $\mathcal{T}_{\mathbb{Q}}$ is the unique faithful irreducible $\mathbb{Q}$-representation of Q_{2^n}, $\mathcal{M}_{\mathbb{Q}}$ contains $\mathcal{T}_{\mathbb{Q}}$. Certainly, $\mathcal{M}_{\mathbb{Q}} \neq \mathcal{T}_{\mathbb{Q}}$, since $H^2(Q_{2^n}, \mathcal{T}) = H^2(\mathbb{Z}_2, \bar{\mathbb{Z}}) = 0$, where $\bar{\mathbb{Z}}$ is a non-trivial $\mathbb{Z}_2$-lattice $\mathbb{Z}$. There are two cases to be considered.

Case 1. Suppose $\mathcal{M}_{\mathbb{Q}}$ contains an irreducible degree 2 dihedral representation coming from the surjection $\mathbb{Q}_{2^n} \twoheadrightarrow D_8$. (If it contains one of larger degree the proof is completed.) But then x^2 acts without fixed points on $\mathcal{M}_{\mathbb{Q}}$ and $H^2(\langle x^2 \rangle, \mathcal{M}) = 0$, a contradiction with Proposition 4.10 (ii).

Case 2. Now we claim that $\mathcal{M}$ must contain three non-trivial linear representations S_i, $i = 1, 2, 3$. Suppose, for example, that S_1 is missing. Because x acts without fixed-points on $\mathcal{T}$ and the other S_i's, $\mathcal{M}^{\langle x \rangle} = 0$, so $H^2(\langle x \rangle, \mathcal{M}) = 0$. This also contradicts Proposition 4.10 (ii).

From these two cases we can conclude that $\operatorname{rank}(\mathcal{M}) \geq 2^{n-1} + 3$ and the proof of Proposition 4.15 is complete. □

In the (semi-)dihedral case we have:

Proposition 4.16 ([64, Theorem 1.7]). $\delta(D_{2^n}) = \delta(SD_{2^n}) = 2^{n-2} + 2$, for $n \geq 2$.

Proof. Since $D_4 = \mathbb{Z}_2 \times \mathbb{Z}_2$, the result follows from the classification in dimension 3, (see previous chapter). Similarly, the tables in dimension 4 (see [16]) show $\delta(D_8) = 4$. Now, we suppose that $n \geq 4$. Both groups admit a unique faithful irreducible $\mathbb{Q}$-representation $\mathcal{T}_{\mathbb{Q}}$ of degree 2^{n-2} obtained by inducing up the faithful irreducible representation of $D_8 \subset D_{2^n}$ (resp. $D_8 \subset SD_{2^n}$) of degree 2, [39, p. 48]. So $\delta \geq 2^{n-2}$, in both cases. We begin by computing $\delta(D_{2^n})$. Consider the infinite dihedral group $D_\infty \simeq \mathbb{Z}_2 * \mathbb{Z}_2$, where $*$ denotes the free product. We choose a presentation $D_\infty = \langle s, t \mid (st)^2 = t^2 = 1 \rangle$, so t and st are the involutions and $\langle s \rangle \simeq \mathbb{Z}$. Applying the Hochschild-Serre spectral sequence to the inclusion of the normal subgroup $\langle s \rangle$ in D_∞, we see that a class in $H^2(D_\infty, N)$ is zero if and only if it restricts trivially to both subgroups $\langle st \rangle$ and $\langle t \rangle$, for any D_∞-module N.

Let K denote the 1-dimensional representation of D_∞ on which s and t act by -1. Then the Hochschild-Serre spectral sequence shows that

$$H^2(D_\infty, K) \simeq H^1(\mathbb{Z}_2, H^1(\langle s \rangle, K)) \simeq K/(s-1)K \simeq \mathbb{Z}_2.$$

The non-trivial class restricts to zero on $\langle t \rangle$ hence it has to restrict non-trivially to $\langle st \rangle$.

Let ζ denote a primitive 2^{n-1}th root of unity. The ring of integers $L = \mathbb{Z}[\zeta]$ is a representation of D_∞, where s acts by multiplication by ζ and t corresponds to complex conjugation. Again we compute: $H^2(D_\infty, L) \simeq \mathbb{Z}_2$. As a representation of $\langle st \rangle$, L is a free module and hence the non-trivial class restricts to zero on $\langle st \rangle$. So this class must restrict non-trivially to $\langle t \rangle$.

Now, it is easy to see that $\mathcal{T} \oplus S_3 \oplus S_1$ is a D_{2^n}-lattice which satisfies the conditions of Proposition 4.10. Hence $\delta(D_{2^n}) \leq 2^{n-2} + 2$. To show that this degree is minimal, we observe that there are three conjugacy classes of elements of order 2 (represented by $y, xy, x^{2^{n-2}}$) and if F is a non-trivial one dimensional D_{2^n}-representation, any non-zero class in $H^2(D_{2^n}, F)$ restricts trivially to at least two of these conjugacy classes.

Let us turn to the calculation of $\delta(SD_{2^n})$. There are two conjugacy classes of elements of order 2. The group SD_{2^n} contains a dihedral subgroup of index 2. Hence we get a faithful irreducible lattice $\mathcal{T}$ of degree 2^{n-2}, where the non-zero class in $H^2(SD_{2^n}, \mathcal{T}) \simeq \mathbb{Z}_2$ restricts non-trivially to the subgroup $\langle y \rangle$.

Let K denote the 1-dimensional lattice with no invariants of the dihedral subgroup of index 2 in SD_{2^n} and let $\mathcal{U}$ be the result of inducing this representation up to SD_{2^n}. The lattice $\mathcal{U}$ has degree 2 and by Shapiro's lemma (Theorem 3.12) $H^2(SD_{2^n}, \mathcal{U})$ has a class that restricts non-trivially to the central involution. Hence $\mathcal{T} \oplus \mathcal{U}$ is a SD_{2^n}-lattice, which satisfies conditions of Proposition 4.10. and has the correct degree $2^{n-2} + 2$. We claim that $\delta(SD_{2^n}) \geq 2^{n-2} + 2$. Let S_1 be the 1-dimensional lattice on which the cyclic subgroup of index 2 acts trivially but the quotient acts by -1. This is the only integral one-dimensional representation with no SD_{2^n}-invariants for which $H^2(\langle x^{2^{n-2}} \rangle, S_1) \neq 0$. But one can check, that the restriction map is zero. This completes the proof. $\qquad\square$

4.3   Exercises

Exercise 4.1. Complete the proof of Proposition 4.1.

Exercise 4.2. Give a proof of the CLAIM from the proof of Theorem 4.4.
Hint: Use [38].

Exercise 4.3. Prove that any finite p-group is primitive if and only if it is not cyclic.
Hint: Use [63, Corollary 3.6].

Exercise 4.4. Let G be a finite group. Prove that G is primitive if and only if $H_1(G, \mathbb{Z}_p) = 0$ whenever the Sylow p-subgroup of G is cyclic.
Hint: Use [63, Theorem 3.9].

Exercise 4.5. Prove that all perfect groups and simple groups are primitive.
Hint: Apply previous exercise.

Exercise 4.6. Prove the second part of Lemma 4.12.

Exercise 4.7. Prove that the faithful Q_{2^n}-lattice $\mathcal{T}$ is isomorphic to $\mathrm{ind}_{\langle y^2 \rangle}^{Q_{2^n}} \bar{\mathbb{Z}}$.

5. Outer Automorphism Groups

This chapter, as the previous one, is related to the Calabi method of classification (see Section 3.2 and Proposition 3.2). This time we are interested in its third part. To classify the short exact sequences of groups

$$0 \to \Gamma' \to \Gamma \to \mathbb{Z} \to 0,$$

where Γ' is a Bieberbach group with a trivial centre, we need to have some information about the outer automorphism group $\mathrm{Out}(\Gamma')$. We give a necessary and sufficient conditions on (in)finiteness of outer automorphism group. We show that it depends only on a holonomy representation. The section 5.1 is mainly based on [23, Chapter V].

5.1 Some representation theory and 9-diagrams

Let Γ be any crystallographic group, which defines a short exact sequence of groups

$$0 \to \mathbb{Z}^n \to \Gamma \xrightarrow{p} G \to 0,$$

where $\mathbb{Z}^n$ is a maximal, free abelian group and G is finite. Let us recall that the above sequence also defines the holonomy representation (see (2.6))

$$h_\Gamma \colon G \to GL(n, \mathbb{Z}).$$

We want to show how many properties of the crystallographic group Γ are saved in the above representation. Since $\mathbb{Z}^n$ is the maximal abelian subgroup, it follows that h_Γ is faithful. Moreover, we have the following obvious inclusions of groups

$$GL(n, \mathbb{Z}) \subset GL(n, \mathbb{Q}) \subset GL(n, \mathbb{R}) \subset GL(n, \mathbb{C}).$$

Hence the holonomy representation can be considered, correspondingly, as rational, real or complex.

We recall a few facts from the representation theory of finite groups [126]. Any representation can be considered as some G-module. A representation is *irreducible*, when the only proper submodule is the 0 module. For complex representations we have the following lemma.

Lemma 5.1. (Schur) *Let V be a finite dimensional irreducible representation of a finite group over the complex numbers. Let $f: V \to V$ be any non-zero G-homomorphism of V. Then f is a homothety (multiplication by a scalar).*

Proof. Let λ be a root of the characteristic polynomial of f. The set

$$\{v \in V \mid f(v) = \lambda v\}$$

is a non-zero G-subspace of V. By assumption it has to be V. Hence f is a homothety.   □

Any finite dimensional representation of a finite group, over a field of characteristic 0 is a direct sum of irreducible representations.

Definition 5.2 ([31], [126, p. 107]). Let T be a $\mathbb{Q}$-irreducible representation of a finite group. Then T is a direct sum of $\mathbb{C}$-irreducible representations. It can be shown that each component has the same multiplicity $m_\mathbb{Q}(T)$. The number $m_\mathbb{Q}(T)$ is defined as the Schur index of the representation T.

By $(Z^n)^G$ we shall denote the G-submodule

$$\{x \in \mathbb{Z}^n \mid gx = h_\Gamma(g)x = x\}.$$

When a representation is defined over a field k, $(k \otimes \mathbb{Z}^n)^G$ is not only a submodule, but also a direct summand of the kG-module $k \otimes \mathbb{Z}^n$. This means that there exists some kG-submodule $N \subset k \otimes \mathbb{Z}^n$, such that $k \otimes \mathbb{Z}^n = (k \otimes \mathbb{Z}^n)^G \oplus N$. Unfortunately, this is no longer true for the G-module $\mathbb{Z}^n$. But the description of crystallographic groups is closely related to the representation theory of finite groups over $\mathbb{Z}$.

Example 5.3. $G = \mathbb{Z}_2$; let us consider $\begin{bmatrix} 0 & 1 \\ 1 & 0 \end{bmatrix}$ the integral representation of G.

One can see that $(\mathbb{Z}^2)^{\mathbb{Z}_2}$ is not a direct summand of $\mathbb{Z}^2$.

Let N be the normalizer of $h_\Gamma(G)$ in $GL(n, \mathbb{Z})$. By definition

$$N = \{X \in GL(n.\mathbb{Z}) \mid \forall f \in h_\Gamma(G) \quad XfX^{-1} \in h_\Gamma(G)\}.$$

It acts on the group $H^2(G, \mathbb{Z}^n) \overset{(2.4)}{\simeq} H^1(G, \mathbb{Q}^n/\mathbb{Z}^n)$ in the following way

$$n * [c(g)] = [n^{-1}c(ngn^{-1})],$$

where $[c] \in H^1(G, \mathbb{Q}^n/\mathbb{Z}^n), n \in \mathbb{N}, g \in G$. One can prove (see [18], [23]) that the action is well defined.

For $\alpha \in H^1(G, \mathbb{Q}^n/\mathbb{Z}^n)$, let

$$N_\alpha = \{n \in N \mid n * \alpha = \alpha\}. \tag{5.1}$$

By $F \colon \operatorname{Aut}(\Gamma) \to \operatorname{Aut}(\mathbb{Z}^n)$ we understand the obvious restriction homomorphism. By the definition of a crystallographic group it is well defined. In fact, it is enough to use the maximality of $\mathbb{Z}^n$. Let $\operatorname{Aut}^0(\Gamma) = \ker F$. The following theorem is our main, technical tool.

Theorem 5.4 ([23, Theorem 1.1, p. 172]). *Let Γ be a crystallographic group, which corresponds to a cocycle $\alpha \in H^2(G, \mathbb{Z}^n)$. Then Diagram 5.1 commutes and all its rows and columns are exact.*

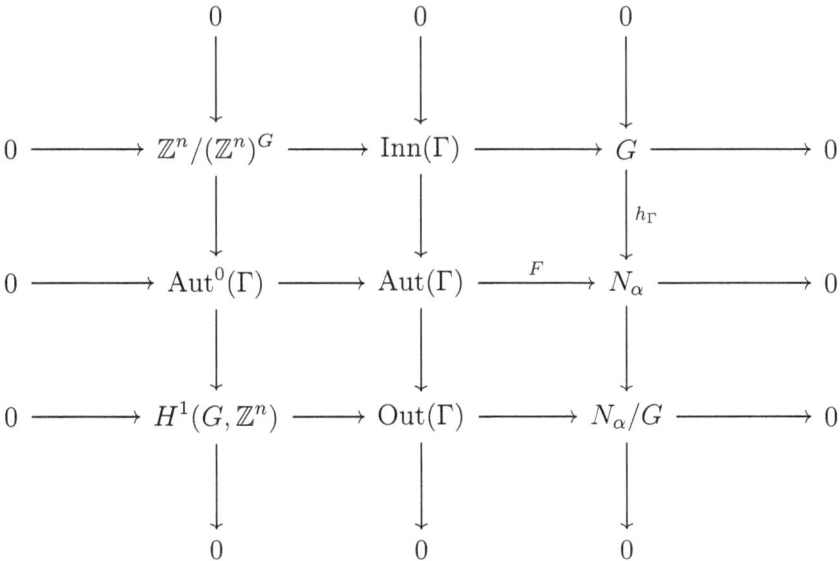

Diagram 5.1

Proof. (We follow [23, p. 172].) Commutativity is obvious. The exactness of the middle column and middle row follow from the definition. We have (see (2.5)) an isomorphism of groups $H^1(G, \mathbb{Z}^n) \simeq Z^1(G, \mathbb{Z}^n)/B^1(G, \mathbb{Z}^n)$ where $Z^1(G, \mathbb{Z}^n) = \mathrm{Der}(G, \mathbb{Z}^n)$ and $B^1(G, \mathbb{Z}^n) = \mathrm{P}(G, \mathbb{Z}^n)$. We claim that $\mathrm{Aut}^0(\Gamma)$ is isomorphic to $Z^1(G, \mathbb{Z}^n)$. In fact, let $c \in H^1(G, \mathbb{Q}^n/\mathbb{Z}^n) \overset{(2.4)}{\simeq} H^2(G, \mathbb{Z}^n)$ be a representative of α. Define a map $H \colon \mathrm{Aut}^0(\Gamma) \to Z^1(G, \mathbb{Z}^n)$ by

$$\psi(g, c(g)) = (g, c(g) + H(\psi)(g)), (\psi \in \mathrm{Aut}^0(\Gamma)).$$

Note that ψ induces the identity on G, so the first component of $\psi(g, c(g))$ is g. Let $H(\psi) = h$ and $g, l \in G, z \in \mathbb{Z}^n$. By expanding $\psi(g, c(g)), (l, c(l))$ first as $\psi(g, c(g)) \cdot \psi(l, c(l))$, and then as $\psi(gl, gc(l) + c(g))$ we get that $h(gl) = gh(l) + h(g)$ or $h \in Z^1(G, \mathbb{Z}^n)$. Cleary, if $H(\psi)(g) = 0, \forall g \in G$, then ψ is the identity, so H is injective. On the other hand, given $h \in Z^1(G, \mathbb{Z}^n)$, define a map $\psi \colon \Gamma \to \Gamma$ by

$$\psi(g, c(g) + z) = (g, c(g) + h(g) + z).$$

It is easily checked that $\psi \in \mathrm{Aut}^0(\Gamma)$, and claim is proved. This also establishes all assertions about the diagram to the left and above $\mathrm{Aut}^0(\Gamma)$. Now, suppose $\psi \in \mathrm{Aut}^0(\Gamma) \cap \mathrm{Inn}(\Gamma)$, i.e. suppose ψ is a conjugation by (g_0, z_0). Since ψ induces the identity on $\mathbb{Z}^n, g_0 = $ the identity and ψ is conjugation by (I, z_0), and one can see that $H(\psi)(g) = z_0 - gz_0$ which implies that $H(\psi) \in B^1(G, \mathbb{Z}^n)$. For a proof that the left column is exact we observe that there is a short exact sequence of groups

$$0 \to (\mathbb{Z}^n)^G \to \mathbb{Z}^n \overset{\Phi}{\to} B^1(G, \mathbb{Z}^n) \to 0,$$

where $\Phi(z)(g) = z - gz$. It only remains to define the maps in the bottom row and to prove that the row is exact. But the maps can be defined to be the unique maps which will make the diagram commute and a diagram chase shows exactness.   □

Let $\mathcal{M}_1, \mathcal{M}_2$ be flat manifolds of dimension n. If the diffeomorphism

$$f \colon \mathcal{M}_1 \to \mathcal{M}_2$$

preserves the Riemann connection then it is called affine. By $\mathrm{Aff}(\mathcal{M}_1)$, we shall denote the group of all affine diffeomorphisms of $\mathcal{M}_1$. That is a Lie group (see [23]). In case $\mathcal{M}_1 = \mathbb{R}^n, \mathrm{Aff}(\mathbb{R}^n) = A(n)$. Hence, we have

$$\mathrm{Aff}(\mathcal{M}_1) = \{f \colon \mathcal{M}_1 \to \mathcal{M}_1 \mid \tilde{f} \colon \mathbb{R}^n \to \mathbb{R}^n \in A(n)\}.$$

Here $\tilde{f}$ is induced by f on the universal covering (see the proof of next proposition). We want to show the relations between the group $\mathrm{Aff}(\mathcal{M}_1)$ and the group $\mathrm{Out}(\pi_1(\mathcal{M}_1))$.

Proposition 5.5. *Let Γ be the fundamental group of the manifold $\mathcal{M}_1$.*

1. Let

$$N(\Gamma) = \{n \in A(n) \mid \forall \gamma \in \Gamma, n\gamma n^{-1} \in \Gamma\}$$

be the normalizer of Γ in $A(n) = GL(n, \mathbb{R}) \ltimes \mathbb{R}^n$. Then the following sequence of groups

$$1 \to \Gamma \to N(\Gamma) \xrightarrow{\bar{f}} \mathrm{Aff}(\mathcal{M}_1) \to 1$$

is exact.

2. Let

$$C(\Gamma) = \{c \in A(n) \mid \forall \gamma \in \Gamma, c\gamma = \gamma c\}$$

be the centralizer of Γ in $A(n)$. Then the following sequence of groups

$$1 \to C(\Gamma) \to N(\Gamma) \xrightarrow{g} \mathrm{Aut}(\Gamma) \to 1$$

is exact.

Proof. We define $\bar{f}$ using the universal covering $p \colon \mathbb{R}^n \to \mathcal{M}_1$.[1] We claim that $\bar{f}$ is a surjection. In fact, choose $\phi \in \mathrm{Aff}(\mathcal{M}_1)$. By definition of the universal covering, we have a map $\bar{\phi} \in A(n)$, which is a lift of ϕ. It is clear, that $\bar{f}(\bar{\phi}) = \phi$. We still have to show that $\bar{\phi} \in N(\Gamma)$. Let $p \colon \mathbb{R}^n \to \mathcal{M}_1$ be the universal covering map, let $\sigma \in \Gamma$, and $x_0 \in \mathbb{R}^n$. We have

$$p[\bar{\phi}(\sigma x_0)] = \phi[p(\sigma x_0)] = \phi[p(x_0)] = p[\bar{\phi}(x_0)].$$

Then $\bar{\phi}(\sigma x_0)$ and $\bar{\phi}(x_0)$ belong to the same orbit. Hence, $\exists \sigma' \in \Gamma$, such that $\bar{\phi}(\sigma x_0) = \sigma' \bar{\phi}(x_0)$. Summing up, we can see that maps $\bar{\phi}\sigma$ and $\sigma'\bar{\phi}$ are lifts of ϕ and coincide at x_0. Hence they are equal and $\bar{\phi} \in N(\Gamma)$. This proves our claim. To finish the proof we have to show that $\ker \bar{f} = \Gamma$. Choose $\bar{\phi} \in N(\Gamma)$ such that it induces the identity on $\mathcal{M}_1$. From properties of a covering, we have $\bar{\phi} \in \Gamma$. The reverse inclusion is obvious. Point 2 follows from the second Bieberbach Theorem (Theorem 2.4). $\qquad \square$

---

[1]Let $\bar{x} \in \mathbb{R}^n$ and $p(\bar{x}) = x$. Then $\bar{x}$ is mapped by some element from $N(\Gamma)$ to $\bar{\bar{x}}$ and $\bar{f}(x) = p(\bar{\bar{x}})$.

We shall need the following lemma.

Lemma 5.6. *Let Γ be the Bieberbach group from the above proposition. Let $\mathcal{Z}$ be its centre, let G be its holonomy group and let $\mathbb{Z}^n$ be its translation subgroup. Then*

1. $C(\Gamma) \subset \mathbb{R}^n$,

2. $C(\Gamma) = (\mathbb{R}^n)^G$,

3. $\mathcal{Z} = C(\Gamma) \cap \Gamma = (\mathbb{Z}^n)^G$.

Proof. Let $(A, t) \in C(\Gamma)$ and $(I, s) \in \mathbb{Z}^n \subset \Gamma$. By definition of the centralizer $C(\Gamma)$, it follows that,

$$(A, As + t) = (A, t)(I, s) = (I, s)(A, t) = (A, t + s).$$

Since $\mathbb{Z}^n$ generates $\mathbb{R}^n$ as a linear space, it follows that A has to be the identity. This proves point 1. Proofs of points 2 and 3, are left as an exercise. $\square$

Let G be any group. By $\mathrm{Out}(G)$ we shall understand the quotient group $\mathrm{Aut}(G)/\mathrm{Inn}(G)$, where $\mathrm{Inn}(G)$ is the subgroup of inner automorphisms.

Theorem 5.7 ([23, Chapter V]). *Let $\mathcal{M}_1$ be a flat manifold of dimension n. Then $\mathrm{Aff}_0(\mathcal{M}_1) = \bar{f}((\mathbb{R}^n)^G)$ and*

$$\mathrm{Aff}(\mathcal{M}_1)/\mathrm{Aff}_0(\mathcal{M}_1) \simeq \mathrm{Out}(\Gamma).$$

Proof. We shall use Diagram 5.2.

$$
\begin{array}{ccccccccc}
 & & 1 & & 1 & & 1 & & \\
 & & \downarrow & & \downarrow & & \downarrow & & \\
1 & \longrightarrow & \mathcal{Z} & \longrightarrow & C(\Gamma) = (\mathbb{R}^n)^G & \to & f\left((\mathbb{R}^n)^G\right) & \longrightarrow & 1 \\
 & & \downarrow & & \downarrow & & \downarrow & & \\
1 & \longrightarrow & \Gamma & \longrightarrow & N & \overset{\bar{f}}{\longrightarrow} & \mathrm{Aff}(\mathcal{M}_1) & \longrightarrow & 1 \\
 & & \downarrow & & \downarrow{g} & & \downarrow & & \\
1 & \longrightarrow & g(\Gamma) & \longrightarrow & \mathrm{Aut}(\Gamma) & \longrightarrow & \mathrm{Out}(\Gamma) & \longrightarrow & 1 \\
 & & \downarrow & & \downarrow & & \downarrow & & \\
 & & 1 & & 1 & & 1 & &
\end{array}
$$

Diagram 5.2

The maps $\bar{f}$ and g were defined in Proposition 5.5. Since $g(\Gamma) = \mathrm{Inn}(\Gamma)$, it is easy to see that Diagram 5.2 commutes and has exact rows and columns. We still have to prove that $\mathrm{Aff}_0(\mathcal{M}_1) = f((\mathbb{R}^n)^G)$. It is clear that $f((\mathbb{R}^n)^G)$ is a connected set which contains the identity. Hence, it contains the connected component of $\mathrm{Aff}_0(\mathcal{M}_1)$. The reverse inclusion follows from the theory of Lie groups [23, p. 214]. □

With support of Theorem 5.4, we start to describe a group of outer automorphisms of crystallographic groups. The crucial fact is the following short exact sequence of groups (see Diagram 5.1)

$$0 \to H^1(G, \mathbb{Z}^n) \to \mathrm{Out}(\Gamma) \to N_\alpha/G \to 0. \qquad (5.2)$$

It follows, that for given Γ, $\mathrm{Out}(\Gamma)$ is an infinite group if and only if the group N_α is infinite. Since α has finite order, we conclude that infiniteness of N_α is equivalent to infiniteness of N.

Proposition 5.8. *If $G \subset GL(n, \mathbb{Z})$ is a finite group then its centralizer*

$$C_{GL(n,\mathbb{Z})}(G) = \{c \in GL(n, \mathbb{Z}) \mid \forall g \in G \ cgc^{-1} = g\}$$

is a normal subgroup of the normalizer

$$N_{GL(n,\mathbb{Z})}(G) = \{d \in GL(n, \mathbb{Z}) \mid \forall g \in G \ dgd^{-1} \in G\}.$$

Moreover, $C_{GL(n,\mathbb{Z})}(G)$ is infinite if and only if, $N_{GL(n,\mathbb{Z})}(G)$ is infinite.

Proof. Since the group G is finite, it follows that for any element a from the normalizer there is a natural number n such that a^n is an element of the centralizer. □

A criterion for (in)finiteness of the centralizer of a finite subgroup of $GL(n, \mathbb{Z})$ was already known to C. S. Siegel more than 60 years ago (see [127]). Then it was reproved again in [17]. In our proof [137] we used the following Dirichlet Unit Theorem. Let L be a finite extension of the rational numbers $\mathbb{Q}$. Let $R \subset L$ be a subring of integral elements of a field L. [2]

Theorem (Dirichlet Unit Theorem [129]). *The group R^* of units of the ring R is isomorphic to $W \times \mathbb{Z}^r$, where W is the finite cyclic group consisting of all the roots of unity in L and r is a natural number. In the case $L = \mathbb{Q}(\sqrt{-d})$, where d is a natural number, $r = 0$. The field $\mathbb{Q}(\sqrt{-d})$ is a complex, quadratic extension of $\mathbb{Q}$.*

---

[2] An element $r \in R \subset L$ is an integral element if and only if it is a root of a polynomial $w \in \mathbb{Z}[X]$, with leading coefficient 1.

5.2 Infinity of outer automorphism group

Theorem 5.9. *Let Γ be a torsion free crystallographic group. The following conditions are equivalent:*

(i) $\mathrm{Out}(\Gamma)$ *is infinite;*

(ii) the $\mathbb{Q}$-decomposition of h_Γ has at least two components of the same isomorphism type or there is a $\mathbb{Q}$-irreducible component which is reducible over $\mathbb{R}$.

Proof. ([137]) Let $S \subset GL(m, \mathbb{Q})$ be a finite group. We shall need the following lemma.

Lemma 5.10. *There is an element $M \in GL(m, \mathbb{Z})$ of infinite order which commutes with all $s \in S$, if and only if there is $M_1 \in GL(m, \mathbb{Q})$, of infinite order, with characteristic polynomial in $\mathbb{Z}[X]$, having constant term ± 1 and which commutes with all $s \in S$.*

Proof of Lemma 5.10. If M exists, it will be suitable for M_1. If M_1 exists, then we have

$$R \in GL(m, \mathbb{Z}), P_1 \in GL(m, \mathbb{Q}),$$

such that $R = P_1^{-1} M_1 P_1$. The integral matrix R is the rational canonical form of M_1, and it exists from the properties of the characteristic polynomial of M_1 and from the structure theorem for finitely generated modules over the principal ideal domain $\mathbb{Q}[X]$ (see [83, Chapter XIV]). Here, the finitely generated $\mathbb{Q}[X]$-module is a m-dimensional linear vector space with the $\mathbb{Q}[X]$-structure defined by the action of M_1. Let h be the product of the denominators of the elements in P_1 and P_1^{-1}. Then there is an integer $l \geq 1$ such that h divides all the entries of $R^l - I$.[3] Hence the matrix M_1^l will be suitable for M. $\qquad\square$

Let us start the proof from the implication $(i) \Rightarrow (ii)$. If $\mathrm{Out}(\Gamma)$ is infinite then $C_{GL(n,\mathbb{Z})}(h_\Gamma(G))$ is infinite. Let

$$h_\Gamma \simeq (h_\Gamma)_1 \oplus (h_\Gamma)_2 \oplus \cdots \oplus (h_\Gamma)_k,$$

where $(h_\Gamma)_i$ are $\mathbb{Q}$-irreducible representations for $i = 1, 2, \ldots, k$. Assume that for each $i = 1, 2, \ldots, k$ representations $(h_\Gamma)_i$ are $\mathbb{R}$-irreducible and non isomorphic to each other. We claim that in this case the group $C_{GL(n,\mathbb{Z})}(h_\Gamma(G))$

---

[3]It is enough to consider the map $q : GL(m, \mathbb{Z}) \to GL(m, \mathbb{Z}_h)$.

is finite. In fact, suppose $H \in C_{GL(n,\mathbb{Z})}(h_\Gamma(G))$. By definition, there exists a new $\mathbb{Q}$-basis such that the matrix $H' = P^{-1}HP$, where P is the matrix of the new basis, has k single blocks H_i on the diagonal. It is clear that $H_i \in C_{GL(n_i,\mathbb{Q})}((h_\Gamma)_i(G))$, for some natural numbers n_i, where $i = 1, 2, \ldots, k$. Hence, from Lemma 5.10 for $M_1 = H_i$, we can assume that h_Γ is $\mathbb{Q}$-irreducible and $\mathbb{R}$-irreducible. We distinguish three cases:

1. h_Γ is absolutely irreducible. Then, by Schur's lemma (Lemma 5.1), the only matrices commuting with h_Γ are scalar matrices λI, where I is the identity matrix. Since the determinant of the matrix is ± 1 and it comes from $GL(n, \mathbb{Z})$, we see that λ is a root of unity and the matrix has a finite order.

2. h_Γ decomposes over $\mathbb{C}$ with Schur index 1. Choosing a suitable basis for K^n, where K is a minimal splitting field for h_Γ, the matrix of h_Γ will be

$$
\begin{pmatrix} (h_\Gamma)_1 & 0 \\ 0 & \overline{(h_\Gamma)_1} \end{pmatrix},
$$

where $(h_\Gamma)_1$ is $\mathbb{C}$-irreducible. Since we assume that the Schur index is 1, we conclude that $(h_\Gamma)_1$ and $\overline{(h_\Gamma)_1}$ are not isomorphic. Again, by Schur's lemma the only commuting matrices which come from $GL(n, \mathbb{Q})$ are of the form

$$
\begin{pmatrix} \lambda I & 0 \\ 0 & \bar{\lambda} I \end{pmatrix},
$$

where $\lambda \in K$. By assumption, the determinant is ± 1. Hence $|\lambda \bar{\lambda}| = 1$ and λ is on the unit circle. But $\lambda \in K$, a complex quadratic extension of $\mathbb{Q}$, and, by **Dirichlet's unit theorem**, λ is a root of unity. Hence our matrix has a finite order.

3. h_Γ decomposes over $\mathbb{C}$ with Schur index 2. Let K be as in the case two. Choosing a suitable basis for K^n, the matrix of h_Γ will be

$$
\begin{pmatrix} (h_\Gamma)_1 & 0 \\ 0 & \overline{(h_\Gamma)_1} \end{pmatrix},
$$

where $(h_\Gamma)_1$ is $\mathbb{C}$-irreducible and $(h_\Gamma)_1, \overline{(h_\Gamma)_1}$ are isomorphic. Let $J \in GL(n/2, K)$ and $J(h_\Gamma)_1 = \overline{(h_\Gamma)_1}J$. Then, since $\bar{J}J$ commutes with $(h_\Gamma)_1$, $\bar{J}J = \kappa I$, with $\kappa \in \mathbb{Q} = K \cap \mathbb{R}$. If $\kappa > 0$ then the matrix

$$
\begin{pmatrix} 0 & \bar{J} \\ J & 0 \end{pmatrix},
$$

whose square is κI, has real eigenvalues and commutes with h_Γ, which is irreducible over $\mathbb{R}$. As it is not a multiple of the identity, this contradicts Schur's lemma, and so $\kappa < 0$. We would like to mention that the matrices [4]

$$\begin{pmatrix} (h_\Gamma)_1 & 0 \\ 0 & \overline{(h_\Gamma)_1} \end{pmatrix}, \begin{pmatrix} 0 & \bar{J} \\ J & 0 \end{pmatrix}.$$

become real after conjugation by the matrix

$$\begin{pmatrix} I & I \\ -iI & iI \end{pmatrix}.$$

Now, by Schur's lemma again every matrix which commutes with h_Γ and comes from $GL(n, \mathbb{Q})$ has to be of the form

$$\begin{pmatrix} \lambda I & \nu \bar{J} \\ \bar{\nu} J & \bar{\lambda} I \end{pmatrix}.$$

Note that the determinant of this matrix is $(\lambda\bar{\lambda} - \kappa\nu\bar{\nu})^{n/2}$. From assumption $\lambda\bar{\lambda} - \kappa\nu\bar{\nu} = 1$. It cannot be -1, because $\kappa < 0$. Then $|\lambda| \leq 1$, so $|\lambda + \bar{\lambda}| \leq 2$. Let us calculate the characteristic polynomial of the above matrix. We have

$$\begin{pmatrix} (\lambda - x)I & \nu\bar{J} \\ \bar{\nu}J & (\bar{\lambda} - x)I \end{pmatrix} =$$

$$\begin{pmatrix} I & \nu(\bar{\lambda} - x)^{-1}\bar{J} \\ (\lambda - x)^{-1}\bar{\nu}J & I \end{pmatrix} \begin{pmatrix} (\lambda - x)I & 0 \\ 0 & (\bar{\lambda} - x)I \end{pmatrix}.$$

With support of Exercise 5.5 we obtain the final formula $(x^2 - (\lambda + \bar{\lambda})x + 1)^{n/2}$. Since for $a = 0, \pm 1, \pm 2$ the roots of the polynomial $x^2 + ax + 1$ are roots of unity, we can assume that $(\lambda + \bar{\lambda}) = 2$. [5] Hence $\lambda = 1$. Because any power of the above matrix has the same properties of the characteristic polynomial and eigenvalues, the only possibility is $\nu = 0$. Hence, again our matrix has finite order and the claim is proved. To finish the first part of the proof it is enough to observe that our centralizer is a group in which any element has an order smaller than some given number r. It follows from the Dirichlet unit theorem. Since the centralizer is a subgroup of $GL(n, \mathbb{Z})$, by the Burnside theorem [58, VIII.19, p. 224] it is finite.

---

[4] We thank K. Dekimpe for this remark.

[5] If λ is an eigenvalue of the matrix A, then λ^n is an eigenvalue of the matrix A^n.

Now, we can start the proof of the implication $(ii) \Rightarrow (i)$. Suppose that h_Γ has at least two components of the same isomorphism type. In the matrix language it means that there is a basis which gives the following decomposition

$$
\begin{pmatrix}
(h_\Gamma)_1 & 0 & \ldots & 0 \\
0\ldots & (h_\Gamma)_1 & 0\ldots & \ldots 0 \\
\ldots & \ldots & (h_\Gamma)_i & 0\ldots 0 \\
0\ldots & \ldots & \ldots & (h_\Gamma)_n
\end{pmatrix}.
$$

It is an exercise to see that

$$
H_1 =
\begin{pmatrix}
I & nI & \ldots & 0 \\
0 & I & 0 & \ldots 0 \\
\ldots & \ldots & I & \ldots \\
0 & \ldots & \ldots & I
\end{pmatrix},
$$

has an infinite order, where I is the identity and $n \in \mathbb{N}$. Moreover, it commutes over $\mathbb{Q}$ with the holonomy representation. Let P be the rational matrix which conjugates the above matrices of the holonomy representation h_Γ onto integral matrices. By definition, we observe that PH_1P^{-1} satisfies the conditions of Lemma 5.10 and the centralizer is infinite.

We still have to prove infiniteness of the centralizer in the case when one of the components $(h_\Gamma)_i = h$ is $\mathbb{R}$-reducible. Suppose $dim_\mathbb{Q}h = m$ and let K be a minimal splitting field of h_Γ whose intersection with $\mathbb{R}$ is non-trivial. Let $n = |G|$. We can assume that K is a subfield of $\mathbb{Q}(\zeta)$, where ζ is a primitive n-th root of unity. Hence the Galois group A of the extension $(K : \mathbb{Q})$ is abelian. Suppose

$$
A = \{\sigma_i\}_{i=1,2,\ldots,l}.
$$

Choose a basis for K^m so that h decomposes as the direct sum of $\mathbb{R}$-irreducible representations

$$
h_i = h_1^{\sigma_i}, i = 1, 2, \ldots, l.
$$

Here $h_1^{\sigma_i}$ are the Galois conjugates of the representation h_1 [32, p. 152]. Suppose $\lambda \in K$, where K is not a complex quadratic extension of $\mathbb{Q}$. Let us denote, the Galois conjugates of λ by $\{\sigma_i(\lambda) = \lambda^{\sigma_i} = \lambda_i\}_{i=1,2,\ldots,l}$. By the Dirichlet's unit theorem we can choose the algebraic unit $\lambda \in K$, such that for any $i = 1, 2, \ldots, l$ λ_i is not on the unit circle. Let $D \in GL(m, K)$ be the diagonal matrix with blocks $\lambda_i I_{dim_\mathbb{Q}h_1}$ on the diagonal. It comes from the $GL(n, \mathbb{Q})$ and its characteristic polynomial

$$
f(X) = ((X - \lambda_1)(X - \lambda_2)\ldots(X - \lambda_l))^{\frac{m}{l}}.
$$

By Galois theory, since λ is an algebraic integer, we have $f(X) \in \mathbb{Z}[X]$. Hence and by arguments of Lemma 5.10, there exists a matrix $C \in GL(m, \mathbb{Z})$ which belongs to the centralizer, see also Exercise 5.7.

Since any $\lambda_i, i = 1, 2, \ldots, l$ is not on the unit circle, C has infinite order. This finishes the proof of the theorem. $\square$

We have the following corollaries and consequences of Theorem 5.9. Let $\mathcal{W}$ be a compact Riemannian manifold. A smooth map

$$f : \mathcal{W} \to \mathcal{W}$$

is called a diffeomorphism, when it is a bijection and f^{-1} is smooth.

Definition 5.11. A function f is an Anosov diffeomorphism, ([95]) when there exist a constant $c > 0$, $0 < \lambda < 1$, such that for any $w \in \mathcal{W}$, $T\mathcal{W}_w = E^s \oplus E^u$, and for any $v \in E^s$, $t \in E^u$ and a positive integral number $r > 0$

$$\|Tf^r v\| \leq c\lambda^r \|v\| \text{ and } \|Tf^{-r} t\| \leq c\lambda^r \|t\|.$$

Example 5.12. Let

$$\varphi = \begin{pmatrix} a & b \\ c & d \end{pmatrix}$$

be an algebraic automorphism of T^2 (i.e. $a, b, c, d \in \mathbb{Z}$, $det\varphi = \pm 1$). If φ has no eigenvalues with norm 1, then it is an Anosov diffeomorphism on T^2.

The main motivation for us was the Porteous's Theorem [117]. It gives description of Anosov diffeomorphisms on a flat manifold M in language of the theory of representations of a holonomy group of the fundamental group Γ. In particular we have:

Theorem 5.13 ([117, Theorem 6.1]). *The following conditions are equivalent:*

(i) M admits the Anosov diffeomorphism;

(ii) any $\mathbb{Q}$-irreducible component of h_Γ with multiplicity one is $\mathbb{R}$-reducible.

The proof is very similar to the proof of the Theorem 5.9 and we leave it as an exercise, (see [117]).

Remark 5.14. For the Anosov diffeomorphisms we have the following conjecture: if a compact aspherical manifold W admits an Anosov diffeomorphism, then W has finite covering by a nilmanifold $\tilde{W}$.[6] (W has finite covering by a nilmanifold if and only if $\pi_1(W)$ is the infinite, torsion free group with a finite index nilpotent subgroup.)

It turns out, that we can give a complete description and characterization of the group of the outer automorphisms as a subgroup of the linear group. In 1972 J. Tits (see [144]) proved that any finitely generated subgroup of $GL(n, \mathbb{Z})$ is either a virtually solvable or has as a subgroup the free group.[7] We want to add that any virtually solvable subgroup of $GL(n, \mathbb{Z})$ is virtually polycyclic [94].

Definition 5.15. A group G is polycyclic if it has a subnormal sequence of subgroups
$$\{e\} = G_0 \trianglelefteq G_1 \trianglelefteq G_2 \trianglelefteq \cdots \trianglelefteq G_{k-1} \trianglelefteq G_k = G,$$
such that, for any $i = 0, 1, \ldots, k-1$, the group G_{i+1}/G_i is cyclic.

Recall (see [32]) that each semisimple finite-dimensional $\mathbb{Q}$-algebra A has a Wedderburn decomposition $A = \oplus_{i=1}^{k} A_i (k \in \mathbb{N})$ into simple algebras A_i. Each component A_i can be identified with a full $(f_i \times f_i)$-matrix ring $M_{f_i}(D_i)$ over a finite-dimensional division algebra D_i over $\mathbb{Q}, 1 \le i \le k$. A subring with the unity Δ of A is a $\mathbb{Z}$-order if it is a finitely generated $\mathbb{Z}$-module such that $\mathbb{Q}\Delta = A$. For some subset X of A, the centralizer $C_A X$ is a subalgebra of A, referred to as the commuting algebra of X in A. We shall need a lemma.

Lemma 5.16 ([79, p. 209], [150, p. 1621]). *The unit group of a $\mathbb{Z}$-order in a semisimple finite-dimensional $\mathbb{Q}$-algebra A is virtually solvable if and only if Wedderburn components of A are number fields or definite quaternions over $\mathbb{Q}$.*

We have:

Theorem 5.17 ([94]). *Let Γ be a torsion free crystallographic group. Then the following conditions are equivalent:*

---

[6]Let N be a simple connected, nilpotent Lie group, let Γ be a discrete, cocompact and torsion free subgroup of Isom(N). The orbit space N/Γ is called a nilmanifold.

[7]A group is virtually solvable (polycyclic), when has a subgroup of a finite index, which is solvable (polycyclic).

(i) Out(Γ) *is virtually polycyclic;*

(ii) *in the $\mathbb{Q}$-decomposition of h_Γ all components occur with multiplicity 1 and the Schur index over $\mathbb{Q}$ of each $\mathbb{R}$-reducible component is equal to 1.*

Proof. From Theorem 5.1 we have a short exact sequence (5.2) of groups

$$0 \to H^1(G, \mathbb{Z}^n) \to \mathrm{Out}(\Gamma) \to N_\alpha/G \to 0,$$

where G is a holonomy group of Γ, $\mathbb{Z}^n \subset \Gamma$ is a maximal abelian subgroup and $\alpha \in H^2(G, \mathbb{Z}^n)$ the cocycle which corresponds to Γ. Moreover, $N_\alpha \subset N = N_{GL(n,\mathbb{Z})}(h_\Gamma(G))$. Hence, it is enough to observe that $\mathrm{Out}(\Gamma), N_\alpha$ and N are all virtually polycyclic if and only if one of them is virtually polycyclic. As a consequence we have to prove that a condition (ii) is equivalent to a condition that a group N is virtually polycyclic. Because G is finite, the centralizer $C_{GL(n,\mathbb{Z})}(h_\Gamma(G))$ is of finite index in N and we shall prove a statement for $C_{GL(n,\mathbb{Z})}(h_\Gamma(G))$. This is exactly the unit group of a commuting ring of $h_\Gamma(G)$ in $M(n, \mathbb{Z})$. Moreover, $C_{M(n,\mathbb{Z})}(h_\Gamma(G))$ is a $\mathbb{Z}$-order in a semisimple finite-dimensional $\mathbb{Q}$-algebra $C_{M(n,\mathbb{Q})}(h_\Gamma(G))$. For a suitable basis of $\mathbb{Q}^n$, h_Γ decomposes into a direct sum of $\mathbb{Q}$-irreducible components $(h_\Gamma)_i$. Without loss of generality, one can assume that in this decomposition equivalent components are identical, then write f_i for a multiplicity of $(h_\Gamma)_i$ and hence

$$h_\Gamma = \bigoplus_{i=1}^{k} f_i (h_\Gamma)_i, \, (k \in \mathbb{N}).$$

Then a Wedderburn decomposition of $C_{M(n,\mathbb{Q})}(h_\Gamma(G))$ is

$$C_{M(n,\mathbb{Q})}(h_\Gamma(G)) \simeq \bigoplus_{i=1}^{k} M_{f_i}(D_i),$$

where D_i can be chosen as the commuting algebra of $(h_\Gamma)_i, 1 \leq i \leq k$. Because of Lemma 5.16, if $f_i > 1$ for some $i, 1 \leq i \leq k$, then $C_{GL(n,\mathbb{Z})}(h_\Gamma(G))$ contains a non-abelian free group. Therefore, assume that all components in the $\mathbb{Q}$-decomposition of h_Γ have multiplicity one (for all $i, 1 \leq i \leq k, f_i = 1$). Then the commuting algebras D_i of the $\mathbb{Q}$-irreducible components $(h_\Gamma)_i$ of h_Γ are exactly the Wedderburn components of $C_{M(n,\mathbb{Q})}(h_\Gamma(G))$. In case when $(h_\Gamma)_i$ is $\mathbb{R}$-reducible, realize that D_i is a number field if and only if $m_{\mathbb{Q}}((h_\Gamma)_i) = 1$.

Moreover, if $(h_\Gamma)_i$ is $\mathbb{R}$-irreducible, then its commuting algebra D_i is either $\mathbb{Q}$, an imaginary quadratic extension of $\mathbb{Q}$ or a definite quaternion algebra over $\mathbb{Q}$. Applying Lemma 5.16 now finishes the proof. $\square$

Example 5.18. The following finite groups have Schur index one over $\mathbb{Q}$ ([94, section 3]):

1) finite abelian groups,

2) finite p-groups with $p \neq 2$,

3) finite solvable groups such that all their Sylow subgroups are elementary abelian,

4) all finite groups of order $2, 3, 4, 5, 9, 11$ and 13.

5.3 $\mathcal{R}_1$-groups

Classification of torsion free crystallographic groups with finite outer automorphisms group is an open problem. The natural first step is the classification of the class of holonomy groups of the Bieberbach groups with finite outer automorphism group. We shall denote this class of groups by $\mathcal{R}_1$. In this section G always denotes a finite group.

Lemma 5.19. *Let M be a faithful $\mathbb{Q}G$-module and p be a prime.*

(a) If G has a cyclic subgroup H of order $p^\alpha, \alpha \geq 1$, then the dimension of M is at least $p^{\alpha-1}(p-1)$. If H is normal, then M has a simple $\mathbb{Q}G$-submodule of dimension divisible by $p^{\alpha-1}(p-1)$.

(b) If G has a non-trivial normal p-subgroup, then M has a simple $\mathbb{Q}G$-submodule of dimension divisible by $p-1$.

Proof. Write $M_H = M_1 \oplus \cdots \oplus M_n$ with simple $\mathbb{Q}H$-modules M_i. By assumption, there is at least one of the M_i's, on which the subgroup of order p of H does not act trivially. By [32, Theorem (74.17)], this M_i has dimension $p^{\alpha-1}(p-1)$. If H is normal in G, let V denote the simple submodule of M such that M_i is a submodule of V_H. The result follows from Clifford's theorem. This proves (a). The assertion in (b) is proved with a similar argument. $\square$

Lemma 5.20. *Let G be the holonomy group of a Bieberbach group Γ with trivial centre and finite outer automorphism group. If H is an $\mathcal{R}_1$-group, then so is the direct product $G \times H$.*

Proof. Let Γ' be a Bieberbach group with holonomy group H and finite outer automorphism group. Then $\Gamma \times \Gamma'$ is a Bieberbach group with holonomy $G \times H$. The $\mathbb{Q}$-representation of $G \times H$ on the translation lattice of $\Gamma \times \Gamma'$ is multiplicity free, since by assumption the fixed point space of G on the translation lattice of Γ is trivial. Clearly, the second condition of the criterion for Bieberbach groups with finite outer automorphism groups is also satisfied. This completes the proof of the lemma.   □

In the case of abelian groups we have a complete classification. Before we shall state the result let us make a few observations. From previous considerations it is clear that any simple $\mathbb{R}G$-module of an abelian G group has dimension ≤ 2, (cf. [68, p. 11]). The second observation is that, if G has an element of prime power order $p^\alpha \geq 5$, then any faithful $\mathbb{Q}G$-module has a simple submodule which does not remain simple after extending scalars to $\mathbb{R}$. In fact by Lemma 5.19(a), there is a simple $\mathbb{Q}G$-submodule of dimension divisible by $p^{\alpha-1}(p-1)$. By assumption, this is larger than 2, and the result follows. We have:

Proposition 5.21 ([68, Theorem 4.2]). *If G is a finite abelian group, then $G \in \mathcal{R}_1$ if and only if G is a direct product of cyclic groups of order at most 4 (in other words, if and only if G has an exponent dividing 12).*

Proof. If G is $\mathcal{R}_1$-group the result is obvious by Lemma 5.19 and the above observation. By Lemma 5.20, it suffices to construct Bieberbach groups with finite outer automorphism group and trivial centre with holonomy group in $\{\mathbb{Z}_2 \times \mathbb{Z}_4, \mathbb{Z}_2^k, (k \geq 2), \mathbb{Z}_3^l, (l \geq 2), \mathbb{Z}_4^m, (m \geq 2)\}$, and Bieberbach groups with finite outer automorphism groups with holonomy group in $\{\mathbb{Z}_2, \mathbb{Z}_3, \mathbb{Z}_4, \mathbb{Z}_2 \times \mathbb{Z}_3, \mathbb{Z}_3 \times \mathbb{Z}_4\}$. For the case $\mathbb{Z}_2 \times \mathbb{Z}_4$ see [134, p. 199]. The construction for the groups $\mathbb{Z}_2^k, (k \geq 2)$ and $\mathbb{Z}_3^l, (l \geq 2)$ is done in [64, Theorem 1.1]. For the group $\mathbb{Z}_4^m$ we use the following construction. Let Γ be Bieberbach group with holonomy group $\mathbb{Z}_2^m$ and finite $\mathrm{Out}(\Gamma)$. Let M be a lattice such that $s \in H^2(\mathbb{Z}_2^m, M)$ defines Γ. Shapiro's lemma (Theorem 3.12) shows that there is an element $s' \in H^2(\mathbb{Z}_4^m, \mathrm{ind}_{\mathbb{Z}_2^m}^{\mathbb{Z}_4^m} M)$ defining a Bieberbach group Γ' with holonomy group $\mathbb{Z}_4^m$. It follows from this construction that $\mathrm{Out}(\Gamma')$ is finite. All the above Bieberbach groups have trivial centre. The groups G in the second set are all cyclic, so if M is the direct sum of the unique simple faithful $\mathbb{Z}G$-lattice and the trivial $\mathbb{Z}G$-lattice, the cohomology group $H^2(G, M)$ contains a special element defining Bieberbach groups with the desired properties. This finishes the proof of the proposition.   □

Example 5.22. For odd $n \geq 3$, let $\Gamma_n \subset E(n)$ be a group generated by

$$\{(B_i, s(B_i) = x_i) : 1 \leq i \leq n - 1\}.$$

Here B_i are the $(n \times n)$ diagonal matrices

$$B_i := \mathrm{diag}(-1, \ldots, -1, \underbrace{1}_{i}, -1, \ldots, -1),$$

and $x_i = s(B_i) = e_i/2 + e_{i+1}/2$, $1 \leq i \leq n-1$. The groups Γ_n are examples of *Hantzsche-Wendt* groups, [124]. From Theorem 5.9 it follows that $\mathrm{Out}(\Gamma_n)$ are finite, (see [137, p. 589] and the last chapter).

If a p-group belongs to $\mathcal{R}_1$, then $p = 2$ or $p = 3$, (see Exercise 5.11 or [68, Cor. 3.2]). Moreover, for $p \in \{2, 3\}$ we have:

Proposition 5.23 ([68, Proposition 3.3]). *Let G be a non-abelian p-group of order $p^\alpha (\alpha \geq 3)$, with cyclic normal subgroup of index p. Then $G \in \mathcal{R}_1$ if and only if G is a group of order 8, 27 or 16 and in the last case is one of the groups with presentation $\langle a, b \mid a^8 = b^2 = 1, bab = a^{4\pm1} \rangle$.*

Proof. We first observe that every simple $\mathbb{C}G$-module has dimension at most p (see [71, Satz V.17.10]). Thus a simple $\mathbb{R}G$-module has dimension at most $2p$, and so, by Lemma 5.19(a), $|G| \leq 27$. Suppose now that $p = 2$. If G has order 8 the result follows from ([64, Theorems 1.6 and 1.7]). So let us assume that $|G| = 16$. Since G is non-abelian, the centre of G is contained in the cyclic normal subgroup of index 2, and so has a unique element of order 2. This element is in the kernel of every non-faithful $\mathbb{Q}G$-module. Hence a faithful $\mathbb{Q}G$-module must have a faithful simple submodule. If G is a generalized quaternion group, then a faithful simple $\mathbb{Q}G$-module has dimension 8. If G is a dihedral group, then the unique faithful simple $\mathbb{Q}G$-module does not remain simple after extending scalars to $\mathbb{R}$ (see [32, Theorem (67.14)]). Let G be one of the remaining two groups of order 16. The matrices given in [16, p. 245] under the headings 32/02/01, 32/03/01 respectively define a faithful $\mathbb{Z}G$-lattice M of rank 4. Let $N = \mathrm{Ind}_{\langle a \rangle}^G(\mathbb{Z})$, where $\langle a \rangle$ acts trivially on $\mathbb{Z}$. We now define $L = M \oplus N$ and construct a special element $s = (s_1, s_2) \in H^2(G, M) \oplus H^2(G, N) \simeq H^2(G, L)$. If s_1 is the cocycle given in [16, p. 245] defining the groups 32/02/01, 32/03/01 respectively then the restriction of s_1 to b is non-trivial. Since $H^2(G, N) \simeq H^2(\langle a \rangle, \mathbb{Z})$, we can find a cocycle $s_2 \in H^2(G, N)$ whose restriction to a is non-trivial. Moreover, since every element of order 2 of G is conjugate to a or b, the cocycle $s = (s_1, s_2)$ is special. Now, let $p = 3$ so that $|G| = 27$. It was shown by Symonds [130, Proposition 4.13] that G is the holonomy group of a Bieberbach group Γ such that the $\mathbb{Q}G$-module arising from the translation lattice of Γ is multiplicity free. The proof is completed by observing that every simple $\mathbb{Q}G$-module remains simple over $\mathbb{R}$.   □

We also have classification in the case of the dihedral groups D_{2n} of order $2n$.

Proposition 5.24 ([68, Proposition 5.1]). *D_{2n} belongs to $\mathcal{R}_1$ if and only if n divides 12.*

Proof. Suppose first D_{2n} is an $\mathcal{R}_1$-group. Let p be a prime and let p^α be the highest power of p dividing n. Since every simple $\mathbb{C}G$-module has dimension at most 2 (see [71, Satz V.17.10]) and is realizable over the real numbers (see [32, Theorem 67.14]), it follows from Lemma 5.19(a) that $p^{\alpha-1}(p-1) \leq 2$. This implies that n divides 12.

It remains to show that D_{2n} is indeed an $\mathcal{R}_1$-group for such n. For $n = 1, 2, 3, 4, 6$ see [64], [134] and [16]. We only have to construct a Bieberbach group with finite outer automorphism group and holonomy group equal to $D_{24} = \{a, b \mid a^{12} = 1, b^2 = 1, bab = a^{11}\}$. We begin by constructing a lattice L. It will be a direct sum of a faithful lattice N and a rank two lattice M. We first describe $M = \mathbb{Z}^2$. We define an action of D_{24} on M by:

$$a \mapsto \begin{pmatrix} 1 & 0 \\ 0 & 1 \end{pmatrix}, b \mapsto \begin{pmatrix} 1 & 0 \\ 0 & -1 \end{pmatrix}.$$

We exhibit an element $s \in H^2(D_{24}, M)$ which restricts non-trivially to the elements: ba^3, b, a^6 and a^4. We describe s as a 1-cocycle using identification $H^2(D_{24}, \mathbb{Z}^2) \simeq H^1(D_{24}, \mathbb{R}^2/\mathbb{Z}^2)$. The 1-cocycle $s \colon D_{24} \to \mathbb{R}^2/\mathbb{Z}^2$ is defined on the given generators by:

$$s(a) = <0, 1/12>, \ s(b) = <1/2, 0>.$$

It is not difficult to check that s extends to an element of $H^1(D_{24}, \mathbb{R}^2/\mathbb{Z}^2)$, and also that $s(ba^3), s(b), s(a^6)$ and $s(a^4)$ are not equal to zero. Hence $(s, 0) \in H^2(D_{24}, M) \oplus H^2(D_{24}, N) \simeq H^2(D_{24}, M \oplus N) = H^2(D_{24}, L)$ defines a

Bieberbach group with the holonomy group D_{24}. We define the faithful $\mathbb{Z}G$-lattice N as a direct sum of two $\mathbb{Z}G$-lattices which give simple $\mathbb{R}G$-modules (one for each of the two factor groups D_8 and D_6). Hence the above Bieberbach group has a finite outer automorphism group. □

Let p and q be odd prime numbers and let $C_{p,q}$ (see [18]), be the non-abelian group defined by the short exact sequence

$$1 \to \mathbb{Z}_p \to C_{p,q} \to \mathbb{Z}_q \to 1.$$

Proposition 5.25 ([68, Proposition 5.2]). *Only* $C_{7,3} \in \mathcal{R}_1$.

Proof. Suppose that $C_{p,q}$ is an $\mathcal{R}_1$-group and let M denote $\mathbb{Q}G$-module defined by the translation lattice of a Bieberbach group with the holonomy group $C_{p,q}$ and finite outer automorphism group. Since q divides $p-1$, the p-adic field $\mathbb{Q}_p$ contains a primitive qth root of unity. By applying [114, Theorem (III.2)] for the prime p, we see that $M \otimes_{\mathbb{Q}} \mathbb{Q}_p$ must contain a non-trivial simple submodule of dimension one. Hence M contains a submodule V of dimension $q-1$, faithfully representing the cyclic complement of C_q. But $V \otimes_{\mathbb{R}} \mathbb{R}$ is simple for $q \le 3$ only. Hence $q = 3$. By Lemma 5.19(b), M must have a simple submodule of the dimension divisible by $p-1$. The dimensions of simple $\mathbb{R}G$-modules are in $\{1, 2, 3, 6\}$ [71, Satz V.17.10] and [32, Theorem 73.9]. This implies that $p = 7$. It is easy to see that $C_{7,3}$ is indeed an $\mathcal{R}_1$-group. □

Definition 5.26. The group $\mathcal{G}_n$ of order 2^{n+1} is extra special, provided $[\mathcal{G}_n, \mathcal{G}_n] = Z(\mathcal{G}_n)$, and $\mathcal{G}_n/Z(\mathcal{G}_n) \simeq (\mathbb{Z}_2)^n$.

It happens that (see [67, Proposition 4.1]) also some infinite family of extra special 2-groups belongs to $\mathcal{R}_1$.

The above statements are mainly consequences of the fact that for $p \ge 5$, any $\mathbb{Q}$-irreducible representation of a finite p-group is $\mathbb{R}$-reducible (see Exercise 5.11). For non-abelian, simple groups we have:

Proposition 5.27 ([68, Proposition 6.1]). *If* $G = PSL(2, p)$, *then* $G \in \mathcal{R}_1$ *if and only if* $p \in \{2, 3, 7\}$.

Proof. It follows from the considerations in [114, Section V], that $PSL(2, p)$ can only be an $\mathcal{R}_1$-group for the primes in the above set. We have already seen that $PSL(2,2) = D_6$ is an $\mathcal{R}_1$-group. $PSL(2,3) = A_4$ is in this class, what follows from the constructions in [16, p. 211]. For example, the space group 24/01/02 004 given there is torsion free. Finally, it follows from the result in [114, Sections IV,V], that $PSL(2,7)$ is an $\mathcal{R}_1$-group in the dimension $15 = 7 + 8$.   □

We want to add (cf. [68, Proposition 6.2]) that the Mathieu sporadic simple group M_{11} belongs to $\mathcal{R}_1$, see also example (5.29) below. We have to mention that the description of the class $\mathcal{R}_1$ is an open problem. We could reduce it to a problem from the pure theory of representations of finite groups, if we proved the following conjecture, (Conjecture 1 in part 10.1).

Conjecture. *Let G be any finite group. Then there exists a Bieberbach group Γ such that:*

1. *G is the holonomy group of Γ,*
2. *the holonomy representation h_Γ is $\mathbb{Q}$ multiplicity free.*

Remark 5.28. For the groups from $\mathcal{R}_1$ the above conjecture is true.

In 2003 there was given an example of the Bieberbach group Γ with trivial centre and trivial Out(Γ). By Theorem 5.7 we can say that the flat manifold with fundamental group Γ has no symmetries. Here it is.

Example 5.29 ([147] and Appendix III). Let G be the Mathieu sporadic simple group M_{11}. Then G has a presentation

$$\langle a, b \mid a^2 = b^4 = (ab)^{11} = (ab^2)^6 = ababab^{-1}abab^2ab^{-1}abab^{-1}ab^{-1} = 1 \rangle, \quad (5.3)$$

and representatives of conjugacy classes of subgroups of order 2, 3 respectively are $\langle a \rangle$, $\langle (ab^2)^2 \rangle$ respectively (see [147]). Let L_1, L_2, L_3 and L_4 be the lattices given in *Appendix III*, and let $\delta_i \in H^1(G, \mathbb{Q} \otimes_{\mathbb{Z}} L_i/L_i)$ for $1 \leq i \leq 4$ be the cocycles given there.

These have the following properties:

(1) The character induced by L_1 is $\chi + \bar{\chi}$, where χ is one of the two non-real irreducible characters of G of degree 10. The order of δ_1 is 6, and we have $res_{\langle a \rangle}^G \delta_1 = 0$, but $res_{\langle (ab^2)^2 \rangle}^G \delta_1 \neq 0$.

(2) The character afforded by L_2 is $\chi + \bar{\chi}$, where χ is one of the two irreducible characters of G of degree 16. The order of δ_2 is 5. Hence the restriction of δ_2 to any subgroup of order 5 is non-zero.

(3) The character afforded by L_3 is the irreducible character of G of degree 44. The order of δ_3 is 6, and we have $res^G_{\langle a \rangle} \delta_3 \neq 0$, but $res^G_{\langle (ab^2)^2 \rangle} \delta_3 = 0$.

(4) The character afforded by L_4 is the irreducible character of G of degree 45. The order of δ_4 is 11. Hence the restriction of δ_4 to any subgroup of order 11 is non-zero.

Thus

$$\delta := \delta_1 + \cdots + \delta_4 \in H^1(G, \mathbb{Q} \otimes_{\mathbb{Z}} L/L), \tag{5.4}$$

where $L := L_1 \oplus \cdots \oplus L_4$, is a special element. Let Γ be an extension of L by G given by δ. Then Γ is torsion free and has trivial centre. Moreover, we have $H^1(G, L) = 0$. This is easily checked by using the fact that, if $L^G = 0$, then a prime p divides $|H^1(G, L)|$ if and only if $(L/pL)^G \neq 0$, (see [67, Lemma 2.1]). Now it remains to check that $N_{\mathrm{Aut}(L)}(G)_\delta = G$. Since G has no outer automorphisms, we have $N_{\mathrm{Aut}(L)}(G) = C_{\mathrm{Aut}(L)}(G)G$, and the centralizer of G in $\mathrm{Aut}(L)$ is $C_{\mathrm{Aut}(L_1)}(G) \times \cdots \times C_{\mathrm{Aut}(L_4)}(G)$. We claim that $C_{\mathrm{Aut}(L_i)}(G) = \{\pm\}$ for $1 \leq i \leq 4$. This is obvious for $i = 3, 4$. Now $_{\mathrm{Aut}(L_i)}(G)$ is the unit group of $\mathrm{End}_{\mathbb{Z}G}(L_i)$, which is a $\mathbb{Z}$-order in $\mathrm{End}_{\mathbb{Q}G}(\mathbb{Q} \otimes_{\mathbb{Z}} L_i)$. For $i = 1, 2$ this endomorphism ring is isomorphic to $\mathbb{Q}(\chi)$, where χ is as above. In the first case, $\mathbb{Q}(\chi) = \mathbb{Q}(\sqrt{-2})$, and in the second case we have $\mathbb{Q}(\chi) = \mathbb{Q}(\sqrt{-11})$. In both cases all $\mathbb{Z}$-orders have the unit group $\{\pm\}$, hence the claim. Now it is clear that $C_{\mathrm{Aut}(L)}(G)_\delta = 1$ and the assertion is completed. It has the dimension 141.

There also are examples of Bieberbach groups with trivial centre and outer automorphism group isomorphic to $\mathbb{Z}_3$, (see [91]) and $\mathbb{Z}_2, \mathbb{Z}_2 \times (\mathbb{Z}_2 \wr F)$, where $F \subset S_{2k+1}$ is a cyclic group generated by the cycle $(1, 2, \ldots, 2k+1)$, $k \geq 2$, (see [67]). Recall that group is directly indecomposable if it cannot be expressed as a direct product of its non-trivial subgroups. Moreover, we have a proposition which we leave as an exercise.

Proposition 5.30. *Let Γ be a directly indecomposable Bieberbach group with trivial centre, $n \in \mathbb{N}$ and $\varphi \in \mathrm{Aut}(\Gamma^n)$, where $\Gamma^n = \underbrace{\Gamma \times \cdots \times \Gamma}_{n}$. Then*

$$\exists \sigma \in S_n \ \forall_{1 \leq i \leq n} \ \varphi(\Gamma_i) = \Gamma_{\sigma(i)},$$

where $\Gamma_i := \{1\}^{i-1} \times \Gamma \times \{1\}^{n-i} \subset \Gamma^n$, for $1 \leq i \leq n$.

$\square$

Corollary I. *Let* Γ *be a directly indecomposable Bieberbach group with trivial centre and* $n \in \mathbb{N}$. *Then*

$$\mathrm{Aut}(\Gamma^n) = \mathrm{Aut}(\Gamma) \wr S_n,$$

hence

$$\mathrm{Out}(\Gamma^n) = \mathrm{Out}(\Gamma) \wr S_n.$$

$\square$

From the above example we have:

Corollary II ([91]). *For every* $n \in \mathbb{N}$ *there exists a flat manifold with group of affine self equivalences isomorphic to the symmetric group* S_n.

$\square$

The above suggests a question about the realization of any finite group as a group of affine self equivalences of some flat manifold. For more about it see Problem 6 at the last chapter.

5.4 Crystallographic groups with trivial center and outer automorphism group

In this part we shall prove the following result. Let Γ be a crystallographic group. From Bieberbach's theorems (see Theorem 2.4) we have a short exact sequence of groups

$$0 \to \mathbb{Z}^n \to \Gamma \xrightarrow{p} G \to 0,$$

where $\mathbb{Z}^n$ is a maximal abelian subgroup of Γ and G is a finite group. Moreover, let $h_\Gamma \colon G \to GL(n, \mathbb{Z})$ be a holonomy representation, see 2.6. Let

$$N = N_{GL(n,\mathbb{Z})}(h_\Gamma(G)) = \{X \in GL(n, \mathbb{Z}) \mid \forall_{f \in h_\Gamma(G)} \ XfX^{-1} \in h_\Gamma(G)\}$$

be the normaliser of $h_\Gamma(G)$ in $GL(n, \mathbb{Z})$. In the case when $Z(\Gamma) = \{e\}$, we have the following commutative Diagram 5.3 (see pp. 65-69) with exact rows and columns:

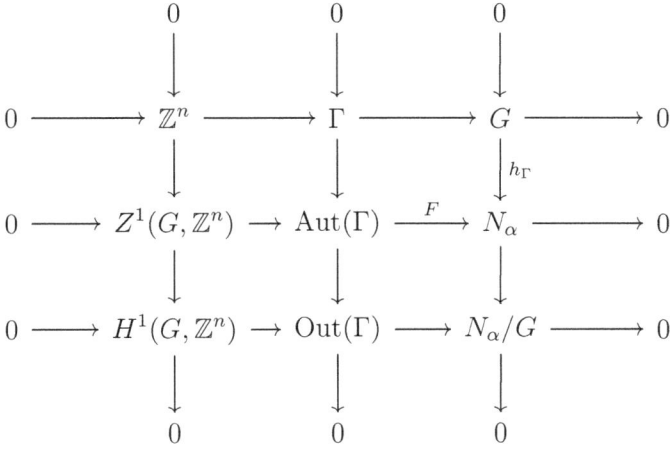

$$
\begin{array}{ccccccccc}
& & 0 & & 0 & & 0 & & \\
& & \downarrow & & \downarrow & & \downarrow & & \\
0 & \longrightarrow & \mathbb{Z}^n & \longrightarrow & \Gamma & \longrightarrow & G & \longrightarrow & 0 \\
& & \downarrow & & \downarrow & & \downarrow{\scriptstyle h_\Gamma} & & \\
0 & \longrightarrow & Z^1(G,\mathbb{Z}^n) & \longrightarrow & \mathrm{Aut}(\Gamma) & \overset{F}{\longrightarrow} & N_\alpha & \longrightarrow & 0 \\
& & \downarrow & & \downarrow & & \downarrow & & \\
0 & \longrightarrow & H^1(G,\mathbb{Z}^n) & \longrightarrow & \mathrm{Out}(\Gamma) & \longrightarrow & N_\alpha/G & \longrightarrow & 0 \\
& & \downarrow & & \downarrow & & \downarrow & & \\
& & 0 & & 0 & & 0 & &
\end{array}
$$

Diagram 5.3

where $Z^1(G,\mathbb{Z}^n)$ is the group of 1-cocycles. Moreover

$$N_\alpha = \{n \in N \mid n * \alpha = \alpha\},$$

and $\alpha \in H^2(G,\mathbb{Z}^n)$ is the cohomology class of the first row of the diagram. The action $*: N \times H^2(G,\mathbb{Z}^n) \to H^2(G,\mathbb{Z}^n)$ is defined by the formula (see 5.1)

$$n * [a] = [n * a],$$

where $n \in N, a \in Z^2(G,\mathbb{Z}^n)$, $[a]$ is the cohomology class of a and

$$\forall_{g_1,g_2 \in G}\ n * a(g_1,g_2) = na(n^{-1}g_1 n, n^{-1}g_2 n).$$

We have the following proposition.

Proposition 5.31. $\mathrm{Aut}(\Gamma)$ *is a crystallographic group if and only if* $\mathrm{Out}(\Gamma)$ *is a finite group.*

Proof. We start with an observation that $Z^1(G,\mathbb{Z}^n)$ is a free abelian group of rank n which is a faithful N_α module. First, assume that $\mathrm{Aut}(\Gamma)$ is a crystallographic group with the maximal abelian subgroup M. From [23, Proposition I.4.1], M is the unique normal maximal abelian subgroup of $\mathrm{Aut}(\Gamma)$. Hence, $M = Z^1(G,\mathbb{Z}^n)$, and $\mathrm{Out}(\Gamma)$ is a finite group. The reverse implication is obvious. This finishes the proof of the proposition. $\square$

Let us formulate the main result of this section 5.4.

Theorem 5.32 ([90, Theorem 2.2]). *For every $n \geq 2$ there exists a crystallographic group Γ of dimension n with $Z(\Gamma) = \mathrm{Out}(\Gamma) = \{e\}$.*

We shall need few lemmas and examples.

Lemma 5.33. *Let G, H be finite groups and $H \subset G \subset GL(n, \mathbb{Z})$. If the group $N_{GL(n,\mathbb{Z})}(H)$ is finite, then $N_{GL(n,\mathbb{Z})}(G)$ is finite.*

Proof of Lemma 5.33. From the assumption, $\mathrm{Aut}(H)$ and $\mathrm{Aut}(G)$ are finite. Moreover, we have monomorphisms:

$$N_{GL(n,\mathbb{Z})}(H)/C_{GL(n,\mathbb{Z})}(H) \overset{\bar{\phi}}{\to} \mathrm{Aut}(H), \quad N_{GL(n,\mathbb{Z})}(G)/C_{GL(n,\mathbb{Z})}(G) \overset{\bar{\phi}}{\to} \mathrm{Aut}(G),$$

where $\bar{\phi}$ is induced by $\phi(s)(g) = sgs^{-1}, g \in G, s \in GL(n, \mathbb{Z})$. Moreover, there is groups inclusion $C_{GL(n,\mathbb{Z})}(G) \subset C_{GL(n,\mathbb{Z})}(H)$. Hence our Lemma is proved. ∎

Using the above lemma for the groups $h_\Gamma(G) \subset N_\alpha$, Proposition 5.31 and [138, theorem 1] we get

Corollary. *If $|\mathrm{Out}(\Gamma)| < \infty$, then $|\mathrm{Out}(\mathrm{Aut}(\Gamma))| < \infty$.* ∎

Lemma 5.34. *Assume $Z(\Gamma) = \{e\}$, then*
  (i) $H^1(G, \mathbb{Z}^n) \simeq (\mathbb{Q}^n/\mathbb{Z}^n)^G = H^0(G, \mathbb{Q}^n/\mathbb{Z}^n)$;
  (ii) $Z^1(G, \mathbb{Z}^n) \simeq \{m \in \mathbb{Q}^n \mid \forall_{g \in G} \; gm - m \in \mathbb{Z}^n\} = A^0(\Gamma)$ as N_α modules;
  (iii) $A(\Gamma) = N_{\mathrm{Aff}(\mathbb{R}^n)}(\Gamma) = \{a \in \mathrm{Aff}(\mathbb{R}^n) \mid \forall_{\gamma \in \Gamma} a\gamma a^{-1} \in \Gamma\} \simeq \mathrm{Aut}(\Gamma)$.

Proof. To the short exact sequence of G-modules

$$0 \to \mathbb{Z}^n \to \mathbb{Q}^n \to \mathbb{Q}^n/\mathbb{Z}^n \to 0$$

we have the following long exact sequence of cohomology groups attached:

$$0 \to H^0(G, \mathbb{Z}^n) \to H^0(G, \mathbb{Q}^n) \to H^0(G, \mathbb{Q}^n/\mathbb{Z}^n) \to H^1(G, \mathbb{Z}^n) \to H^1(G, \mathbb{Q}^n) \to \ldots$$

Since $H^1(G, \mathbb{Q}^n) = 0$ and by assumption $Z(G) \simeq (\mathbb{Z}^n)^G = H^0(G, \mathbb{Z}^n) = 0$ we also get $H^0(G, \mathbb{Q}^n) = (\mathbb{Q}^n)^G = 0$ and part (i) follows.

Now consider a homomorphism $\Phi\colon A^0(\Gamma) \to Z^1(G, \mathbb{Z}^n)$ of N_α-modules given by the formula

$$\forall_{m \in A^0(\Gamma)} \forall_{g \in G} \;\; \Phi(m)(g) = gm - m.$$

Note that the action $'*'$ defined by equation (5.1) can be extended to any cocycle group (see [23, page 168]) and this is the N_α module structure on $Z^1(G, \mathbb{Z}^n)$ that we use. Recall that $H^1(G, \mathbb{Q}^n) = 0$, hence every cocycle from the group $Z^1(G, \mathbb{Q}^n)$ is coboundary and Φ is onto. Easy calculation shows that

$$\ker \Phi = (\mathbb{Q}^n)^G = 0$$

and by the isomorphism theorem we prove part (ii) of the lemma. By theorem 5.7 we have the following short exact sequence

$$0 \to (\mathbb{R}^n)^G \to A(\Gamma) \to \mathrm{Aut}(\Gamma) \to 1.$$

Using again the triviality of the centre of Γ we get that $(\mathbb{R}^n)^G = 0$ and the group $A(\Gamma)$ and $\mathrm{Aut}(\Gamma)$ are isomorphic. This finishes the proof of part (iii) Lemma 5.34. □

We have the following modification of the Diagram 5.3.

$$\begin{array}{ccccccccc}
 & & 0 & & 0 & & 0 & & \\
 & & \downarrow & & \downarrow & & \downarrow & & \\
0 & \longrightarrow & \mathbb{Z}^n & \longrightarrow & \Gamma & \longrightarrow & G & \longrightarrow & 0 \\
 & & \downarrow & & \downarrow & & \downarrow{\scriptstyle h_\Gamma} & & \\
0 & \longrightarrow & A^0(\Gamma) & \longrightarrow & A(\Gamma) & \xrightarrow{F} & N_\alpha & \longrightarrow & 0 \\
 & & \downarrow & & \downarrow & & \downarrow & & \\
0 & \longrightarrow & (\mathbb{Q}^n/\mathbb{Z}^n)^G & \longrightarrow & \mathrm{Out}(\Gamma) & \longrightarrow & N_\alpha/G & \longrightarrow & 0 \\
 & & \downarrow & & \downarrow & & \downarrow & & \\
 & & 0 & & 0 & & 0 & &
\end{array}$$

Diagram 5.4

Let Γ be a crystallographic group of rank n with trivial centre and holonomy group G. Moreover, assume that the group $H^1(G, \mathbb{Z}^n) = \{e\}$, and the group $\text{Out}(\Gamma)$ is finite. Inductively, put $\Gamma_0 = \Gamma$ and $\Gamma_{i+1} = A(\Gamma_i)$, for $i \geq 0$.

Lemma 5.35. *$\exists N$ such that $\Gamma_{N+1} = \Gamma_N$.*

Proof. We start from observations that for $i > 0$, Γ_i is a crystallographic group, $Z(\Gamma_i) = \{e\}$ and $M_0 = M_i$, where $M_i = A^0(\Gamma_{i-1}) \subset \Gamma_i$ is the maximal abelian normal subgroup (a subgroup of translations). Let $G_i = \Gamma_i / M_i$. From definition we can consider (G_i) as a non-decreasing sequence of finite subgroups of $GL(n, \mathbb{Z})$. From Bieberbach theorems, (see Theorem 2.4) and from Diagrams 5.3 and 5.4, there is only a finite number of possibilities for G_i. Hence $\exists N \in \mathbb{N}$ such that $\forall_{i>N} \ G_i = G_N$. This finishes the proof. $\qquad\square$

Example 5.36. Let $\Gamma_1 = G_1 \ltimes \mathbb{Z}^2$ be the crystallographic group of dimension 2 with holonomy group $G_1 = D_{12}$, where

$$D_{12} = \text{gen} \left\{ \begin{bmatrix} 0 & 1 \\ -1 & 1 \end{bmatrix}, \begin{bmatrix} 0 & 1 \\ 1 & 0 \end{bmatrix} \right\}$$

is the dihedral group of order 12. Moreover, let $\Gamma_2 = G_2 \ltimes \mathbb{Z}^3$ be the crystallographic group of dimension 3, with holonomy group $G_2 = S_4 \times \mathbb{Z}_2$ generated by matrices

$$B = \begin{bmatrix} 1 & 1 & 0 \\ 0 & -1 & 0 \\ 0 & 1 & 1 \end{bmatrix}, \quad C \begin{bmatrix} 0 & 0 & 1 \\ 0 & -1 & -1 \\ -1 & 0 & 1 \end{bmatrix}.$$

Here S_4 denotes the symmetric group on four letters.

Lemma 5.37. *For $i = 1, 2 \Gamma_i$ is centerless and we have*

$$N_{GL(n_i, \mathbb{Z})}(G_i) = G_i$$

and

$$H^1(G_i, \mathbb{Z}^{n_i}) = 0,$$

where n_i is the rank of Γ_i.

Proof. First of all note that the representations of both groups by the identity maps are absolutely irreducible (and non-trivial). Hence the centre

$$Z(\Gamma_i) \simeq (\mathbb{Z}^{n_i})^{G_i}$$

is trivial and using Schur's Lemma [126, proposition 4, page 13] one gets

$$C_{GL(n_i,\mathbb{Z})}(G_i) = \{\pm I_{n_i}\} \subset G_i,$$

where I_{n_i} is the identity matrix of degree n_i, for $i = 1, 2$.

Now $\text{Out}(G_i)$ is a cyclic group of order two for $i = 1, 2$. Consider non-inner automorphism of G_1 defined as follows

$$\begin{bmatrix} 0 & 1 \\ -1 & 1 \end{bmatrix} \mapsto \begin{bmatrix} 0 & 1 \\ -1 & 1 \end{bmatrix}, \begin{bmatrix} 0 & 1 \\ 1 & 0 \end{bmatrix} \mapsto -\begin{bmatrix} 0 & 1 \\ 1 & 0 \end{bmatrix}.$$

An easy calculation shows that this automorphism cannot be realized as a conjugation by an element of $GL(2, \mathbb{Z})$.

As for the group G_2, if you identify it with the group generated by the cycles $(12), (1234)$ and (56) you'll get

$$B \leftrightarrow (12) \quad \text{and} \quad C \leftrightarrow (234)(56).$$

Consider an automorphism of G_2 which corresponds to the automorphism of permutation group defined by

$$(12) \mapsto (12)(56), (1234) \mapsto (1234)(56), (56) \mapsto (56).$$

In that case B is mapped to $-B$ and hence traces of those matrices differ, the automorphism cannot be realized by a conjugation inside $GL(3, \mathbb{Z})$. The cohomology groups can be calculated using Lemma 5.34.

For $i = 1, 2$ we have

$$N_{GL(n_i,\mathbb{Z})}(G_i) = G_i$$

and

$$H^1(G_i, \mathbb{Z}^{n_i}) = 0,$$

where n_i is the rank of Γ_i. Hence $A(\Gamma_i) = \Gamma_i$, and $\text{Out}(\Gamma_i) = \{e\}$, for $i = 1, 2$. □

Now we are ready to finish the proof of Theorem 5.32. The cases $n = 2, 3$ are done in the above example. Assume $n \geq 4$. Let $n = 2k + 3i$, where $i \in \{0, 1\}$. Put $\Gamma' = \Gamma_1^k \times \Gamma_2^i$. Then Γ' is centerless and by [91, Theorem 3.4] the bottom exact sequence of the Diagram 5.1 looks as follows

$$0 \to 0 \to \mathrm{Out}(\Gamma') \to S_k \to 0.$$

Hence, Γ' satisfies the assumption of Lemma 5.35 and the sequence $\Gamma_0 = \Gamma'$, $\Gamma_{i+1} = \mathrm{A}(\Gamma_i)$ stabilizes, i.e., $\exists N$ such that $\forall_{i \geq N}$ $\Gamma_i = \Gamma_N$. Moreover, $\mathrm{Out}(\Gamma_N) = \{e\}$ and $Z(\Gamma_N) = \{e\}$.

$\square$

5.5 Exercises

Exercise 5.1. Prove the assertion in Example 5.3.

Exercise 5.2. Prove parts 2 and 3, from Lemma 5.6.

Exercise 5.3. Let $\mathcal{M}$ be a flat manifold. Prove that

$$\mathrm{Aff}_0(\mathcal{M}) \simeq (S^1)^{b_1},$$

where b_1 is the first Betti number of $\mathcal{M}$.

Exercise 5.4. Prove that any finite subgroup of $GL(m, \mathbb{Q})$ is conjugate to a finite subgroup of $GL(m, \mathbb{Z})$.

Exercise 5.5. Let I be the identity matrix of degree n. Prove that for any matrices $A, B \in GL(n, K)$, the determinant of the matrix

$$\begin{pmatrix} I & B \\ A & I \end{pmatrix}$$

is equal to the determinant of the matrix $(I - AB)$. K is any field of the characteristic zero.

Exercise 5.6. Prove that the centralizer $C_{GL(n,\mathbb{Z})}(F)$ of a finite group $F \subset GL(n, \mathbb{Z})$ is finitely generated.

Exercise 5.7. Let $f \in \mathbb{Z}[X]$ be any polynomial whose splitting field is a subfield of $\mathbb{Q}(\zeta)$, where ζ is a primitive root of unity. Moreover, assume that the leading coefficient is ± 1. Prove that there exists an integral matrix whose characteristic polynomial is f.

Exercise 5.8. With support of Example 5.2, describe all Anosov diffeomorphisms on the torus.

Exercise 5.9. Prove Theorem 5.13 by means of the proof of Theorem 5.9.

Exercise 5.10. Define the non-abelian free group as a subgroup of $GL(2, \mathbb{Z})$.

Exercise 5.11. Prove that for any finite p-group G with $p \geq 5$, G is not an $\mathcal{R}_1$-group.
Hint: Use [68].

Exercise 5.12. Give a proof of Proposition 5.30.
Hint: Use [52].

6. Spin Structures and Dirac Operator

This chapter is mainly concerned with the existence of spin structures on a flat manifold. We want to convince the reader that such structure completely depends on some *condition* on the fundamental group of a manifold (see Corollary before Proposition 6.15). We have learned it from an article by F. Pfaffle, [112]. We start with an elementary definition of Spin(n) group and then introduce some basic information about vector bundles. It is necessary for a complete explanation of the above *condition*. Next, by using only algebraic techniques, we prove some classification results about the existence or non-existence of spin structures on flat manifolds. The last, most technical subsection is about a spectrum of the Dirac operator on flat manifolds. Here the approach essentially requires some knowledge of differential geometry which can be found, for example, in [46].

6.1 Spin(n) group

Let C_n be Clifford's algebra over the real numbers. By definition it is an associative algebra with unity, generated by elements

$$\{e_1, e_2, \ldots, e_n\}$$

and with relations

$$\forall i, e_i^2 = -1,$$

$$\forall i, j, e_i e_j = -e_j e_i,$$

where $1 \leq i, j \leq n$. We define $C_0 = \mathbb{R}$. It is easy to see that $C_1 = \mathbb{C}$ and $C_2 = \mathbb{H}$, where $\mathbb{H}$ is the four-dimensional quaternion algebra. Moreover, $\mathbb{R}^n \subset C_n$ and $\dim_{\mathbb{R}} C_n = 2^n$, where $\mathbb{R}^n$ is n-dimensional $\mathbb{R}$-vector space with the basis $e_1, e_2, \ldots, e_n$.

We have the following homomorphisms (involutions) on C_n :

(i) $*: e_{i_1} e_{i_2} \ldots e_{i_k} \mapsto e_{i_k} e_{i_{k-1}} \ldots e_{i_2} e_{i_1}$,

(ii) $': e_i \mapsto -e_i$,

(iii) $^-\colon a \mapsto (a')^*, a \in C_n$.

Suppose $C_n^0 = \{x \in C_n \mid x' = x\}$. It is easy to observe that

$$\forall a\ b \in C_n, (ab)^* = b^* a^*.$$

We define subgroups of C_n,

$$Pin(n) = \{x_1 x_2 \ldots x_k \mid x_i \in S^{n-1} \subset \mathbb{R}^n \subset C_n, i = 1, 2, \ldots k\},$$

$$Spin(n) = Pin(n) \cap C_n^0.$$

Moreover, we have:

Proposition 6.1. *Suppose* $u \in Spin(n), y \in \mathbb{R}^n$. *The map*

$$\lambda_n \colon Spin(n) \to SO(n),$$

where

$$\lambda_n(u)(y) = uyu^* \tag{6.1}$$

is a continuous epimorphism of groups.

Proof. Let $u = u_1 u_2 \ldots u_{2k} \in Spin(n)$, where $u_i \in S^{n-1}$ for $i = 1, 2, \ldots, 2k$ and $k \in \mathbb{N}$. We claim, that the map

$$\lambda_n(u) \colon \mathbb{R}^n \to \mathbb{R}^n$$

is a linear isometry. In fact, by definition we can assume that $u \in S^{n-1} \subset \mathbb{R}^n$. Let $x_1, x_2 \in \mathbb{R}^n$. From the above we have

$$\lambda_n(u)(\alpha x_1) = \alpha \lambda_n(u)(x_1)$$

and

$$\lambda_n(u)(x_1 + x_2) = \lambda_n(u)(x_1) + \lambda_n(u)(x_2).$$

On the other hand

$$\|\lambda_n(u)x_1\|^2 = -(ux_1 u^*)(ux_1 u^*) = -ux_1^2 u^* =$$

$$u\|x_1\|^2 u^* = \|x_1\|^2.$$

Finally, for $u, v \in Spin(n)$,

$$\lambda_n(uv)(x_1) = uvx_1(uv)^* = u(vx_1v^*)u^* = \lambda_n(u)(\lambda_n(v)(x_1)).$$

We still have to prove that λ_n is a surjection. For this, we show that for any $u \in S^{n-1}$

$$\lambda_n(u) \colon \mathbb{R}^n \to \mathbb{R}^n$$

is a reflection in the hyperplane orthogonal to u. In fact, it is well known, that such a reflection is defined by the formula

$$x \mapsto x - 2\langle x, u \rangle \frac{u}{\|u\|^2}. \tag{6.2}$$

Let $x_1 \in \mathbb{R}^n$. There exist $t \in \mathbb{R}$ and $u_1 \in \mathbb{R}^n$ such that $\langle u, u_1 \rangle = 0$ and

$$x_1 = tu + u_1.$$

Hence, by (6.2)

$$x_1 = tu + u_1 \mapsto (tu + u_1) - 2\langle tu + u_1, u \rangle u = -tu + u_1.$$

Moreover,

$$\lambda_n(u)(x_1) = ux_1u^* = ux_1u = -u\bar{x}_1u = x_1 - x_1\bar{u}u - u\bar{x}_1u =$$

$$x_1 - (x_1\bar{u} + u\bar{x}_1)u \overset{(6.14)}{=} x_1 - 2\langle x_1, u \rangle u = -tu + u_1.$$

Since $SO(n)$ is generated by reflections, λ_n is an epimorphism, and the claim is proved. □

We define again [1] the 2-extraspecial group.

Example 6.2 ([121]). Let $H \subset SO(n)$ be a subgroup of all diagonal matrices with ± 1 only on the diagonal. One can show, that the set $(\lambda_n)^{-1}(H) \subset Spin(n)$ is a 2-extraspecial group.

---

[1]See Definition 5.26.

6.2   Vector bundles

For the formulation of a definition of the spin structure (Definition 6.12) we have to introduce vector bundles. Let E, X, F be topological spaces.

Definition 6.3. A map $p \colon E \to X$ is a locally trivial fibration (E, p, X, F), with a fibre F, when the following condition is satisfied: For any $x_0 \in X$ there exists an open set $U(x_0) \subset X$ and a homeomorphism

$$\Phi_{U(x_0)} \colon p^{-1}(U(x_0)) \to U(x_0) \times F,$$

such that Diagram 6.1

$$
\begin{array}{ccc}
p^{-1}(U(x_0)) & \xrightarrow{\ \Phi_{U(x_0)}\ } & U(x_0) \times F \\
& {\scriptstyle p}\searrow \quad \swarrow {\scriptstyle pr_1} & \\
& U(x_0) &
\end{array}
$$

Diagram 6.1

commutes. Here pr_1 is a projection on the first factor.

Example 6.4.

1. Let $E = X \times F$ and $p \colon E \to X$ be a projection. Then (E, p, X, F) is a locally trivial fibration.

2. Let $X = W^n$ be a smooth manifold and $E = \cup_{x \in W^n} T_x(W^n)$, when $p \colon E \to X$ is defined by $p(w) = x, \forall w \in T_x(W^n)$. Here, $T_x(W^n)$ is the tangent space W^n at x. Then $(E, p, W^n, \mathbb{R}^n)$ is a locally trivial fibration with the fibre $\mathbb{R}^n$.

Two fibrations, (E, p, X, F), (E', p', X, F') are equivalent when there exists a homeomorphism $f \colon E \to E'$, such that Diagram 6.2 commutes. The local fibration (E, X, p, F) is trivial, when it is equivalent to $(X \times F, X, pr, F)$. Suppose $p \colon E \to X$ is a fibration. The map $s \colon X \to E$ is a *section* of p, when $ps = 1_X$. If s is defined locally, that is, on the open sets of a topological basis, then it is a *local section*.

Diagram 6.2

Let $p\colon E \to X$ be a fibration and let $f\colon Y \to X$ be a continuous function. We define the topological space

$$f^*E = \{(y, e) \in Y \times E : f(y) = p(e)\}.$$

The projection on the first factor

$$p^*\colon f^*E \to Y,$$

is called the *induced fibration*.

Definition 6.5. Let G be a topological group. (P, π, X, G) is the principal G-bundle when the following conditions are satisfied:

1. P is a topological space and G acts freely from the right on P;

2. $\pi\colon P \to X$ is a continuous surjection and $\pi(p_1) = \pi(p_2)$, for $p_1, p_2 \in P$, if and only if, $\exists g \in G$, such that $p_1 g = p_2$;

3. for any $x \in X$ there exists a neighbourhood $U \subset X$ of x, and a map $\Phi_U\colon \pi^{-1}(U) \to U \times G$, such that

$$\Phi_U(p) = (\pi(p), \bar\phi_U(p)).$$

Moreover, $\bar\phi_U\colon \pi^{-1}(U) \to G$ has the property

$$\bar\phi_U(pg) = \bar\phi_U(p)g.$$

Remark 6.6.

1. The analogous definition is for the left, free action of G on P.

2. Any principal G-bundle is the locally trivial fibration with a fibre $F = G$.

3. Suppose (E, p, X, G) is the principal G-bundle and $f\colon Y \to X$ is the continuous map, then (f^*E, p^*, Y, G) is a principal bundle.

Definition 6.7. Let (E, p, X, G) and (E_1, p_1, X, G) be principal G-bundles with the same base X. They are isomorphic, provided there exists a homeomorphism $f \colon E \to E_1$ with the following properties:

1. Diagram 6.3 commutes;
2. $\forall e \in E, g \in G, f(eg) = f(e)g$.

Diagram 6.3

Immediately from the definition we have:

Proposition 6.8. *If the principal G-bundle (E, p, X, G) has a section, then it is isomorphic to a trivial principal G-bundle $(E \times G, pr, X, G)$.*

Example 6.9. Let G be a Lie group and H a closed subgroup of G. Suppose $p \colon G \to G/H$ is a projection. Then

$$(G, p, G/H, H)$$

is the principal H-bundle.

Example 6.10. Suppose

$$X = S^2 = \mathbb{C}P^1$$

and

$$E = S^3 = \{(z_1, z_2) \in \mathbb{C}^2 \mid |z_1|^2 + |z_2|^2 = 1\}.$$

We define

$$p \colon S^3 \to \mathbb{C}P^1$$

such that $p(z_1, z_2) = [z_1 : z_2]$. Moreover, we have two possibilities of the action S^1 on S^3 :

$$(z_1, z_2)z = (z_1 z, z_2 z), \tag{6.3}$$

$$(z_1, z_2)z = (z_1 z^{-1}, z_2 z^{-1}), \forall z, z_1, z_2 \in \mathbb{C}. \tag{6.4}$$

This gives us two principal S^1-bundles

$$\psi_1 = (S^3, p, \mathbb{C}P^1, S^1)$$

and
$$\psi_2 = (S^3, p, \mathbb{C}P^1, S^1),$$

with different actions (6.3) and (6.4) of S^1 on S^3. We leave as an exercise to prove that they are not isomorphic. ψ_1 is called the Hopf bundle.

Let us consider an example of a principal bundle which is crucial for the definition of the Dirac operator.

Example 6.11. Let M^n be a smooth manifold of dimension n. For any $x \in M^n$, let

$$L_x(M^n) = \{(v_1, v_2, \ldots, v_n) \in T_x(M^n) \mid det(v_1, v_2, \ldots, v_n) \neq 0\}.$$

Suppose
$$L(M^n) = \cup_{x \in M^n} L_x(M^n).$$

We define the *right* action of $GL(n, \mathbb{R})$ on $L(M^n)$ as the matrix multiplication. Hence we have the principal $GL(n, \mathbb{R})$-bundle

$$(L(M^n), p, M^n, GL(n, \mathbb{R})).$$

The above bundle, can be considered over the orthogonal group. In fact, let (P, π, X, G) be the principal G-bundle, and let $\lambda : G_1 \to G$ be a continuous homomorphism of groups. The principal G_1-bundle (Q, π, X, G_1) is a λ-extension of the bundle (P, π, X, G), when there exists a map $f : Q \to P$, such that Diagram 6.4 commutes and $f(qg_1) = f(q)\lambda(g_1), \forall q \in Q, g_1 \in G_1$.

$$\begin{array}{ccc} Q & \xrightarrow{f} & P \\ {\scriptstyle \pi}\downarrow & \swarrow {\scriptstyle \pi} & \\ X & & \end{array}$$

Diagram 6.4

6.3   Spin structure

We are ready to formulate the main definition of this chapter.

Definition 6.12. Let W^n be an orientable, compact manifold of dimension n. We define a spin-structure on W^n as a choice of the λ_n-extension of the principal $SO(n)$-tangent bundle to a $Spin(n)$-bundle.

Remark 6.13. Existence of a spin structure is equivalent to the second Stiefel-Whitney class $w_2 \in H^2(W^n, \mathbb{Z}_2)$ of the manifold W^n being zero, [46, p. 40].

Example 6.14. Since the tangent bundle of any oriented and compact 3-manifold W^3 is trivial *(see [104, §12, Exercise 12.B])*, W^3 admits a spin-structure.

Now, we shall describe the spin-structures on the oriented, flat manifold M^n of dimension n with the fundamental group $\Gamma \subset E(n)$. Next, consider the trivial principal bundle

$$(L(\mathbb{R}^n) = \mathbb{R}^n \times SO(n), p, \mathbb{R}^n, SO(n)),$$

which has the λ_n-extension to the principal $Spin(n)$-bundle

$$(\mathbb{R}^n \times Spin(n), p, \mathbb{R}^n, Spin(n)).$$

This defines the spin-structure on $\mathbb{R}^n$. Since it is a 1-connected space, there is only one spin-structure, (see [46]). We can extend an action of Γ on $\mathbb{R}^n$ to $L(\mathbb{R}^n) = \mathbb{R}^n \times SO(n)$, in the following way. For any $\gamma \in \Gamma$ and $(x, v) \in \mathbb{R}^n \times SO(n)$, we have

$$\gamma(x, v) = (\gamma x, p_1(\gamma)v).$$

Recall (see Lemma 2.13) that, $p_1 \colon E(n) \to O(n)$ is a projection onto the first factor. By definition $p_1(\gamma)$ is also the linear component (a derivative) of an isometry. We have $L(M^n) = (\mathbb{R}^n \times SO(n))/\Gamma$. From the elementary properties of covering, any $\gamma \in \Gamma$ has two lifts $\gamma^\pm$, such that the Diagram 6.5 commutes.

$$
\begin{array}{ccc}
\mathbb{R}^n \times Spin(n) & \xrightarrow{\;\gamma^\pm\;} & \mathbb{R}^n \times Spin(n) \\
\downarrow{\scriptstyle id\times\lambda_n} & & \downarrow{\scriptstyle id\times\lambda_n} \\
\mathbb{R}^n \times SO(n) & \xrightarrow{\;\gamma\;} & \mathbb{R}^n \times SO(n)
\end{array}
$$

Diagram 6.5

Summing up, there is a bijection between the spin-structures on M^n and actions α of Γ on $\mathbb{R}^n \times Spin(n)$ with: $\alpha(\gamma) \in \{\gamma^{\pm}\}$ for all $\gamma \in \Gamma$.

In fact, the spin structure associated to such α is given by

$$(\mathbb{R}^n \times Spin(n))/\Gamma \to (\mathbb{R}^n \times SO(n))/\Gamma \simeq L(M^n).$$

We can observe, that $\gamma^{\pm}$ determines $A^{\pm} \in \lambda_n^{-1}(p_1(\gamma))$, such that

$$\forall (x, s) \in \mathbb{R}^n \times Spin(n), \gamma^{\pm}(x, s) = (\gamma x, A^{\pm}s),$$

(see [112] and [46, p. 42-45]).

A next Corollary translates an existence of a spin structure on flat manifolds into a pure algebraic language.

Corollary ([112, Proposition 3.2]). *M^n has a spin-structure if and only if there exists a homomorphism $\epsilon \colon \Gamma \to Spin(n)$, such that $\lambda_n \circ \epsilon = p_1 \mid_{\Gamma}$.*

Proof. Given a homomorphism ϵ with $p_1 \mid_{\Gamma} = \lambda_n \circ \epsilon$ one defines an action α on $\mathbb{R}^n \times Spin(n)$ via $\alpha(\gamma) \colon (x, s) \mapsto (\gamma s, \epsilon(\gamma)s)$, and one gets a spin structure as described above. For the proof of other direction we shall use the Remark 6.13. Since $H^2(SO(n), \mathbb{Z}_2) = \mathbb{Z}_2$, the second Stiefel-Whitney class $w_2 \in H^2(M^n, \mathbb{Z}_2) = H^2(\Gamma, \mathbb{Z}_2)$ defines the upper row of the following diagram.

$$
\begin{array}{ccccccccc}
1 & \longrightarrow & \mathbb{Z}_2 & \longrightarrow & \bar{\Gamma} & \longrightarrow & \Gamma & \longrightarrow & 1 \\
& & \| & & \downarrow & & \downarrow{\scriptstyle p_1|_\Gamma} & & \\
1 & \longrightarrow & \mathbb{Z}_2 & \longrightarrow & Spin(n) & \xrightarrow{\ \lambda_n\ } & SO(n) & \longrightarrow & 1
\end{array}
$$

Summing up, if M^n has a spin-structure then $w_2 = 0$ and the first row of the above diagram splits. Hence (see Theorem 3.1), there exists a homomorphism $\epsilon \colon \Gamma \to Spin(n)$, such that $\lambda_n \circ \epsilon = p_1 \mid_{\Gamma}$. $\qquad \square$

Since for $n \geq 3$, $\ker \lambda_n = \mathbb{Z}_2$, any flat manifold with an odd order holonomy G, has a spin-structure. It follows from the fact that any short, exact sequence of the finite groups

$$0 \to \mathbb{Z}_2 \to H \to G \to 0$$

splits.

The following proposition was told to me by G. Hiss.

Proposition 6.15 ([69, Proposition 2.1]). *Suppose $n \geq 2$. Let M_i be flat oriented n-manifolds with fundamental groups $\Gamma_i, i = 1, 2$. Suppose that Γ_1 is isomorphic to Γ_2 (as abstract groups). Then M_1 has a spin structure if and only if M_2 has one.*

Proof. It is well known ([104, Lemma 11.13]) that having a spin structure is a homotopy invariant. However, we shall give an elementary, algebraic and direct proof. We may assume that Γ_1 and Γ_2 are subgroups of $E(n)$, and $M_i = \mathbb{R}^n / \Gamma_i, i = 1, 2$. By Bieberbach's second theorem there is an element $\alpha \in A(n)$ such that $\Gamma_2 = \alpha \Gamma_1 \alpha^{-1}$. Let $\rho_i \colon \Gamma_i \to SO(n)$ denote the holonomy representations of $\Gamma_i, i = 1, 2$. Suppose that $\alpha = (X, v)$ with $X \in GL(n, \mathbb{R})$ and $v \in \mathbb{R}^n$. Then

$$\rho_2(\alpha \gamma \alpha^{-1}) = X \rho_1(\gamma) X^{-1},$$

for all $\gamma \in \Gamma_1$. Let $ML(n)$ denote the metalinear group, a two fold covering of the general linear group $GL(n, \mathbb{R})$, and let $\lambda \colon ML(n) \to GL(n, \mathbb{R})$ denote the covering homomorphism. Then $\lambda^{-1}(SO(n)) = Spin(n)$. Choose $\tilde{X} \in ML(n)$ with $\lambda(\tilde{X}) = X$. Suppose $\epsilon_1 \colon \Gamma_1 \to Spin(n)$ is a spin structure. Define $\epsilon_2 \colon \Gamma_2 \to Spin(n)$ by $\epsilon_2(\alpha \gamma \alpha^{-1}) := \tilde{X} \epsilon_1(\gamma) \tilde{X}^{-1}$ for $\gamma \in \Gamma_1$. Then ϵ_2 is a spin structure on M_2. By symmetry, this completes the proof. $\qquad \square$

As we have mentioned above (Example 6.14) all oriented three flat manifolds have a spin structure. In the dimension 4 we have the following result.

Theorem 6.16 ([120]). *All but 3 of the 27 closed, oriented, flat, four-dimensional manifolds have a spin structure. The holonomy groups of manifolds which do not admit a spin structure are equal to $\mathbb{Z}_2 \oplus \mathbb{Z}_2$ and D_8.*

For the proof and complete classification we refer to [120]. Let us give a list of all four-dimensional orientable flat manifolds without spin structure.

Example 6.17. We shall prove that the group *(22.1.4 in [109] or 05/01/04/006 in [16])*

$$\Gamma_7 = \text{gen}\{\gamma_1, \gamma_2, \bar{t}_1, \bar{t}_2, \bar{t}_3, \bar{t}_4\},$$

where

$$\gamma_1 = \left(\begin{bmatrix} 1 & 0 & 0 & 0 \\ 0 & -1 & 0 & 0 \\ 0 & 0 & 0 & -1 \\ 0 & 0 & -1 & 0 \end{bmatrix}, (1/2, 1/2, 0, 0) \right) \text{ and}$$

$$\gamma_2 = \left(\begin{bmatrix} -1 & 0 & 0 & 0 \\ 0 & -1 & 0 & 0 \\ 0 & 0 & 1 & 0 \\ 0 & 0 & 0 & 1 \end{bmatrix}, (0,0,-1/2,1/2) \right) \text{ has no spin structure.}$$

After conjugation by an element $(A_6, 0)$ where

$$A_6 = \begin{bmatrix} 1 & 0 & 0 & 0 \\ 0 & 1 & 0 & 0 \\ 0 & 0 & -1 & 1 \\ 0 & 0 & 1 & 1 \end{bmatrix},$$

we get the Bieberbach group

$$\Gamma'_7 = (A_6, 0)\Gamma_7(A_6^{-1}, 0) = \text{gen}\{A_6\gamma_1 A_6^{-1} = \gamma'_1, A_6\gamma_2 A_6^{-1} = \gamma'_2, g_1, g_2, g_3, g_4\}$$

where, $\gamma'_1 = ([1,-1,1,-1], (1/2,1/2,0,0)), \gamma'_2 = ([-1,-1,1,1], (0,0,1,0))$ and $\gamma'_1\gamma'_2 = ([-1,1,1,-1], (1/2,1/2,1,0))$. Here $[a,b,c,d]$ is a diagonal 4×4-matrix with a,b,c,d on the diagonal. Assume that there exists a homomorphism $\epsilon: \Gamma'_7 \to Spin(4)$ of groups, such that $\lambda_4\epsilon = h$. We have $\epsilon(g_1) = -1, \epsilon(g_4 - g_3) = -1, \epsilon(g_2 + g_4 - g_3) = -1$. Hence $\epsilon(\gamma'_1 g_4\gamma'_1) = \epsilon(g_1 - g_3) = \epsilon((\gamma'_1)^2)\epsilon(g_3) = -\epsilon((\gamma'_1)^2)\epsilon(g_4) = -\epsilon(\gamma'_1)\epsilon(g_4)\epsilon(\gamma'_1) = -\epsilon(\gamma'_1 g_4\gamma'_1)$ and the spin structure does not exist.

The second group, which has no spin structure, is the group *(22.1.8 in [109] or 05/01/06/006 in [16])*

$$\Gamma_9 = \text{gen}\{\gamma_1, \gamma_2, \bar{t}_1, \bar{t}_2, \bar{t}_3, \bar{t}_4\}$$

where

$$\gamma_1 = \left(\begin{bmatrix} -1 & 0 & 0 & 0 \\ 0 & 1 & 0 & 0 \\ 0 & -1 & 0 & -1 \\ 0 & -1 & -1 & 0 \end{bmatrix}, (1/2,-1/2,1/2,0) \right) \text{ and}$$

$$\gamma_2 = \left(\begin{bmatrix} -1 & 0 & 0 & 0 \\ 0 & 0 & 1 & -1 \\ 0 & 1 & 0 & 1 \\ 0 & 0 & 0 & 1 \end{bmatrix}, (0,-1/2,0,1/2) \right).$$

After conjugation by an element $(A_9, 0)$ where

$$A_9 = \begin{bmatrix} 1 & 0 & 0 & 0 \\ 0 & 1 & 1 & -1 \\ 0 & 1 & 1 & 1 \\ 0 & 1 & -1 & 1 \end{bmatrix},$$

we get the Bieberbach group

$$\Gamma'_9 = (A_9, 0)\Gamma_9(A_9^{-1}, 0) = \text{gen}\{A_9\gamma_1 A_9^{-1} = \gamma'_1, A_9\gamma_2 A_9^{-1} = \gamma'_2, k_1 = (I, A_9 t_1), k_2 = (I, A_9 t_2), k_3 = (I, A_9 t_3), k_4 = (I, A_9 t_4)\}$$

where

$$\gamma'_1 = ([-1, 1, -1, 1], (1/2.0, 0, 0, -1)), \gamma'_2 = ([-1, 1, 1, -1], (0, -1, 0, 0))$$

and $\gamma'_1\gamma'_2 = ([1, 1, -1, -1], (1/2, -1, 0, -1))$.

According to the definition $\epsilon(k_3 - k_2) = -1, \epsilon(k_4 - k_2) = -1$ and $\epsilon(k_1 - k_2 + k_4) = -1$. Hence $\epsilon(k_1) = 1, \epsilon(k_2) = 1, \epsilon(k_3) = -1$ and $\epsilon(k_4) = -1$ or $\epsilon(k_1) = 1, \epsilon(k_2) = -1, \epsilon(k_3) = 1$ and $\epsilon(k_4) = 1$. In any case we have $\epsilon(\gamma'_2 k_2 \gamma'_2) = \epsilon(k_3 + k_4 - k_2) = -\epsilon(\gamma'_2)\epsilon(k_2)\epsilon(\gamma'_2)$ and it proves our statement.

The last Bieberbach group *(29.1.2) in [109] or 13/04/04/011 in [16])* of rank 4, with holonomy group D_8, has the set of generators.

$$\Delta_4 = \text{gen}\{(F_1, (0, -1/2, 0, 1/2)), (F_2, (1/4, 0, 1/2, 0)), \bar{t}_1, \bar{t}_2, \bar{t}_3, \bar{t}_4\},$$

where $F_1 = \begin{bmatrix} -1 & 0 & 0 & 0 \\ 0 & 0 & 1 & -1 \\ 0 & 1 & 0 & 1 \\ 0 & 0 & 0 & 1 \end{bmatrix}$, $F_2 = \begin{bmatrix} -1 & 0 & 0 & 0 \\ 0 & 0 & 0 & -1 \\ 0 & 0 & 1 & 0 \\ 0 & -1 & 0 & 0 \end{bmatrix}$.

Let us conjugate the group Δ_4 by the element $(A, 0) \in GL(4, \mathbb{R}) \ltimes \mathbb{R}^4$, where

$$A = \begin{bmatrix} 0 & 1 & 1 & 1 \\ 0 & 1 & -1 & 1 \\ 0 & 1 & 1 & -1 \\ 1 & 0 & 0 & 0 \end{bmatrix}.$$

We get $AF_1 A^{-1} = D_1$ and $AF_2 A^{-1} = D_2$. Hence

$$\Delta'_4 = \{\gamma_1 = (D_1, (0, 0, -1, 0)), \gamma_2 = (D_2, (1/2, -1/2, 1/2, 1/4)), l_1 = (I, At_1) = (I, (0, 0, 0, 1)), l_2 = (I, At_2) = (I, (1, 1, 1, 0)), l_3 = (I, At_3) = (I, (1, -1, 1, 0)) \, l_4 = (I, At_4) = (I, (1, 1, -1, 0))\}.$$

We claim that this group has no spin structure. By contradiction, assume that $\epsilon \colon \Delta'_4 \to Spin(4)$ defines a spin structure.

Similarly to the previous cases, we have:

$$\epsilon(\gamma_1) = \pm e_2 e_4, \epsilon(\gamma_2) = \pm \frac{1}{\sqrt{2}}(e_1 + e_2)e_4.$$

Hence

$$\epsilon(l_3) = \epsilon(\gamma_2^2) = -1,$$

$$\epsilon(l_3 + l_4) = \epsilon((\gamma_1\gamma_2^2)^2) = -1 \Rightarrow \epsilon(l_4) = 1.$$

Moreover,

$$\epsilon(l_3 - l_2) = \epsilon((\gamma_2\gamma_1\gamma_2)^2) = -1 \Rightarrow \epsilon(l_2) = 1.$$

Finally,

$$1 = \epsilon(l_4)\epsilon(l_2) = \epsilon(l_4 - l_2) = \epsilon(\gamma_1^2) = -1.$$

This is impossible.

Now, we want to present some general results about spin structures.

Proposition 6.18 ([69, Corollary 2.2]). *Let M be a flat oriented n-manifold, $n \geq 4$, with the fundamental group Γ and holonomy group G. Suppose that the translation lattice Λ of Γ is of the form $\Lambda = \Lambda_1 \oplus \Lambda_2$ with G-invariant sublattices Λ_1 and Λ_2, such that the $\mathbb{R}$-representations ρ'_i of G afforded Λ_i map into $SO(m), i = 1, 2$, and that ρ'_1 and ρ'_2 are equivalent (over $\mathbb{R}$). Then M has a spin structure.*

Proof. We may assume that $\Gamma \subset E(n)$ and $M = \mathbb{R}^n/\Gamma$. Our hypothesis implies that Γ is conjugate in $A(n)$ to a Bieberbach group $\tilde{\Gamma} \subset E(n)$ with translation lattice $\tilde{\Lambda} = \tilde{\Lambda}'_1 \oplus \tilde{\Gamma}'_2$, such that the representations $\tilde{\rho}'_i : G \to SO(m)$ afforded by $\tilde{\Lambda}_i$ are in fact equal. The "double construction" in the proof of [37, Theorem 1] implies that $\tilde{M} = \mathbb{R}^n/\tilde{\Gamma}$ has a spin structure. By Proposition 6.15, the manifold M also has a spin structure. $\square$

By a spin structure of a pair (ρ, F), where F is a finite group and $\rho: F \to SO(n)$ a homomorphism we shall understand any homomorphism $\epsilon: F \to Spin(n)$ such that $\rho = \lambda_n \circ \epsilon$. We recall a well known criterion, due to Griess [53] and Gagola, and Garrison [48], for the non-existence of a spin structure in this case. The criterion is based on the following lemma. By $|x|$ we denote the order of the element x of some group.

Lemma 6.19 ([48], [53]). *Let $A \in SO(n)$ be of order 2 and let $a \in \lambda_n^{-1}(A)$. Then $|a| = 4$ if and only if*

$$\tfrac{1}{2}(n - Trace(A)) \equiv 2 \pmod 4.$$

Proof of Lemma 6.19. Let d denote the dimension of the (-1)-eigenspace of A. Then d is even. By [48, Corollary], there is an inverse image $a \in Spin(n)$ of A with $a^2 = (-1)^{d(d-1)/2}$. Now $d = (n - \text{Trace}(A))/2$ and the result follows. $\square$

Proposition 6.20 ([48], [53]). *Let F be a finite group and $\rho \colon F \to SO(n)$ a homomorphism with character χ. Let $g \in F$ have order 2. If*

$$\frac{1}{2}(\chi(1) - \chi(2)) \equiv 2(mod\ 4),$$

then there is no $\epsilon \colon F \to Spin(n)$ such that $\rho = \lambda_n \circ \epsilon$.

Proof. This is a direct consequence of the previous lemma. $\square$

The following example shows that the hypothesis of Proposition 6.18, namely that ρ'_i $(i = 1, 2)$ map into $SO(m)$ cannot be omitted.

Example 6.21. Consider the flat oriented manifold $\tilde{M}_1$ of [99, Table 1 p. 327]. Its holonomy representation is a "double" one (each irreducible component has multiplicity two), but the sum of the three distinct irreducible components yields a representation of $\mathbb{Z}_2 \times \mathbb{Z}_2$ into $O(n)$ but not into $SO(n)$. The fundamental group $\pi_1(\tilde{M}_1)$ is generated by its translation subgroup $\mathbb{Z}^6$ and two generators (copied from [99, Table 1, p. 327])

$$\gamma_1 = \left(\begin{bmatrix} 1 & 0 & 0 & 0 & 0 & 0 \\ 0 & 1 & 0 & 0 & 0 & 0 \\ 0 & 0 & 1 & 0 & 0 & 0 \\ 0 & 0 & 0 & -1 & 0 & 0 \\ 0 & 0 & 0 & 0 & -1 & 0 \\ 0 & 0 & 0 & 0 & 0 & 1 \end{bmatrix}, (0,0,1/2,0,0,0) \right)$$

and

$$\gamma_2 = \left(\begin{bmatrix} 1 & 0 & 0 & 0 & 0 & 0 \\ 0 & 1 & 0 & 0 & 0 & 0 \\ 0 & 0 & -1 & 0 & 0 & 0 \\ 0 & 0 & 0 & 1 & 0 & 0 \\ 0 & 0 & 0 & 0 & 1 & 0 \\ 0 & 0 & 0 & 0 & 0 & -1 \end{bmatrix}, (1/2,1/2,0,0,0,0) \right).$$

Let A_2 and A_3 denote the linear parts of γ_2 and $\gamma_1\gamma_2$, respectively. Then (6-Trace(A_2))/2 = 2 and (6-Trace(A_3))/2 = 4. Thus, by Lemma 6.19, inverse images of A_2 and A_3 in $Spin(6)$ have order 4 and 2, respectively. Hence a spin structure $\epsilon \colon \pi_1(\tilde{M}_1) \to Spin(6)$ would map γ_2 and $\gamma_1\gamma_2$ to elements of order 4 and 2, respectively. But

$$
\gamma_1\gamma_2 = \left(\begin{bmatrix} 1 & 0 & 0 & 0 & 0 & 0 \\ 0 & 1 & 0 & 0 & 0 & 0 \\ 0 & 0 & -1 & 0 & 0 & 0 \\ 0 & 0 & 0 & -1 & 0 & 0 \\ 0 & 0 & 0 & 0 & -1 & 0 \\ 0 & 0 & 0 & 0 & 0 & -1 \end{bmatrix}, (1/2, 1/2, 1/2, 0, 0, 0)\right)
$$

and $(\gamma_2)^2 = (\gamma_1\gamma_2)^2$. Thus, there is no spin structure on $\tilde{M}_1$.

6.3.1   Case of cyclic holonomy

In this part we shall study spin structures on manifolds with cyclic holonomy groups of 2-power order. Let $A \in \mathrm{SL}(n, \mathbb{Z})$ have order $2^m, m \geq 1$, and let $G := \langle A \rangle$. Choose an embedding $\rho' \colon G \to SO(n)$. We want to investigate Bieberbach groups with holonomy G and translation lattice $\Lambda = \{(\mathrm{id}, v) \mid v \in \mathbb{Z}^n\} \subset E(n)$. We identify the (multiplicatively written) group Λ with the natural $\mathbb{Z}G$-module $\mathbb{Z}^n$. Any element $\delta \in \mathbb{Q}^n$ gives rise to a derivation $\hat{\delta} \colon G \to \mathbb{Q}^n/\mathbb{Z}^n$ by sending A to $\delta + \mathbb{Z}^n$. We identify $\mathbb{Q}^n/\mathbb{Z}^n$ with $\mathbb{Z}G$-module $\mathbb{Q} \otimes_\mathbb{Z} \Lambda/\Lambda$. If the cohomology class of $\hat{\delta}$ in $H^1(G, \mathbb{Q} \otimes_\mathbb{Z} \Lambda/\Lambda)$ is special, then δ defines a Bieberbach group

$$
\Gamma = \langle (A, \delta), \Lambda \rangle \subset A(n)
$$

with holonomy group G and translation lattice Λ. Composing δ' with the canonical map $\pi \colon \Gamma \to G$, we obtain the holonomy representation $\rho := \rho' \circ \pi \colon \Gamma \to SO(n)$. The natural action of G on $\mathbb{Z}^n$ equips $\mathbb{Z}^n/2\mathbb{Z}^n$ with the structure of an $\mathbb{F}_2G$-module, isomorphic to the $\mathbb{F}_2G$-module $\bar{\Lambda} = \Lambda/\Lambda^2$ arising from the conjugation action of Γ on Λ. We write $\bar{v}$ for the image of $v \in \mathbb{Z}^n$ in $\mathbb{Z}^n/2\mathbb{Z}^n$, which is identified with $\bar{\Lambda}$. Recall that the radical of a module is the intersection of its maximal submodules.

Theorem 6.22 ([69, Theorem 3.1]). *Let Γ and δ be as above.   Then the following holds.*

(1) *If $\lambda_n^{-1}(\rho(A))$ contains an element of order $2^m = |A|$, then (ρ, Γ) has a spin structure.*

(2) *If $\lambda_n^{-1}(\rho(A))$ contains an element of order $2^m = |A|$ if and only if*

$$\frac{1}{2}(n - Trace(A^{2^{m-1}})) \equiv 0 \ (mod\, 4).$$

(3) *If $m = 1$ or $m \le 3$, then (ρ, Γ) has a spin structure.*

(4) *Suppose that $m = 2$ and that $\lambda_n^{-1}(\rho(A))$ does not contain elements of order 4. Put $\delta' := A^3\delta + A^2\delta + A\delta + \delta$. Then $\delta' \in \Lambda$ and (ρ, Γ) has a spin structure if and only if $\bar{\delta}'$ is not contained in the radical of $\bar{\Lambda}$.*

Proof. Let $a \in \lambda_n^{-1}(A)$. (1) Since $|a| = |A|$, the homomorphism $\epsilon \colon \Gamma \to Spin(n)$ defined by sending (A, δ) to a and Λ to 1 defines the required spin structure.

(2) Clearly $\lambda_n^{-1}(\rho(A))$ contains an element of order $2^m = |A|$, if and only if $\lambda_n^{-1}(\rho(A^{2^{m-1}}))$ contains an involution. This is equivalent to the given condition by Lemma 6.19. (3) First, consider the case $m = 1$. Using the well known integral representation theory of groups of order 2, we may assume that $\Lambda = \Lambda_1 \oplus \Lambda_2$ with G-invariant sublattices Λ_1 and Λ_2 such that A acts diagonally on Λ_1 and $H^1(G, \mathbb{Q} \otimes_{\mathbb{Z}} \Lambda_2/\Lambda_2) = 0$. Write $\delta = \delta_1 + \delta_2$ with $\delta_i \in \mathbb{Q} \otimes_{\mathbb{Z}} \Lambda_i, i = 1, 2$. Then $\delta_2 \in \Lambda_2$ and we may assume that $\delta_2 = 0$, i.e., $\delta \in \Lambda_1$.

Put $\delta' := A\delta + \delta$. The fact that $\delta \mapsto \delta + \Lambda$ defines a non-zero element of $H^1(G, \mathbb{Q} \otimes_{\mathbb{Z}} \Lambda/\Lambda)$ implies that $\bar{\delta}' \in \bar{\Lambda}$ is non-zero, i.e., $\delta' \notin \Lambda_1^2$. Let Λ_0 be a sublattice of index 2 in Λ with $\Lambda_1^2 + \Lambda_2 \subset \Lambda_0$ and $\delta' \notin \Lambda_0$. Then Λ_0 is normal in Γ since G acts trivially on Λ_1/Λ_1^2. Hence Γ/Λ_0 is a cyclic group of order 4, generated by the image of (A, δ). It follows that (ρ, Γ) has a spin structure. Now, suppose that $m \ge 3$. Let χ denote the character of the embedding $G \to GL(n, \mathbb{C})$. Let s denote the multiplicity of an irreducible complex faithful character ζ of G in χ. Then the multiplicity of ζ^j in χ equals s for all odd integers $1 \le j \le 2^m - 1$ since χ is rational valued.

The matrix A is equivalent to a block diagonal matrix with (1×1)-blocks containing the entries ± 1, and (2×2)-blocks of the form

$$\begin{pmatrix} \cos\alpha & -\sin\alpha \\ \sin\alpha & \cos\alpha \end{pmatrix},$$

where $0 < \alpha < \pi$ with $2^m\alpha \in 2\pi i\mathbb{Z}$. Such a matrix contributes the value -2 to the trace of $A^{2^{m-1}}$, if and only if $2^{m-1}\alpha \notin 2\pi i\mathbb{Z}$. By the above, the matrix contains exactly $s2^{m-2}$ such blocks. Hence $n - Trace(A^{2^{m-1}}) = 2^m s$. The result follows from (1) and (2). (4) In this case, $|a| = 2, |A| = 8$ and thus $a^4 = -1$. Consider the $\mathbb{F}G$-module $\bar{\Lambda} = \Lambda/\Lambda^2$. Suppose first, that (ρ, Γ) has a spin structure ϵ. Then $\epsilon(\text{id}, \delta') = -1$ since $(A, \delta)^4 = (\text{id}, \delta')$. Hence $(\text{id}, \delta') \notin K := \ker(\epsilon) \subset \Lambda$. On the other hand, $\Lambda_2 \subset K$ and $\bar{K}$ is a maximal $\mathbb{F}_2 G$-submodule of $\bar{\Lambda}$. Thus $\bar{\delta}'$ is not contained in the radical of $\bar{\Lambda}$.

Suppose now that $\bar{\delta}'$ is not contained in the radical of $\bar{\Lambda}$. Then there is a normal subgroup K of Γ with $\Lambda_2 \subset K \subset \Lambda$ and $|\Lambda : K| = 2$ such that $(\text{id}, \delta') \notin K$. Clearly, Γ/K is cyclic of order 8, and we obtain a spin structure by sending (A, δ) to a and K to 1. □

Using the description of flat $\mathbb{Z}_4$-manifolds by H. Hiller [62], we further investigate this case. The classification of the (finitely many) isomorphism classes of indecomposable integral representations of a cyclic group G of order 4 is due to Heller and Reiner, and is reproduced in [62, Theorem 1.3]. The corresponding low degree cohomology is computed in [62, Propositions 2.2, 2.3]. For the convenience of the reader, these results are presented in Table 6.1. The notation for the $\mathbb{Z}G$-lattices is taken from Hiller's paper. On M_1, M_2, and M_4 the subgroup of order 2 of G acts trivially, the other modules afford faithful representations of G. If X is an indecomposable $\mathbb{Z}G$-lattice, then either $H^2(G, X)$ is trivial or has two elements. In the latter case, the last column of Table 6.1 describes the non-trivial element in $H^2(G, X)$. These elements are special only for $X = M_1$, M_4 or $M_9(0)$ (see [62, Proposition 2.7]).

Proposition 6.23 ([69, Theorem 3.2]). *Let $A \in SL(n, \mathbb{R})$ be of order 4, satisfying $(n - \text{Trace}(A^2))/2 \equiv 2(\text{mod } 4)$. Put $G = \langle A \rangle$, and let δ and Γ be as above. Choose an embedding $\rho \colon \Gamma \to E(n)$.*

Write $\Lambda = \oplus_{i=1}^m \Lambda_i$ with indecomposable $\mathbb{Z}G$-lattices Λ_i, $i = 1, \ldots, m$. Decompose δ accordingly as $\delta = \sum_{i=1}^m \delta_i$ with $\delta_i \in H^1(G, \mathbb{Q} \otimes_{\mathbb{Z}} \Lambda_i/\Lambda_i)$. Then (ρ, Γ) has a spin structure if and only if, for some $1 \leq i \leq m$, (Λ_i, δ_i) is equivalent to $(M_1, (1/4))$ or $(M_9(0), (1/4, 0, 0, 0))$.

Proof. By Theorem 6.2 (4) we have to investigate when $\bar{\delta}'$ lies in the radical of $\bar{\Lambda}$ (viewed as an $\mathbb{F}_2 G$-module). The radical of a direct sum of modules is the direct sum of their radicals. Hence $\bar{\delta}'$ lies in the radical of $\bar{\Lambda}$, if and only if $\bar{\delta}_i'$ lies in the radical of $\bar{\Lambda}_i := \Lambda_i/\Lambda_i^2$ for all $1 \leq i \leq m$.

Note that $\delta_i' \in \Lambda_i$ is a fixed vector of A. Hence if there are no non-zero fixed vectors in Λ_i, we have $\bar{\delta}_i' = 0$ and hence it lies in the radical of $\bar{\Lambda}_i$. This is the case for the lattices M_2, M_3 and $M_8(0)$. Next suppose that $\bar{\Lambda}_i$ has a unique non-zero fixed vector v_i. Then v_i lies in the radical of $\bar{\Lambda}_i$. Moreover,

Table 6.1: Indecomposable integral representations of $\mathbb{Z}_4$

Name	Matrix	Cocycle
M_1	(1)	$\left(\frac{1}{4}\right)$
M_2	(-1)	
M_3	$\begin{pmatrix} 0 & -1 \\ 1 & 0 \end{pmatrix}$	
M_4	$\begin{pmatrix} 0 & 1 \\ 1 & 0 \end{pmatrix}$	$\begin{pmatrix} \frac{1}{2} \\ 0 \end{pmatrix}$
M_5	$\begin{pmatrix} 1 & 0 & 1 \\ 0 & 0 & -1 \\ 0 & 1 & 0 \end{pmatrix}$	$\begin{pmatrix} \frac{1}{4} \\ 0 \\ 0 \end{pmatrix}$
$M_6(0)$	$\begin{pmatrix} 0 & 1 & 0 & 0 \\ 1 & 0 & 0 & 1 \\ 0 & 0 & 0 & -1 \\ 0 & 0 & 1 & 0 \end{pmatrix}$	
$M_6(1)$	$\begin{pmatrix} 0 & 1 & 0 & 1 \\ 1 & 0 & 0 & -1 \\ 0 & 0 & 0 & -1 \\ 0 & 0 & 1 & 0 \end{pmatrix}$	$\begin{pmatrix} \frac{1}{2} \\ 0 \\ 0 \\ 0 \end{pmatrix}$
$M_8(0)$	$\begin{pmatrix} -1 & 0 & 1 \\ 0 & 0 & -1 \\ 0 & 1 & 0 \end{pmatrix}$	
$M_9(0)$	$\begin{pmatrix} 1 & 0 & 0 & 1 \\ 0 & -1 & 0 & 1 \\ 0 & 0 & 0 & -1 \\ 0 & 0 & 1 & 0 \end{pmatrix}$	$\begin{pmatrix} \frac{1}{4} \\ 0 \\ 0 \\ 0 \end{pmatrix}$

either $\bar{\delta}'_i = 0$ or $\bar{\delta}'_i = v_i$. This is the case for the lattices M_1, M_4, M_5, and $M_6(0)$.

Suppose next that Λ_i is equivalent to $M_6(1)$. In this case, $\delta' = (1, 1, 0, 0)$. Observe that $\bar{\Lambda}_i = U_1 \oplus U_2$ as $\mathbb{F}_2 G$-modules with

$$U_1 = \langle (0,0,1,0), (0,0,0,1), (1,1,1,0) \rangle \quad \text{and} \quad U_2 = \langle (1,0,1,1) \rangle.$$

Hence the radical of $\bar{\Lambda}_i$ equals the radical of U_1 which is spanned by the fixed vector $(1, 1, 0, 0)$. Thus, again, either $\bar{\delta}'_i = 0$ or $\bar{\delta}'_i$ lies in the radical of $\bar{\Lambda}_i$.

It remains to consider the case that Λ_i is equivalent to $M_9(0)$. Here, as in the $M_6(1)$ case, $\bar{\delta}'_i \in \langle (1,1,0,0) \rangle$. But contrary to that case, $\bar{\Lambda}_i = \langle (1,1,0,0) \rangle \oplus \langle (0,0,1,0), (0,0,0,1), (1,1,1,0) \rangle$ as $\mathbb{F}_2 G$-modules. Hence if $\bar{\delta}'_i \neq 0$, it does not lie in the radical of $\bar{\Lambda}_i$. This completes the proof. $\square$

Example 6.24. (a) Let

$$A := \left(\begin{array}{cc|cccc} 0 & 1 & & & & \\ 1 & 0 & & & & \\ \hline & & 1 & 0 & 0 & 1 \\ & & 0 & 1 & 0 & 1 \\ & & 0 & 0 & 0 & -1 \\ & & 0 & 0 & 1 & 0 \end{array} \right) \in SL(6, \mathbb{Z}).$$

Then A has order 4 and $(6 - \text{Trace}(A^2))/2 = 2$. (Observe that $\mathbb{Z}^6$ is a direct sum of the $\mathbb{Z}\langle A \rangle$-modules M_4 and $M_9(0)$ of Table 6.1.)

(1) Let $\delta := (1/2, 0, 0, 0, 0, 0) \in \mathbb{Q}^6$. Then δ gives rise to a special cocycle, and we let Γ be defined as above. By Proposition 6.23, there is no spin structure on Γ.

(2) Now let $\delta := (0, 0, 1/4, 0, 0, 0) \in \mathbb{Q}^6$. Then, again, δ gives rise to a special cocycle, and we let Γ be defined as above. By Proposition 6.23, Γ has a spin structure. This yields another example of two Bieberbach groups with the same holonomy representation, one with and one without spin structure. The first example appears in [99, Table 1, p. 327].

(b) A 5-dimensional $\mathbb{Z}_4$-manifold without a spin structure is provided by the module $M_2 \oplus M_3 \oplus M_4$ and the special cocyle arising from M_4. In this case we may take

$$A := \left(\begin{array}{ccccc} -1 & 0 & 0 & 0 & 0 \\ 0 & 0 & -1 & 0 & 0 \\ 0 & 1 & 0 & 0 & 0 \\ 0 & 0 & 0 & 0 & 1 \\ 0 & 0 & 0 & 1 & 0 \end{array} \right) \in SL(5, \mathbb{Z}),$$

and $\delta = (0, 0, 0, 1/2, 0)$.

Let G be a finite group. As in [37, Definition 2], we let $s(G)$ denote the minimal dimension of a flat oriented spin manifold with holonomy group G.

Corollary. *Let m be a positive integer. Then $s(\mathbb{Z}_{2^m}) = 2^{m-1} + 1$ if $m > 1$, and $s(\mathbb{Z}_2) = 3$.*

Proof. The smallest degree of a flat oriented $\mathbb{Z}_2$-manifold equals 3, so that the result for $m = 1$ follows from Theorem 6.22. Assume that $m > 1$. Let r be minimal such that there is an $(r \times r)$-matrix B over $\mathbb{Z}$ of order 2^m. Then $r = 2^{m-1}$ and there is no Bieberbach group of dimension r with holonomy group isomorphic to $\langle B \rangle$ (see [61]). The minimal polynomial of B is the 2^mth cyclotomic polynomial $X^{2^{m-1}} + 1$, and thus B has determinant 1. Now put $n := r + 1$ and

$$A = \begin{pmatrix} 1 & 0 \\ 0 & B \end{pmatrix},$$

and let $\delta \in \mathbb{Q}^n$ have entry $1/2^{m-1}$ in the first coordinate, and zeroes elsewhere. Then δ gives rise to a special cocycle and we define Γ as above. In this case δ' is the first standard basis vector. Clearly, $\bar{\delta}'$ is not contained in the radical of the $\mathbb{F}_2 G$-module $\mathbb{Z}_2^n$, so that Γ has a spin structure by Theorem 6.22. $\qquad \square$

6.4   The Dirac operator

We finish this chapter with some application of a spin structure. Using a material from previous sections we are going to define the Dirac operator. Let $C_n^{\mathbb{C}} = C_n \otimes \mathbb{C}$ be the complexification of Clifford's algebra C_n. We shall define the space of complex n-spinors Σ_n, as the linear space $\mathbb{C}^K$, of complex dimension $K = 2^{[\frac{n}{2}]}$, where

$$[\frac{2k}{2}] = k \text{ and } [\frac{2k+1}{2}] = k. \tag{6.5}$$

Hence $\Sigma_{2k} = \Sigma_{2k+1}$. It is often called the irreducible representation of the algebra $C_n^{\mathbb{C}}$ (see [46, p. 13, 20]).

Let $\Sigma_3 = \Sigma_2 = C_2^{\mathbb{C}} = M_2(\mathbb{C})$ with the basis, the identity matrix I, and matrices

$$g_1 = \begin{pmatrix} i & 0 \\ 0 & -i \end{pmatrix}, g_2 = \begin{pmatrix} 0 & i \\ i & 0 \end{pmatrix}, T = \begin{pmatrix} 0 & -i \\ i & 0 \end{pmatrix}.$$

We can observe, that [46, p. 13] if $n = 2k$, then

$$C_n^{\mathbb{C}} = \underbrace{M_2(\mathbb{C}) \otimes \cdots \otimes M_2(\mathbb{C})}_{k \text{ times}} = End(\underbrace{\mathbb{C}^2 \otimes \cdots \otimes \mathbb{C}^2}_{k \text{ times}}) = End(\Sigma_n). \quad (6.6)$$

Let j be a natural number. Put

$$\alpha(j) = \begin{cases} 1, & \text{when } j \text{ is odd,} \\ 2, & \text{when } j \text{ is even.} \end{cases}$$

Suppose $u = u_1 \otimes \cdots \otimes u_k \in \Sigma_n$, and $e_1, \ldots, e_n$ is the basis of $\mathbb{C}_n^{\mathbb{C}}$. We define a structure of $C_n^{\mathbb{C}}$-module by

$$e_j u = (I \otimes \cdots \otimes I \otimes g_{\alpha(j)} \otimes \underbrace{T \otimes \cdots \otimes T}_{[\frac{j-1}{2}] \text{ times}})(u),$$

for $j \leq n$.
When $n = 2k + 1$,

$$C_n^{\mathbb{C}} = \{M_2(\mathbb{C}) \otimes \cdots \otimes M_2(\mathbb{C})\} \oplus \{M_2(\mathbb{C}) \otimes \cdots \otimes M_2(\mathbb{C})\} = End(\Sigma_n) \oplus End(\Sigma_n). \quad (6.7)$$

For $j \leq 2k$,

$$e_j u = (I \otimes \cdots \otimes I \otimes g_{\alpha(j)} \otimes \underbrace{T \otimes \cdots \otimes T}_{[\frac{j-1}{2}] \text{ times}})(u), (I \otimes \cdots \otimes I \otimes g_{\alpha(j)} \otimes \underbrace{T \otimes \cdots \otimes T}_{[\frac{j-1}{2}] \text{ times}})(u)).$$

Finally,

$$e_n u = (iT \otimes \cdots \otimes T(u), -iT \otimes \cdots \otimes T(u)).$$

Using a projection $pr_1 \colon End(\Sigma_n) \oplus End(\Sigma_n) \to End(\Sigma_n)$ on the first factor, we define a structure of $C_n^{\mathbb{C}}$-module on Σ_n. Summing up, for even n we have an isomorphism (6.6) $\kappa_n \colon C_n^{\mathbb{C}} \to End(\Sigma_n)$ and for odd n, κ_n is an isomorphism (6.7) followed by the projection pr_1. Hence, the structure of module over the Clifford algebra $C_n^{\mathbb{C}}$ on the vector space of n-spinors is established by κ_n.

Let $M^n = \mathbb{R}^n/\Gamma$ be an oriented, flat manifold with the spin structure $\epsilon \colon \Gamma \to Spin(n)$. Let us consider the trivial spinor bundle

$$(\Sigma\mathbb{R}^n = \mathbb{R}^n \times \Sigma_n, \pi, \mathbb{R}^n, \Sigma_n)$$

together with the spinor bundle on M^n

$$(\Sigma M^n, \pi, M^n, \Sigma_n).$$

Here $\Sigma M^n = \Sigma\mathbb{R}^n/\Gamma$, where Γ acts on $\Sigma\mathbb{R}^n$ in the following way

$$\forall (x, \sigma) \in \Sigma\mathbb{R}^n, \gamma \in \Gamma, \gamma(x, \sigma) = (\gamma x, \epsilon(\gamma)\sigma).$$

By spinors on M^n we understand the maps

$$\Psi \colon \mathbb{R}^n \to \Sigma_n,$$

which satisfy the condition

$$\Psi = \epsilon(\gamma)\Psi \circ (\gamma)^{-1}, \forall \gamma \in \Gamma. \tag{6.8}$$

The Levi-Civita connection on the bundle $L(M^n)$ lifts to the bundle ΣM^n and induces the covariant derivative ∇. Since κ_n establishes on Σ_n a structure of $C_n^{\mathbb{C}}$-module and $\mathbb{R}^n \subset C_n \subset C_n^{\mathbb{C}}$, this leads to the so-called *Clifford multiplication of vectors and spinors*. In the simplest case it is a linear map

$$\mu \colon \mathbb{R}^n \otimes_{\mathbb{R}} \Sigma_n \to \Sigma_n, \tag{6.9}$$

where $\mu(x \otimes \psi) = \kappa_n(x)(\psi)$, for $x \in \mathbb{R}^n$ and $\psi \in \Sigma_n$.[2] Hence in case of the vector bundle we have

$$\mu \colon T_p M^n \otimes \Sigma_p M^n \to \Sigma_p M^n,$$

where $p \in M^n$ (see [46, p. 21]).

We define the Dirac operator locally, with the formula

$$(D\Psi)(p) = \mu(\Sigma_{i=1}^n e_i \otimes \nabla_{e_i}\Psi),$$

where $e_1, e_2, \ldots, e_n \in T_p M$ is the orthonormal basis, $p \in M^n$ and $\Psi \in \Sigma_p M^n$.

---

[2]Instead of $\mu(x \otimes \psi)$, we will sometimes write $x \cdot \psi$.

Suppose $\Gamma(S)$ denotes all sections of spinor bundles on the manifold M^n, and suppose $\Gamma(T \otimes S)$ denotes all sections of the tensor product of the tangent bundle with the spinor bundle S, then (see [46, p. 68])

$$D = \mu \circ \nabla \colon \Gamma(S) \to \Gamma(T \otimes S) \to \Gamma(S).$$

Here μ is the *Clifford multiplication*, (6.9).

By definition, it is a first order, self-adjoint, elliptic operator.[3] The following example, which considers the Dirac operator on the torus can be an easy introduction.

Example 6.25. Let $T^n = \mathbb{R}^n/\mathbb{Z}^n$ be a flat torus. The spin-structures on T^n are defined with homomorphisms $\epsilon \colon \mathbb{Z}^n \to \{\pm 1\} \subset Spin(n)$. Let $a_1^*, a_2^*, \ldots, a_n^*$ be the basis of $(\mathbb{Z}^n)^* = \mathrm{Hom}(\mathbb{Z}^n, \mathbb{Z})$. Suppose

$$a_\epsilon := \frac{1}{2} \sum_{\epsilon(a_l) = -1} a_l^*.$$

Let $b \in (\mathbb{Z}^n)^* + a_\epsilon$ and let

$$\{\sigma^j \mid j = 1, 2, 3, \ldots, 2^{[\frac{n}{2}]}\}$$

be the standard basis of Σ_n, see (6.5). Consider spinors $\Psi_b^j \colon \mathbb{R}^n \to \Sigma_n$ sending x to $e^{2\pi i \langle b, x \rangle} \sigma^j$. To deal with the spectrum of the Dirac operator it is convenient to describe it in terms of the spectrum of D^2. By immediate calculation or following [45, p. 61, Theorem 1]) we have

$$D^2(\Psi_b) = 4\pi^2 |b|^2 (\Psi_b). \tag{6.10}$$

Hence the space $E_b(D^2)$, generated by Ψ_b^j is D^2-eigenspace. Moreover, we have ([100, page 4576])

$$D(\Phi) = 2\pi i b \cdot \Phi,$$

where $\Phi \in E_b(D^2), b \neq 0$ and $\cdot$ is Clifford multiplication (6.9).

---

[3]We refer the reader for historical remarks and the motives of the Dirac operator to the introduction of T. Friedrich's book, [46]. For more information about the theory of the Dirac operator the reader is referred to the monograph of N. Berline, E. Getzler and M. Vergne [9].

In that case we are interested in the set of eigenvalues of the Dirac operator. Let

$$E_{b\pm}(D) = \{\Psi \in E_b(D^2) \mid D\Psi = \pm 2\pi|b|\Psi\}.$$

Hence $E_b(D^2) = E_{b+}(D) \oplus E_{b-}(D)$. We define the operator $F^\pm: E_b(D^2) \to E_{b\pm}(D)$ such that

$$F^\pm(\Psi) = (1 \pm \frac{1}{2\pi|b|}D)\Psi = (1 \pm \frac{2\pi ib}{2\pi|b|})\Psi = (1 \pm i\frac{b}{|b|})\Psi.$$

Since $F^\pm$ is surjection, it follows that spinors

$$\Phi_{b\pm}^j = F^\pm\Psi_b^j = (1 \pm i\frac{b}{|b|})\Psi_b^j$$

generate $E_{b\pm}(D)$, where $j = 1, 2, \ldots, 2^{[\frac{n}{2}]}$. In the case $b = 0 \in (\mathbb{Z}^n)^* + a_\epsilon$ one gets $E_0(D^2) = E_0(D) = \Sigma_n$. For $b \neq 0$, we have an isomorphism $E_{b+}(D) \simeq E_{b-}(D)$. In fact, let $c \in \mathbb{R}^n$ be the vector perpendicular to b with norm one. Let $M_c: E_b(D^2) \to E_b(D^2)$ be the *Clifford's multiplication* (cf. (6.9)) by c. By definition $M_c D\Phi = -DM_c\Phi$, for $\Phi \in E_b(D^2)$. Hence $M_c: E_{b\pm} \to E_{b\mp}$, $(M_c)^2 = id$ and M_c is an isomorphism. Moreover, $\dim(E_{b\pm}) = \frac{1}{2}2^{[\frac{n}{2}]}$.

Let us to consider the general case. Denote by $\Gamma \subset E(n)$ the fundamental group of the orientable, flat manifold M^n, with the spin structure $\epsilon: \Gamma \to Spin(n)$. From the first Bieberbach Theorem 2.4 we known that M^n has a finite covering $T^n = \mathbb{R}^n/\mathbb{Z}^n$, where $\mathbb{Z}^n \subset \Gamma$ is the subgroup of all translations. In connection with the condition 6.8, define

$$A\Phi := \epsilon(\gamma)\Phi \circ \gamma^{-1},$$

where $A = p_1(\gamma)$ is a linear part of $\gamma \in \Gamma$, and Φ is a spinor on T^n. The spinors of the torus, which are invariant with respect to the action of a holonomy group, are spinors of the flat manifold $M^n = \mathbb{R}^n/\Gamma$.

Lemma 6.26 ([112, Lemma 4.1]). *Let A be the linear part of the isometry $\gamma = (A, a) \in \Gamma$. Then for $b \neq 0$ we have*

$$A\Phi_{b\pm}^j = e^{-2\pi i\langle Ab,a\rangle}(1 \pm i\frac{Ab}{|Ab|})(\epsilon(\gamma)\Psi_{Ab}^j).$$

Proof. For $x \in \mathbb{R}^n$ we have

$$(\Psi_b^j \circ \gamma^{-1})(x) = e^{2\pi i\langle b, A^{-1}(x-a)\rangle}\sigma^j$$

$$= e^{-2\pi i \langle Ab, a \rangle} \Psi^j_{Ab}(x).$$

By definition

$$A\Phi^j_{b\pm} = \epsilon(\gamma)\Phi^j_{b\pm} \circ \gamma^{-1} = \epsilon(\gamma)(1 \pm i\frac{b}{|b|})\Psi^j_b \circ \gamma^{-1}$$

$$= (1 \pm i\frac{1}{|b|}\epsilon(\gamma)b\epsilon(\gamma)^{-1})\epsilon(\gamma)\Psi^j_b \circ \gamma^{-1}.$$

From definition of the spin-structure, see Corollary 6.3, we have $\lambda_n\epsilon = p_1$. Hence and from equation (6.1)

$$\epsilon(\gamma)b\epsilon(\gamma)^{-1} = \epsilon(\gamma)b\epsilon(\gamma)^* = (\lambda_n\epsilon(\gamma))(b) = A(b).$$

Moreover, $|b| = |Ab|$ and hence

$$A\Phi^j_{b\pm} = (1 \pm i\frac{Ab}{|Ab|})(\epsilon(\gamma)e^{-2\pi i \langle Ab, a \rangle}\Psi^j_{Ab}).$$

$\square$

We can write $A\Phi^j_{b\pm} = F^{\pm}(e^{-2\pi i \langle Ab, a \rangle}\epsilon(\gamma)\Psi^j_{Ab})$. Hence, for all $\Phi \in E_{b\pm}(D)$, we have

$$A\Phi \in E_{Ab\pm}(D). \tag{6.11}$$

Let H be the holonomy group of M^n. By definition it can be considered as a subgroup of $SO(n)$.

Proposition 6.27 ([112, Theorem 4.2]). *Suppose that for $b \in (\mathbb{Z}^n)^* + a_\epsilon, b \neq 0$ one has $\#H = \#(Hb)$. Consider*

$$V = \oplus_{A \in H} E_{Ab}(D^2).$$

Then, the subspace K of D-eigenspinors of the manifold M^n, with respect to the eigenvalue $\pm 2\pi|b|$, has dimension $\frac{1}{2}2^{[\frac{n}{2}]}$.

Proof. The assumption $k = \#H = \#(Hb)$ denotes that H acts on H-orbits b without fixed points. Hence, for $A_i \in H$, all elements $\{A_ib\}_{i=1,2,\ldots,k}$ are distinct, the subspaces $E_{A_jb}(D^2)$ are orthogonal and

$$V^{\pm} = \oplus_{A \in H} E_{Ab\pm}(D).$$

The holonomy representation defines the representation

$$\rho^{\pm} \colon H \to GL(V^{\pm}).$$

Let $\chi^{\pm}$ denote the associated characters. From definition of the space V and Lemma 6.26 together with formula 6.11 it follows that $\chi^{\pm}(A) = tr(\rho^{\pm}(A)) = 0$, for $I \neq A \in H$. Moreover, the subspace K of D-eigenspinors is equal to the subspace $(V^{\pm})^H$. Finally, using a formula on the dimension of the fix points space from the theory of characters (see [126, part 2.3]), we have

$$\dim K = \langle \chi^{\pm}, 1 \rangle = \tfrac{1}{\#H} \sum_{A \in H} \chi^{\pm}(A) = \tfrac{1}{k} \chi^{\pm}(id) =$$
$$= \tfrac{1}{k} \dim(V^{\pm}) = \tfrac{1}{k} \cdot \tfrac{1}{2} \cdot k \cdot 2^{[\frac{2}{n}]}. \tag{6.12}$$

$\square$

Corollary ([112, Corollary 4.3]). *If the action of the holonomy group on the set $(\mathbb{Z}^n)^* + a_\epsilon$ is free then the spectrum of the Dirac operator on the manifold M^n is symmetric.*

$\square$

From the general theory of elliptic operators [4] for compact spin manifolds it follows that the spectrum of the Dirac operator is discrete, consisting of real eigenvalues λ with finite multiplicity d_λ. The spectrum is said to be asymmetric if for some eigenvalue λ, one has that $d_\lambda \neq d_{-\lambda}$. To study this phenomenon Atiyah, Patodi and Singer introduced the η series

$$\eta(z) = \sum_{\lambda \neq 0} sgn\lambda |\lambda|^{-z} \tag{6.13}$$

This series converges for $z \in \mathbb{C}$ with $Re(z)$ sufficiently large ([4, Theorem 3.10]). Here summation is taken over all non-zero eigenvalues λ of D, each eigenvalue being repeated according to its multiplicity. It can be proved [4, Theorem 3.10] that $\eta(z)$ extends to a meromorphic function on the complex plane with no pole at 0.

Definition 6.28. The number $\eta(0)$ is called the η-invariant of the Dirac operator D.

The η-invariant of the elliptic operator measures the symmetry of its spectrum. Hence an operator with symmetric spectrum has η-invariant zero. The table, with values of η-invariants for 6 orientable, flat 3-manifolds can be found in [112].

Remark 6.29. The above few facts about the Dirac operator are very closely related to Problem 11.22 and Question 11.27 from the last Section of Problems. There are also interesting facts about the spectra of the Dirac operator on flat manifolds in R. J. Miatello and R. A. Podestá's article [100].

6.5 Exercises

Exercise 6.1. Prove that for any $x \in \mathbb{R}^n$, $xx = -\|x\|^2$ and hence, $x^{-1} = \frac{-x}{\|x\|^2} = \frac{\bar{x}}{\|x\|^2}$.

Exercise 6.2. Let $y \in \mathbb{R}^n \subset C_n$ and let $x \in Pin(n)$. Prove, that $xyx^* \in \mathbb{R}^n$.

Exercise 6.3. Let $x \in Spin(n)$. Prove, that $x^{-1} = x^*$.

Exercise 6.4. Let $x, y \in \mathbb{R}^n \subset C_n$. Prove, that

$$\langle x, y \rangle = \frac{1}{2}(x\bar{y} + y\bar{x}). \tag{6.14}$$

Exercise 6.5. Prove that for $n \geq 3$ the group Spin(n) is simply connected and λ_n is a covering map.

Exercise 6.6. Give an example of non isomorphic principal bundles which are isomorphic as fibrations (consider the last example).

Exercise 6.7. Prove Proposition 6.8.

Exercise 6.8. Prove that flat manifolds with the Hantzsche-Wendt fundamental groups have no spin structure.

Exercise 6.9. Find all spin-structures on the torus. In particular, prove that all spin-structures on T^n are related to the set of sequences $(\delta_{i_1}, \delta_{i_2}, \ldots, \delta_{i_n})$, where $\delta_{i_j} \in \{0, 1\}$ for $j = 1, 2, \ldots, n$.

Exercise 6.10. Describe an isomorphism of $C_2^{\mathbb{C}}$ with $M_2(\mathbb{C})$. In particular, show that the matrices I, g_1, g_2, T are the basis of the complex algebra $M_2(\mathbb{C})$.

Exercise 6.11. Show that spinors Ψ_b^j are eigenspinors of D^2 (cf. [45]).

Exercise 6.12. Prove that the map M_c is an isomorphism.

Exercise 6.13. Calculate the η-invariant of the Dirac operator of 6 orientable flat manifolds (see [112, Theorem 5.6]).

7. Flat Manifolds with Complex Structures

In this part we are interested in the classification of six dimensional flat manifolds, admitting the complex structure, which in this case is equivalent to the Kähler structure. At the beginning we give necessary and sufficient conditions on existence of a complex structure on the flat manifold. We show that they depend on properties of a holonomy representation. At the end of this chapter we shall describe some homology properties of the flat Kähler manifolds.

The text below is based on [36]. Assume that n is a positive, even natural number and Γ is the torsion free crystallographic group with holonomy group G of dimension n.

Definition 7.1. The holonomy representation $\Phi \colon G \to GL(n, \mathbb{Z})$ is essentially complex if there is a matrix $A \in GL(n, \mathbb{R})$, such that

$$\forall g \in G, A\Phi(g)A^{-1} \in GL(\frac{n}{2}, \mathbb{C}).$$

It is an exercise to show that the above is equivalent to the existence of a linear, Φ-invariant map $t \colon \mathbb{R}^n \to \mathbb{R}^n$, such that $t^2 = -id$. Recall a definition of the unitary group,

$$U(n) = \{A = [a_{ij}] \in GL(n, \mathbb{C}) \mid A\bar{A}^T = I\},$$

where $\bar{A} = [\bar{a}_{ij}], a_{ij} \in \mathbb{C}, i, j = 1, 2, \ldots, n$.

Proposition 7.2. *The following conditions are equivalent* [76, Theorem 3.1].
  (i) Γ *is the fundamental group of a Kähler flat manifold;*
  (ii) *the holonomy representation of* Γ *is essentially complex;*
  (iii) Γ *is a discrete, cocompact and torsion-free subgroup of* $U(\frac{n}{2}) \ltimes \mathbb{C}^{\frac{n}{2}}$.

Proof. We prove $(i) \Rightarrow (ii) \Rightarrow (iii) \Rightarrow (i)$. Let Γ be a fundamental group of Kähler flat manifold M of complex dimension $\frac{n}{2}$. Relative to the Hermitian connection (see [81, chapter 9, corollary 4.4, p. 149]) of M, a parallel translation around a closed curve induces a unitary transformation on the (complex) tangent space at the base point. Hence the holonomy representation $h_\Gamma\colon G \to GL(n,\mathbb{Z}) \subset GL(n,\mathbb{R})$ is essentially complex.

Let us start to prove the implication $(ii) \Rightarrow (iii)$. Suppose that $h_\Gamma\colon G \to GL(n,\mathbb{Z})$ is essentially complex that is, there exists an invertible real $n \times n$ matrix B such that $Bh_\Gamma(G)B^{-1} \subset GL(\frac{n}{2},\mathbb{C})$. By averaging over the finite group G, (cf. Exercise 2.15) we see that $Dh_\Gamma(G)D^{-1} \subset U(\frac{n}{2})$ for some $D \in GL(\frac{n}{2},\mathbb{C})$.

Repeating the argument of Section 2.3, we get a morphism of exact sequences

$$
\begin{array}{ccccccccc}
0 & \longrightarrow & \mathbb{Z}^n & \longrightarrow & \Gamma & \longrightarrow & G & \longrightarrow & 0 \\
 & & \downarrow{\scriptstyle P} & & \downarrow{\scriptstyle \phi} & & \downarrow{\scriptstyle h_\Gamma^P} & & \\
0 & \longrightarrow & \mathbb{C}^{\frac{n}{2}} & \longrightarrow & E_{\mathbb{C}}\left(\frac{n}{2}\right) & \longrightarrow & U\left(\frac{n}{2}\right) & \longrightarrow & 0,
\end{array}
$$

where $E_{\mathbb{C}}(\frac{n}{2})$ is the complex Euclidean group $U(\frac{n}{2}) \ltimes \mathbb{C}^{\frac{n}{2}}$, and $\phi(x,y) = (Px, h_\Gamma^P(y))$ with $P = DB$; ϕ embeds Γ as a discrete cocompact subgroup of $U(\frac{n}{2}) \ltimes \mathbb{C}^{\frac{n}{2}}$, as required.

For the proof of the last implication $(iii) \Rightarrow (i)$, let us see that for any torsion-free discrete, cocompact subgroup Γ of $E_{\mathbb{C}}(\frac{n}{2})$, $(E_{\mathbb{C}}(\frac{n}{2})/U(\frac{n}{2}))/\Gamma = M$ is a compact flat Kähler manifold with $\pi_1(M) = \Gamma$. Since $U(\frac{n}{2}) \subset SO(n)$, we can use the Bieberbach theorems (Theorem 2.4) and it finishes the proof. $\square$

Proposition 7.3 ([76]). *Let $h\colon G \to GL(m,\mathbb{R})$ be a faithful representation. Then h is essentially complex if and only if m is even and each $\mathbb{R}$-irreducible summand of h which is also $\mathbb{C}$-irreducible occurs with an even multiplicity.*

Proof. Let W be a left $\mathbb{R}[G]$ module which is defined by the representation h. We have a direct sum $W = V_1 \oplus V_2 \oplus \cdots \oplus V_k$ of $\mathbb{R}[G]$ irreducible modules V_i for $i = 1,2,...,k$. It follows from (see [32, Theorem (73.9)]) that there are three kinds of summand: absolutely irreducible, "complex" and "quaternionic". A simple $\mathbb{R}[G]$ module V is "complex" if $End_{\mathbb{R}[G])}(V) = \mathbb{C}$ and is "quaternionic" if $End_{\mathbb{R}[G]}(V) = \mathbb{H}$. If $End_{\mathbb{R}[G])}(V) = \mathbb{R}$ then V is absolutely irreducible. Hence, we have a decomposition $W = W_{\mathbb{R}} \oplus W_{\mathbb{C}} \oplus W_{\mathbb{H}}$ into absolutely irreducible, "complex" and "quaternionic" summands. Suppose that

$\mathbb{R}[G]$-module W is essentially complex. Then, m is even and the space

$$End_{\mathbb{R}[G]}(W_{\mathbb{R}}) \simeq M(n, \mathbb{R})$$

is also essentially complex of dimension n^2, for some even natural number n. In fact, a complex structure t on W induces a complex structure t_* on the real vector space $End_{\mathbb{R}[G]}(W_{\mathbb{R}})$. This means $t_*(f) = t \circ f$, for $f \in End_{\mathbb{R}[G]}(W_{\mathbb{R}})$. Hence $n^2 = dim_{\mathbb{R}}(End_{\mathbb{R}[G]}(W_{\mathbb{R}}))$ is even and also the number n of the absolutely irreducible components of W is even. This proves the necessary condition. On the other hand, if we consider the $\mathbb{C}[G]$-module $V \otimes \mathbb{C}$, then in the first case it is simple, in the "complex" case it is the direct sum of two non-isomorphic simple, complex conjugate modules, and in the "quaternionic" case is the direct sum of two isomorphic simple modules. Hence it is clear that "complex" and "quaternionic" summands of W are essentially complex. Finally, from the assumption about the even number of the absolutely irreducible components we get the essentially complex structure from an isomorphism of $V \oplus V$ with $V \otimes \mathbb{C}$. $\qquad \square$

Definition 7.4 ([84, page 495]). A hyperelliptic variety is a complex projective variety, not isomorphic to an abelian variety, but admitting an abelian variety as a finite covering.

It is proved in [77] that the class of fundamental groups of complex flat manifolds (with exception of the complex torus) and hyperelliptic varieties coincide. However, in dimension three there are nonalgebraic Kähler flat manifolds. An example of such manifold is given in [84, page 495, page 501 Remark 3.9].

In the next part, using the above results, we shall give a list of the Kähler flat manifolds of $\mathbb{R}$-dimension 4 and 6.

7.1 Kähler flat manifolds in low dimensions

Before we start our investigation of Kähler flat manifolds in low dimensions, we first prove a lemma providing constraints on the possible holonomy groups of such manifolds.

Lemma 7.5. Let $\mathbb{Z}_2^k$ be a holonomy group of a n-dimensional, complex flat Kähler manifold. Then $k \leq n - 1$.

Proof. Let $\varphi \colon \mathbb{Z}_2^k \to GL(2n, \mathbb{R})$ denote the realization of the holonomy representation. Then, $\mathbb{R}^{2n}$ seen as a $\mathbb{Z}_2^k$-module, can be written as a direct sum

$$\mathbb{R}^{2n} = V_1^{l_1} \oplus V_2^{l_2} \oplus \cdots \oplus V_m^{l_m},$$

where each V_i is an $\mathbb{R}$-irreducible $\mathbb{Z}_2^k$-module and V_i and V_j are not equivalent if $i \neq j$. Of course, as a vector space, each of the $V_i = \mathbb{R}$ and the corresponding representation $\mathbb{Z}_2^k \to GL(\mathbb{R})$ has its image lying inside $\{1, -1\} \cong \mathbb{Z}_2$. Note that the $\mathbb{R}$-irreducible components are also $\mathbb{C}$-irreducible and hence all l_i are even numbers (Proposition 7.3).

It follows that, because the holonomy representation φ is faithful, m must be at least k. We can also exclude the case where $m = k$. For, if $m = k$, the case where V_i is the trivial module does not occur. It follows that for any i we can find an element $a_i \in \mathbb{Z}_2^k$ acting as -1 on V_i and as $+1$ on the other components. This would imply that the element $a_1 a_2 \cdots a_m$ acts as -1 on the total space, which is impossible. So we have that $k + 1 \leq m$.

Finally, as each of l_i is even, we have that the real dimension of the manifold is at least $2m$ or the complex dimension $n \geq m$, which finishes the proof.

$\square$

Lemma 7.6. *In complex dimension 2, the only groups appearing as a holonomy group of a Kähler flat manifold are 1, $\mathbb{Z}_2$, $\mathbb{Z}_3$, $\mathbb{Z}_4$ and $\mathbb{Z}_6$. In fact there are exactly eight Kähler flat manifolds in dimension 2.*

Proof. Looking at the classification (cf. [16], [24] and [109]) of flat manifolds in real dimension 4, one sees that the groups occurring as a holonomy group of a 4-dimensional flat manifold are

$$1, \; \mathbb{Z}_2, \; \mathbb{Z}_2^2, \; \mathbb{Z}_2^3, \; \mathbb{Z}_3, \; \mathbb{Z}_6, \; \mathbb{Z}_2 \times \mathbb{Z}_6, \; \mathbb{Z}_4, \; \mathbb{Z}_2 \times \mathbb{Z}_4, \; D_8, \; D_6, \; \mathbb{Z}_2 \times D_6,$$

where D_n is a finite dihedral group of order n. As all of the groups $\mathbb{Z}_2^2$, $\mathbb{Z}_2^3$, $\mathbb{Z}_2 \times \mathbb{Z}_6$, $\mathbb{Z}_2 \times \mathbb{Z}_4$, D_8 and $\mathbb{Z}_2 \times D_6$ contain a subgroup which is isomorphic to $\mathbb{Z}_2^2$, we deduce from Lemma 7.5 that those groups cannot occur as the holonomy groups of a 2-dimensional Kähler flat manifold. There are three flat manifolds in dimension 4 having D_6 as their holonomy group. It is however easy to see that all of them have the first Betti number one, so that we can exclude this group too. For the rest of the possible holonomy groups, there are flat manifolds supporting a Kähler structure. Going through the list of all such groups, we find the following table of 2-dimensional Kähler flat manifolds.

holonomy	CARAT symbols
1	15.1.1
$\mathbb{Z}_2$	18.1.1; 18.1.2
$\mathbb{Z}_3$	35.1.1; 35.1.2
$\mathbb{Z}_4$	25.1.2; 27.1.1
$\mathbb{Z}_6$	70.1.1

□

In the case of $\mathbb{R}$-dimension 6 we have more possibilities.

Lemma 7.7. *The following finite groups occur as holonomy groups of a three-dimensional Kähler flat manifold:* 1, $\mathbb{Z}_n$, *for* $n = 2, 3, 4, 5, 6, 8, 10, 12$, $\mathbb{Z}_2 \times \mathbb{Z}_2$, $\mathbb{Z}_2 \times \mathbb{Z}_4$, $\mathbb{Z}_3 \times \mathbb{Z}_3$, $\mathbb{Z}_6 \times \mathbb{Z}_2$, $\mathbb{Z}_4 \times \mathbb{Z}_4$, $\mathbb{Z}_6 \times \mathbb{Z}_3$, $\mathbb{Z}_6 \times \mathbb{Z}_4$, $\mathbb{Z}_6 \times \mathbb{Z}_6$, D_8.

Proof. In [109], [24] a list of all holonomy groups of six-dimensional flat manifolds is given. We use the notation of [24].

We shall now go through the list of all finite groups appearing as the holonomy groups of a 6-dimensional flat manifold. We first remark that all the groups

$$[64, 250], [32, 47], [32, 46], [32, 36], [32, 33], [24, 11], \mathbb{Z}_2 \times [16, 9], [16, 9],$$

$$\mathbb{Z}_2 \times D_8, \mathbb{Z}_2 \times \mathbb{Z}_2 \times D_8, \mathbb{Z}_4 \times D_8, \mathbb{Z}_2 \times \mathbb{Z}_2 \times A_4, \mathbb{Z}_2 \times A_4,$$

$$\mathbb{Z}_2^2 \times \mathbb{Z}_4, \mathbb{Z}_2^3 \times \mathbb{Z}_4, \mathbb{Z}_2^3 \times \mathbb{Z}_3, \mathbb{Z}_2^4 \times \mathbb{Z}_3, \mathbb{Z}_2^n, n = 3, 4, 5,$$

have a group $(\mathbb{Z}_2)^3$ as a subgroup, hence by Lemma 7.5 they can be eliminated.

Moreover, the finite groups

$$\mathbb{Z}_2 \times [80, 52], [80, 52], [32, 31], [16, 11], \mathbb{Z}_2 \times D_{10}, D_{10}, \mathbb{Z}_6 \times D_8, \mathbb{Z}_3 \times D_8,$$

$$\mathbb{Z}_2 \times (\mathbb{Z}_3^2 \rtimes \mathbb{Z}_2), \mathbb{Z}_3^2 \rtimes \mathbb{Z}_2, \mathbb{Z}_3 \rtimes \mathbb{Z}_8, \mathbb{Z}_3 \times (\mathbb{Z}_3 \rtimes \mathbb{Z}_4), \mathbb{Z}_2 \times (\mathbb{Z}_3 \rtimes \mathbb{Z}_4),$$

$$\mathbb{Z}_3 \rtimes \mathbb{Z}_4, \mathbb{Z}_3 \times Q_8, Q_8, \mathbb{Z}_2 \times \mathbb{Z}_4 \times A_4, \mathbb{Z}_6 \times A_4, \mathbb{Z}_3 \times A_4, \mathbb{Z}_2^3 \times A_4,$$

$$\mathbb{Z}_3 \times S_4, \mathbb{Z}_2 \times S_4, D_6^2, \mathbb{Z}_2 \times \mathbb{Z}_4 \times D_6, \mathbb{Z}_4 \times D_6, \mathbb{Z}_2^3 \times D_6, \mathbb{Z}_3 \times D_8,$$

$$\mathbb{Z}_6 \times D_8, \mathbb{Z}_2 \times \mathbb{Z}_8, \mathbb{Z}_2 \times \mathbb{Z}_{10}, \mathbb{Z}_2 \times \mathbb{Z}_4 \times \mathbb{Z}_6, \mathbb{Z}_4 \times A_4, \mathbb{Z}_2 \times D_{24}$$

only occur as holonomy groups of flat manifolds with the first Betti number one. By Proposition 7.3 these can be eliminated, too.

Let $M \cong \mathbb{Z}^6$ be any faithful D_6-module which is essentially complex and where the D_6-sublattice M^{D_6} is of rank 2. Then using similar methods as in the proof of Proposition ⸬ in [134, p. 192] and properties of the group $H^2(D_6, \mathbb{Z})$ we can show that any cocycle $\alpha \in H^2(D_6, M)$ is mapped to zero by the homomorphism $res_{\langle x \rangle}^{D_6}: H^2(D_6, M) \to H^2(\langle x \rangle, M)$, where x is an element of order three. Hence we can eliminate the groups:

$$D_6, \mathbb{Z}_2 \times D_6, \mathbb{Z}_3 \times D_6, D_{24}, S_4, \mathbb{Z}_6 \times D_6, \mathbb{Z}_2 \times \mathbb{Z}_2 \times D_6.$$

There exists one Kähler flat manifold with holonomy group D_8 and the first Betti number equal to zero. It has the symbol[1] 207.1.1. As a subgroup of $E(6)$, it is generated by the following elements

$$(I, (0,0,1,0,0,0)), \; (I, (0,0,0,1,0,0)), \; (I, (0,0,0,0,1,0)),$$

$$(I, (0,0,0,0,0,1)), \; (A_1, (1/2,0,0,0,1/4,0)), \; (A_2, (0,1/2,0,0,0,0)),$$

where I denotes the identity 6×6 matrix and

$$A_1 = \begin{bmatrix} 1 & 0 & 0 & 0 & 0 & 0 \\ 0 & 0 & 0 & -1 & 0 & 0 \\ 0 & 0 & -1 & 0 & 0 & 0 \\ 0 & -1 & 0 & 0 & 0 & 0 \\ 0 & 0 & 0 & 0 & -1 & 0 \\ 0 & 0 & 0 & 0 & 0 & -1 \end{bmatrix}, A_2 = \begin{bmatrix} 0 & 0 & 1 & 0 & 0 & 0 \\ 0 & 1 & 0 & 0 & 0 & 0 \\ 1 & 0 & 0 & 0 & 0 & 0 \\ 0 & 0 & 0 & -1 & 0 & 0 \\ 0 & 0 & 0 & 0 & -1 & 0 \\ 0 & 0 & 0 & 0 & 0 & -1 \end{bmatrix}.$$

Moreover, it is known [63] that any flat manifold with holonomy group D_6 has a non zero first Betti number. The group A_4 has one absolutely irreducible faithful representation of rank 3. Hence, by Proposition 7.3 it cannot be on our list (there are no 6-dimensional flat manifolds with holonomy group A_4 and with the first Betti number 0). Let us now consider the group of order sixteen which is referred to as [16, 10] in the notations of [24] and which we have not considered, yet. It is the holonomy group of 31 Bieberbach groups of rank 6 with non trivial centre. We can prove that all of them have the first Betti number one (see [134, Lemma 1, page 194].) Moreover, it is also the

---

[1] We use notations from CARAT, (cf. [109]).

holonomy group of 3 Bieberbach groups of rank 6 with trivial centre. In this case it is easy to see, for example from elementary representation theory, that the conditions of Proposition 7.3 are not satisfied. By an analogous procedure we can eliminate the groups [16.8], [16, 13] and D_{16} which completes the proof. □

Finally, we have:

Theorem 7.8. *There are 174 3-dimensional Kähler flat manifolds.*

Proof. We shall use the results about the holonomy groups proved in Lemma 7.7 and the list of six dimensional Bieberbach groups from CARAT, [109], [24]. To prepare the final list we shall mainly use Proposition 7.3. □

Let us present the final Table 7.1.

Table 7.1

Holonomy group	Number of groups	Notations in CARAT and Betti number β_1
$\mathbb{Z}_2$	5	$\beta_1 = 2$, 174.1.1, 174.1.2, $\beta_1 = 4$, 173.1.1, 173.1.2, 173.1.3,
$\mathbb{Z}_3$	4	$\beta_1 = 2$, 291.1.1, 291.1.2, $\beta_1 = 4$, 311.1.1, 311.1.2,
$\mathbb{Z}_4$	22	$\beta_1 = 2$, 201.1.1, 202.1.2, 225.1.1., 225.1.10, 225.1.11, 225.1.12 (2 groups), 225.1.13, 225.1.2, 225.1.3, 225.1.4 (2 groups), 225.1.5, 225.1.6 (2 groups), 225.1.7 (2 groups), 225.1.8 (2 groups), 225.1.9, $\beta_1 = 4$, 219.1.1, 219.1.2,
$\mathbb{Z}_5$	2	$\beta_1 = 2$, 626.1.1, 626.1.2,
$\mathbb{Z}_6$	14	$\beta_1 = 2$, 1611.1.1, 318.1.1, 318.1.2, 318.1.3, 318.1.5, 319.1.1, 319.1.2, 319.1.3, 319.1.5, 404.1.1, 404.1.2, 404.1.3, 404.1.4, $\beta_1 = 4$, 1694.1.1,
$\mathbb{Z}_8$	1	$\beta_1 = 2$, 468.1.1,
$\mathbb{Z}_{10}$	1	$\beta_1 = 2$, 7093.1.1,
$\mathbb{Z}_{12}$	6	$\beta_1 = 2$, 359.1.1, 359.1.3, 359.1.4, 361.1.1, 361.1.2, 554.1.1,

$\mathbb{Z}_2 \times \mathbb{Z}_2$	33	$\beta_1 = 0$, 185.1 (4 groups),
		$\beta_1 = 2$, 186.1 (29 groups),
$\mathbb{Z}_2 \times \mathbb{Z}_4$	45	$\beta_1 = 2$, 257.1 (19 groups), 1135.1 (26 groups),
$\mathbb{Z}_3 \times \mathbb{Z}_3$	13	$\beta_1 = 2$, 405.1.1 (5 groups), 405.1.2 (3 groups),
		405.1.3 (3 groups), 405.1.4, 405.1.5,
$\mathbb{Z}_6 \times \mathbb{Z}_2$	7	$\beta_1 = 2$, 1732.1 (5 groups), 2701.1 (2 groups),
$\mathbb{Z}_4 \times \mathbb{Z}_4$	8	$\beta_1 = 2$, 1264.1 (8 groups),
$\mathbb{Z}_6 \times \mathbb{Z}_3$	6	$\beta_1 = 2$, 2719.1 (2 groups), 2720.1 (4 groups),
$\mathbb{Z}_6 \times \mathbb{Z}_4$	4	$\beta_1 = 2$, 1920.1 (4 groups),
$\mathbb{Z}_6 \times \mathbb{Z}_6$	1	$\beta_1 = 2$, 2752.1.

Finally, we would like to mention that in [77] it has been observed that using "the double" construction it is possible to construct for any finite group G, a Kähler flat manifold with holonomy group G.

7.2   The Hodge diamond for Kähler flat manifolds

In this section we shall show how to compute the real cohomology and Hodge numbers for any flat Kähler manifold. We shall explicitly list all possible Hodge diamonds up to the complex dimension 3. We shall continue this study in the next section where we shall be dealing with a general class of flat Kähler manifolds in arbitrarily high dimensions. Any flat Kähler complex n-dimensional manifold M is a quotient of the form T^{2n}/H, where T^{2n} is a real $2n$-dimensional torus and $H \subset U(n)$ is a finite group. From the standard observations we have:

$$H^{p,q}(M) = (\Lambda^{p,q}(\mathbb{C}^n \oplus (\mathbb{C}^n)^*))^H,$$

where $H^{*,*}$ denotes the Hodge cohomology. Recall that $\Lambda^{p,q}(\mathbb{C}^n \oplus (\mathbb{C}^n)^*)$ is the vector space with basis elements

$$dz_{i_1} \wedge dz_{i_2} \wedge \cdots \wedge dz_{i_p} \wedge d\bar{z}_{j_1} \wedge d\bar{z}_{j_2} \wedge \cdots \wedge d\bar{z}_{j_q},$$
$$1 \leq i_1 < i_2 < \cdots < i_p \leq n, 1 \leq j_1 < j_2 < \cdots < j_q \leq n \qquad (7.1)$$

on which the action of H is induced by the holonomy representation $H \to U(n)$. For a given flat manifold M, such a representation is unique up to conjugation in $GL(2n, \mathbb{R})$, however, sometimes it is not unique up to conjugation in $GL(n, \mathbb{C})$. When this is the case, such a flat Riemannian manifold M carries different kinds of complex structures, with possibly different Hodge numbers. For example, we will see that every complex 3-dimensional Kähler flat manifold with $\beta_1 = 2$ and $\beta_2 = 5$ has two different complex structures leading to different Hodge numbers, (see the example below). We can also calculate the Betti numbers directly from the holonomy representation $G \to GL(2n, \mathbb{R})$, using the equation: $\beta_i(M) = \dim(\Lambda^i(\mathbb{R}^{2n}))^G$. Let us present the Table 7.2 of Kähler flat manifolds from the previous section with their Betti numbers. [2]

Table 7.2

β_1	β_2	β_3	Holonomy	CARAT symbol
0	3	8	$\mathbb{Z}_2 \times \mathbb{Z}_2$	185.1
0	2	6	D_8	207.1
2	3	4	$\mathbb{Z}_4$	225.1
			$\mathbb{Z}_5$	all
			$\mathbb{Z}_6$	318.1, 319.1, 404.1
			$\mathbb{Z}_8$	all
			$\mathbb{Z}_{10}$	all
			$\mathbb{Z}_{12}$	all
			$\mathbb{Z}_2 \times \mathbb{Z}_2$	186.1
			$\mathbb{Z}_2 \times \mathbb{Z}_4$	all
			$\mathbb{Z}_3 \times \mathbb{Z}_3$	all
			$\mathbb{Z}_6 \times \mathbb{Z}_2$	all
			$\mathbb{Z}_4 \times \mathbb{Z}_4$	all
			$\mathbb{Z}_6 \times \mathbb{Z}_3$	all
			$\mathbb{Z}_6 \times \mathbb{Z}_4$	all
			$\mathbb{Z}_6 \times \mathbb{Z}_6$	all
2	5	8	$\mathbb{Z}_3$	291.1
			$\mathbb{Z}_4$	202.1
			$\mathbb{Z}_6$	1611.1
2	7	12	$\mathbb{Z}_2$	174.1
4	7	8	$\mathbb{Z}_2$	173.1
			$\mathbb{Z}_3$	311.1
			$\mathbb{Z}_4$	219.1
			$\mathbb{Z}_6$	1694.1
6	15	20	1	170.1.1 $(M = T^6)$

[2]Note that $\beta_4 = \beta_2$, $\beta_5 = \beta_1$, $\beta_6 = \beta_0 = 1$. Moreover, as the Euler characteristic of such a manifold is 0, we also have the relation $\beta_3 = 2 - 2\beta_1 + 2\beta_2$.

In what follows we present some calculations of the Hodge numbers $\{h^{p,q}\}$, for some of the manifolds above. The case $\beta_1 = 0$ and holonomy $\mathbb{Z}_2 \times \mathbb{Z}_2$ is also being considered in the next section. As we are working in complex dimension 3, we have that $p, q \in \{0, 1, 2, 3\}$.

Example 7.9. To illustrate several possibilities, we consider as an example what happens in case the holonomy group G is isomorphic to $\mathbb{Z}_6$. If t denotes the generator of $\mathbb{Z}_6$, we can distinguish, up to conjugation inside $GL(3, \mathbb{C})$, 4 possibilities for the representation $\mathbb{Z}_6 \to U(3)$ given by the following possible images for t:

$$
t \mapsto \begin{pmatrix} 1 & 0 & 0 \\ 0 & z & 0 \\ 0 & 0 & w \end{pmatrix}, \quad \begin{pmatrix} 1 & 0 & 0 \\ 0 & z & 0 \\ 0 & 0 & z \end{pmatrix}, \quad \begin{pmatrix} 1 & 0 & 0 \\ 0 & z & 0 \\ 0 & 0 & \bar{z} \end{pmatrix},
$$

or

$$
\begin{pmatrix} 1 & 0 & 0 \\ 0 & 1 & 0 \\ 0 & 0 & z \end{pmatrix},
$$

where z denotes a primitive 6-th root of unity and $1 \neq w$ denotes a non-primitive 6-th root of unity. One can easily check that the second and third possibilities, when regarded as representations in $GL(6, \mathbb{R})$, are conjugate to each other.

When we perform these computations for all possible holonomy groups, we find Table 7.3 of Hodge diamonds, where in the case of manifolds with $\beta_1 = 2$ and $\beta_2 = 5$, there are always 2 possibilities, depending on the choice of the complex structure.

In the same way we can compute the Hodge diamond of all hyperelliptic surfaces in which case we always find:

$$
\begin{array}{ccccc}
 & & 1 & & \\
 & 1 & & 1 & \\
0 & & 2 & & 0 \\
 & 1 & & 1 & \\
 & & 1 & &
\end{array}
$$

In the last 20 years, after the solution of Calabi's conjecture by S. T. Yau, studies connected with complex manifolds have grown. One of the most important classes are the so called **Calabi-Yau manifolds**, that is Kähler manifolds with holonomy group contained in $SU(n)$, (see [36]).

There are also other definitions of Calabi-Yau manifolds. In [78] the author mentions five non-equivalent definitions. For example one definition requires such a manifold to be projective. Moreover, there are two definitions which are not interesting in the case of flat manifolds: the first defines Calabi-Yau manifolds as Ricci-flat Kähler manifolds, while the second requires that the holonomy group be the full $SU(n)$. These two definitions are beyond our interest, since any flat manifold is Ricci-flat and the holonomy group of a flat manifold is always finite. We have:

Proposition 7.10 ([78, Corollary 6.2.5]). *Let M be a flat Kähler manifold of complex dimension n with induced holonomy representation $\varphi \colon H \to U(n)$. Then $h^{n,0} = 1$ if and only if $\varphi(H) \subseteq SU(n)$.*

Proof. $h^{n,0}$ is the dimension of $(\Lambda^{n,0}(\mathbb{C}^n \oplus (\mathbb{C}^n)^*))^H$. Therefore, $h^{n,0}$ is 1 if and only if $dz_1 \wedge dz_2 \wedge \cdots \wedge dz_n$ is fixed under the action of any element $h \in H$. However, as the action of h on this basis vector is given by

$$^h(dz_1 \wedge dz_2 \wedge \cdots \wedge dz_n) = \mathrm{Det}(\varphi(h))dz_1 \wedge dz_2 \wedge \cdots \wedge dz_n,$$

we have that h fixes this basis vector if and only if $\mathrm{Det}(\varphi(h)) = 1$. Therefore, $h^{n,0} = 1 \Leftrightarrow \varphi(H) \subseteq SU(n)$. $\square$

Corollary. *There are no Calabi-Yau hyperelliptic surfaces. In complex dimension three, there are twelve Calabi-Yau flat Kähler manifolds with non-trivial holonomy:*

1. *five manifolds with the first Betti number equal to zero, where four manifolds have holonomy $\mathbb{Z}_2 \times \mathbb{Z}_2$ and one has holonomy D_8;*

2. *two manifolds with the first Betti number equal to 2 and holonomy $\mathbb{Z}_2$;*

3. *five manifolds with the following Betti numbers: $\beta_1 = 2$, $\beta_2 = 5$, where two manifolds have holonomy $\mathbb{Z}_3$, two have holonomy $\mathbb{Z}_4$ and one has holonomy $\mathbb{Z}_6$.*

7.3 Exercises

Exercise 7.1. Prove that if $A \in U(n)$ then $\det(A)$ is a complex number of absolute value 1, and deduce that $U(n)/SU(n) \simeq S^1$.

Exercise 7.2. Prove that $U(\frac{n}{2}) \subset SO(n)$.

Exercise 7.3. Prove that any Bieberbach group with the holonomy group $(\mathbb{Z}_3)^n$, $n \geq 2$ is the fundamental group of a Kähler flat manifold.

Exercise 7.4. Find an example of the fundamental group Γ of a Kähler flat manifold such that the group $\mathrm{Out}(\Gamma)$ is finite.
Hint: Consider a group from the previous exercise.

Exercise 7.5. Prove that any finite group is a holonomy group of a Kähler flat manifold.

Table 7.3

Manifold with $\beta_1 = 6$ (T^6):	Manifolds with $\beta_1 = 2$ and $\beta_2 = 3$:

```
              1                                              1
          3       3                                      1       1
      3       9       3                              0       3       0
  1       9       9       1                      0       2       2       0
      3       9       3                              0       3       0
          3       3                                      1       1
              1                                              1
```

Manifolds with $\beta_1 = 2$ and $\beta_2 = 5$

```
          1                                          1
      1       1                                   1       1
  0       5       0                            1       3       1
0       4       4       0        or        1       3       3       1
  0       5       0                            1       3       1
      1       1                                   1       1
          1                                          1
```

Manifolds with $\beta_1 = 2$ and $\beta_2 = 7$:	Manifolds with $\beta_1 = 4$

```
              1                                              1
          1       1                                      2       2
      1       5       1                              1       5       1
  1       5       5       1                      0       4       4       0
      1       5       1                              1       5       1
          1       1                                      2       2
              1                                              1
```

Manifold with holonomy $\mathbb{Z}_2 \times \mathbb{Z}_2$:	Manifolds with holonomy D_8:

```
              1                                              1
          0       0                                      0       0
      0       3       0                              0       2       0
  1       3       3       1                      1       2       2       1
      0       3       0                              0       2       0
          0       0                                      0       0
              1                                              1
```

8. Crystallographic Groups as Isometries of $\mathbb{H}^n$

In this chapter we consider a phenomenon of relations between hyperbolic and Euclidean geometries. It is well known that the boundary of hyperbolic space $\mathbb{H}^n$ is equal to $\mathbb{R}^{n-1}$. Using this observation, we show that any Bieberbach group of dimension $n - 1$ is also a discrete subgroup of the group $\text{Isom}(\mathbb{H}^n)$. Most facts of this chapter are based on [86] and [87].

8.1 Hyperbolic space $\mathbb{H}^n$

We start with a simple description of the hyperbolic space $\mathbb{H}^n$. Let us consider the Euclidean space $\mathbb{R}^n$ with the quadratic form

$$f_n = \langle -1, 1, \ldots, 1, 1 \rangle$$

of signature $(n, 1)$ with $(n + 1) \times (n + 1)$ symmetric matrix X. Let

$$O(f_n, \mathbb{R}) = \{ Y \in GL(n + 1, \mathbb{R}) \mid Y^t X Y = X \}$$

be the orthogonal group of f_n, and

$$SO(f_n, \mathbb{R}) = O(f_n, \mathbb{R}) \cap SL(n + 1, \mathbb{R}),$$

the special orthogonal group of f_n. The hyperboloid model of hyperbolic n-space is the metric space

$$\mathbb{H}^n = \{ x \in \mathbb{R}^{n+1} \mid x \circ x = -1 \text{ and } x_{n+1} > 0 \}$$

with metric d defined by

$$\cosh d(x, y) = -x \circ y. \tag{8.1}$$

A matrix $A \in O(f_n, \mathbb{R})$ is said to be either positive ($\in O_0(f_n, \mathbb{R})$) or negative ($\notin O_0(f_n, \mathbb{R})$) according to whether A maps $\mathbb{H}^n$ to $\mathbb{H}^n$ or $-\mathbb{H}^n$, where

$O_0(f_n, \mathbb{R})$ is the connected component of the identity in $O(f_n, \mathbb{R})$. We identify $O_0(f_n, \mathbb{R})$ with $\mathrm{Isom}(\mathbb{H}^n)$. When passing to the connected component of the identity in $SO(f_n, \mathbb{R})$, denoted $SO_0(f_n, \mathbb{R})$ (which has index 4 in $O(f_n, \mathbb{R})$), we obtain a group which may be identified with $Isom_+(\mathbb{H}^n)$; it preserves the upper sheet of the hyperboloid $f_n(x) = -1$ and the orientation. An element of $O_0(f_n, \mathbb{R})$ is parabolic (resp. elliptic) if it has a unique fixed point which lies on hyperbolic boundary $\partial\mathbb{H}^n = S_\infty^{n-1}$ (resp. has a fixed point in $\mathbb{H}^n$). Let Δ be a discrete and (torsion-free) subgroup of $O_0(f_{n+1}, \mathbb{R})$, such that $\mathbb{H}^{n+1}/\Delta = V^{n+1}$ has a finite volume. It is well known (cf. [122]) that V^{n+1} has finite number of cusps and each cusp topologically is equivalent to $M^n \times \mathbb{R}^{\geq 0}$, where M^n is a n-dimensional flat manifold (orbifold). An example of such a group is $O_0(f_{n+1}; \mathbb{Z})$. For any flat manifold (orbifold) M we can ask if there is some finite volume hyperbolic manifold (orbifold) V with one cusp homeomorphic to $M \times \mathbb{R}^{\geq 0}$?

We have:

Theorem 8.1 ([87, Theorem 1.2]). *Let Γ be the fundamental group of a flat manifold of dimension n. Then there is a quadratic form q_{n+2} defined over $\mathbb{Q}$, of signature $(n + 1, 1)$ for which Γ embeds as a subgroup of $O_0(q_{n+2}; \mathbb{Z})$.*

Proof. [1] We have the following short exact sequence

$$0 \to \mathbb{Z}^n \to \Gamma \xrightarrow{p} G \to 0.$$

Let $e_i \in \mathbb{Z}^n, i = 1, 2, ..., n$ be a standard basis. Moreover, from the correspondence (2.7) considered in Chapter III and Proposition 1.12 (see also [87, page 292]), there exists the faithful representation $\Phi^* : \Gamma \to GL(n + 1, \mathbb{Q})$ given by

$$\Phi^*(\gamma) = \begin{pmatrix} h_\Gamma(p(\gamma)) & t_\gamma \\ 0 & 1 \end{pmatrix}, \tag{8.2}$$

where h_Γ is a holonomy representation and $\gamma \in \Gamma$. By conjugating the representation, we may rescale the last basis vector e_{k+1} (cf. Exercise 5.4) and thus arrange that the representation Φ^* is into the group $GL(n + 1, \mathbb{Z})$. We shall need another equivalent form of this representation.

$$\Phi(\gamma) = (\Phi^*(\gamma)^T)^{-1} = \begin{pmatrix} h_\Gamma(p(\gamma^{-1}))^T & 0 \\ (-h_\Gamma(p(\gamma^{-1}))(t_\gamma))^T & 1 \end{pmatrix}. \tag{8.3}$$

---

[1]The idea of proof is taken from [87].

Then we can extend this representation to $\mathbb{R}^{n+2}$ by mapping g to

$$\hat{\Phi} = \begin{pmatrix} \Phi(\gamma) & \mathrm{v}_\gamma \\ 0 & 1 \end{pmatrix},$$

where the column vector v_γ is to be determined.

Now, let $\langle, \rangle$ be any h_Γ^T invariant positive definite inner product on the maximal abelian subgroup $\mathbb{Z}^n$ of Γ; such an inner product exists by taking a random inner product and forming the h_Γ-average, see Example 2.15. Let D be the symmetric rational matrix associated to this form in the $\{e_i\}_{i=1,2,\dots,n}$ basis. Extend this form to $\mathbb{R}^{n+2}$ by summing on a subspace H_2, which in the language of quadratic forms is a hyperbolic plane. More precisely, we let H_2 denote the 2-dimensional form $2XY$, with associated symmetric matrix

$$\begin{pmatrix} 0 & 1 \\ 1 & 0 \end{pmatrix}$$

(see [83, p. 590]). The form $D \oplus H_2$ now has the signature $(n+1, 1)$.

Denoting vectors lying in $\mathbb{R}^n$ by w and the last two coordinates by v_1 and v_2, we will show that $\mathrm{v}_\gamma = (W_\gamma, \tau_\gamma) \in \mathbb{R}^n \oplus \langle v_1 \rangle$ may be chosen so that each $\hat{\Phi}(\gamma)$ is an isometry of the form $D \oplus H_2$.

Fix some generator $\gamma \in \Gamma$. We will continue to use $\langle, \rangle$ to denote the bilinear form associated to the quadratic form $D \oplus H_2$. It suffices to show that $\hat{\Phi}(\gamma)$ can be chosen to be an isometry of the bilinear form. There are various cases, for convenience of notation we will suppress $\hat{\Phi}$ in the following:

- By choice of D, $\langle \gamma(w), \gamma(w') \rangle = \langle h_\Gamma(\gamma)(w) + \xi_1 v_1, h_\Gamma(\gamma)(w') + \xi_2 v_1 \rangle = \langle h_\Gamma(\gamma)(w), h_\Gamma(\gamma)(w') \rangle = \langle w, w' \rangle$ since v_1 is orthogonal to the $\mathbb{R}^n$ subspace and D is $h_\Gamma(\Gamma)$ invariant.

- $\langle v_1, v_1 \rangle (= 0)$ is preserved since $\gamma v_1 = v_1$, cf. (8.3).

- $\langle w, v_1 \rangle = 0$, but $\langle \gamma w, \gamma v_1 \rangle = \langle w' + \xi v_1, v_1 \rangle = 0$, so the inner product is preserved in this case.

- $\langle v_1, v_2 \rangle = 1$. But $\langle \gamma v_1, \gamma v_2 \rangle = \langle v_1, \mathrm{v}_\gamma + v_2 \rangle = \langle v_1, v_2 \rangle$.

- $\langle w, v_2 \rangle = 0$. Now $\langle \gamma(w), \gamma(v_2) \rangle = \langle w' + a_\gamma(w)v_1, W_\gamma + \tau_\gamma v_1 + v_2 \rangle$ where $\mathrm{v}_\gamma = (W_\gamma, \tau_\gamma)$ so that W_γ and τ_γ are ours to choose. We need that right hand side of the equality to be zero, that is we must arrange

$$\langle w', W_\gamma \rangle + a_\gamma(w) = 0. \tag{8.4}$$

Now, as w runs over some basis of $\mathbb{R}^n$, the vectors w' also run over a basis so that there is a rational vector W_γ for which all the equations (8.4) hold by the nondegeneracy of the form defined by D.

- $\langle v_2, v_2 \rangle = 0$ and then we need $0 = \langle \gamma v_2, \gamma v_2 \rangle = \langle W_\gamma + \tau_\gamma v_1 + v_2, W_\gamma + \tau_\gamma v_1 + v_2 \rangle = \langle W_\gamma, W_\gamma \rangle + 2\tau_\gamma$ which determines how we should choose τ_γ.

To sum up, for each generator of Γ, we may choose v_γ so that the resulting matrix is an isometry of the form $\langle , \rangle$. These matrices are potentially rational in the last column, since the vectors v_γ need only be rational. However, by conjugating by a matrix of the form

$$\begin{pmatrix} I & 0 \\ 0 & k \end{pmatrix},$$

$k \in \mathbb{N}$ we may find a new collection of matrices which are the same in the last column, and which have v_γ replaced by $k \cdot v_\gamma$. In particular, for suitable k we may arrange that the conjugated representation is integral. After this conjugation, the new representation now leaves invariant a different form, but this new form is rationally equivalent to $D \oplus H_2$; in particular, it continues to have signature $(n+1, 1)$. To finish the proof we have to show that with this choice, we get a faithful integral representation of the group Γ. In fact, we need only show that the relations in Γ hold, faithfulness will follow, because if a product of these matrices is the identity, then it must at least be the identity in the $(n+1)$ representation which is already a faithful representation of Γ. Moreover, we shall show that any isometry, say δ, which is an identity on upper $(n+1) \times (n+1)$ block is in fact the identity.

For this pick a random $w \in \mathbb{R}^n$. Then $0 = \langle w, v_2 \rangle = \langle \delta(w), \delta(v_2) \rangle = \langle w, w' + \xi v_1 + v_2 \rangle = \langle w, w' \rangle$. This holds for all w, so that $w' = 0$. Also $0 = \langle v_2, v_2 \rangle = \langle \delta(v_2), \delta(v_2) \rangle = \langle \xi v_1 + v_2, \xi v_1 + v_2 \rangle = 2\xi$, so that $\xi = 0$, implying $\delta(v_2) = v_2$ as required. Finally, note that the construction exhibits Γ explicitly as a subgroup of the stabiliser of the light-like vector v_1.   □

The above construction yields Γ as a subgroup of the stabiliser $\mathrm{Stab}(v_1) \subset O_0(q_{n+2}, \mathbb{Z})$ of the light-like vector v_1, where q_{n+2} is the form considered above. However, we do not know whether Γ is actually equal to $\mathrm{Stab}(v_1)$. If we use a subgroup separability argument (see [87, Lemma 2.6 and Proposition 3.2]) to pass to a subgroup of finite index in $O_0(q_{n+2}, \mathbb{Z})$ for which the group Γ is a maximal peripheral subgroup, then we obtain:

Corollary ([87, Theorem 1.1]). *For every $n \geq 2$, the diffeomorphism class of every flat n-manifold has a representative W which arises as some cusp cross-section of a finite volume cusped hyperbolic $(n+1)$-orbifold.*

□

Remark 8.2. In [98] the above Corollary is proved for manifolds. See also Problem 11.24.

We shall illustrate the above theorem (8.1) by an example.

Example 8.3 ([87]). Let Γ be the fundamental group of the 3-dimensional Hantzsche-Wendt manifold. It is generated by two elements:

$$a = (\mathrm{diag}(1,-1,-1),(1/2,1/2,0)) \in E(3),$$

$$b = (\mathrm{diag}(-1,1,-1),(0,1/2,1/2)) \in E(3).$$

Then we have:

$$\Phi^*(a) = \begin{pmatrix} 1 & 0 & 0 & 1/2 \\ 0 & -1 & 0 & 1/2 \\ 0 & 0 & -1 & 0 \\ 0 & 0 & 0 & 1 \end{pmatrix},$$

and

$$\Phi^*(b) = \begin{pmatrix} -1 & 0 & 0 & 0 \\ 0 & 1 & 0 & 1/2 \\ 0 & 0 & -1 & 1/2 \\ 0 & 0 & 0 & 1 \end{pmatrix}.$$

We conjugate the above to get rid of fractions and form the representation Φ and then $\hat{\Phi}$. In this case an invariant form for the finite holonomy is $D = X^2 + Y^2 + Z^2$ and when we solve the equations for the group to be an isometry of the form $D \oplus H_2$ we obtain:

$$\hat{\Phi}(a) = \begin{pmatrix} 1 & 0 & 0 & 0 & 1 \\ 0 & -1 & 0 & 0 & 1 \\ 0 & 0 & -1 & 0 & 0 \\ -1 & 1 & 0 & 1 & -1 \\ 0 & 0 & 0 & 0 & 1 \end{pmatrix},$$

and

$$\hat{\Phi}(b) = \begin{pmatrix} -1 & 0 & 0 & 0 & 0 \\ 0 & 1 & 0 & 0 & 1 \\ 0 & 0 & -1 & 0 & 1 \\ 0 & -1 & 1 & 1 & -1 \\ 0 & 0 & 0 & 0 & 1 \end{pmatrix}.$$

The above matrices preserve the rationally equivalent form $X^2 + Y^2 + Z^2 + 2WT$. Note that the above form is equivalent over $\mathbb{Q}$ to $f_4 = X^2 + Y^2 + Z^2 + W^2 - T^2$.

In [123] the authors obtain the above flat manifold as a cusp cross-section of a hyperbolic 4-manifold arising from a torsion-free subgroup in $O_0(f_4; \mathbb{Z})$.

8.2   Exercises

Exercise 8.1. Prove that the function $d \colon \mathbb{H}^n \times \mathbb{H}^n \to \mathbb{R}^{\geq 0}$ defined by formula 8.1 is a metric.

Exercise 8.2. Prove that $SO_0(f_n, \mathbb{R})$ has index 4 in $O(f_n, \mathbb{R})$.

Exercise 8.3. Prove that for a discrete subgroup Δ of $O(f_n, \mathbb{R}), \Delta \cap SO_0(f_n, \mathbb{R})$ has index ≤ 4 in Δ.

Exercise 8.4. Give examples of parabolic (resp. elliptic) elements of $O_0(f_n, \mathbb{R})$.

Exercise 8.5. Prove that $O_0(f_{n+1}; \mathbb{Z})$ is arithmetic and $\mathbb{H}^n / O_0(f_{n+1}; \mathbb{Z})$ has a finite volume.

Exercise 8.6. Give a proof of the corollary before Example 8.3.
Hint: See [87].

Exercise 8.7. Let Γ_n be an oriented generalized Hantzsche-Wendt group (cf. Example 5.22 and 9.13). Find $\hat{\Phi}(\Gamma_n) \subset O_0(f_{n+1}; \mathbb{Z})$.

9. Hantzsche-Wendt Groups

Only one 3-dimensional Bieberbach group Γ_9 has a non-cyclic holonomy group $\mathbb{Z}_2 \times \mathbb{Z}_2$. Γ_9 is a finitely presented group and has a cyclic presentation, see [74]. In particular it belongs to the family of Fibonacci groups, see [75]. We emphasize that Γ_9 is a member of another infinite class of groups, which we shall consider in this part. In fact, in the theory of torsion free crystallographic groups we define an infinite class of generalized Hantzsche-Wendt groups. Roughly speaking, a Bieberbach group of dimension n is the generalized Hanztsche-Wendt if and only if it has a holonomy group $(\mathbb{Z}_2)^{n-1}$. Since this 3-dimensional group Γ_9 was first discovered by German mathematicians W. Hantzsche and H. Wendt, these groups are called Hantzsche-Wendt groups. In the following Chapter we present a few interesting and exceptional properties of the above groups.

9.1 Definitions

We have already considered these groups in Section 5.3, (see Example 5.22). Let us give a more general definition.

Definition 9.1 ([124, Definition 2.1]). Let Γ be a Bieberbach group of dimension n. A flat manifold $\mathbb{R}^n/\Gamma$ is called a *generalized Hantzsche-Wendt* (GHW for short) manifold if its holonomy group is $\mathbb{Z}_2^{n-1}$. In this case Γ is called a *generalized Hantzsche-Wendt* group.

Example 9.2. In dimension two the Klein bottle belongs to this family and in dimension three there are three such manifolds: the classical flat manifold first considered by Hantzsche and Wendt [56] (also called didicosm, see Section 3.3) and two non-orientable ones; the first and second amphidicosm (see Section 3.3).

These manifolds (groups) have already been studied in [85], [97], [101], [102], [119], [131], mainly in the oriented case, where we call them *Hantzsche-Wendt manifolds* (HW manifolds for short; respectively HW (Bieberbach) groups).

By a theorem of Cartan-Ambrose-Singer (see [148, Corollary 3.4.7]) any closed Riemannian manifold with a finite holonomy group must be flat. Hence, with the support of the Bieberbach Theorem we have:

Proposition 9.3. *Let Γ be a Bieberbach group of dimension n. Then the following conditions are equivalent.*

(i) Γ is the GHW group of dimension n;

(ii) Γ is the fundamental group of a closed Riemannian manifold of dimension n with holonomy group $\mathbb{Z}_2^{n-1}$;

(iii) Γ is the fundamental group of a flat manifold of dimension n with holonomy group $\mathbb{Z}_2^{n-1}$.

$\square$

Definition 9.4. We say that a Bieberbach group Γ is *diagonal* or of *diagonal type* if its corresponding lattice of translations $\mathbb{Z}^n$ has an orthogonal basis in which the integral holonomy representation h_Γ diagonalizes. The corresponding flat manifold $\mathbb{R}^n/\Gamma$ will also be called diagonal.

Consider

$$
C_i := \begin{bmatrix}
1 & 0 & \cdots & 0 & 0 & 0 & \cdots & 0 \\
0 & 1 & \cdots & 0 & 0 & 0 & \cdots & 0 \\
\vdots & \vdots & \ddots & & \vdots & \vdots & & \vdots \\
0 & 0 & & 1 & 0 & 0 & \cdots & 0 \\
0 & 0 & \cdots & 0 & -1 & 0 & \cdots & 0 \\
0 & 0 & \cdots & 0 & 0 & 1 & & 0 \\
\vdots & \vdots & & \vdots & \vdots & & \ddots & \vdots \\
0 & 0 & \cdots & 0 & 0 & 0 & \cdots & 1
\end{bmatrix}, \quad 1 \leq i \leq n, \qquad (9.1)
$$

where -1 is placed in the (i,i) entry. For brevity, we will denote $C_i = \mathrm{diag}(1,\ldots,1,\underbrace{-1}_{i},1,\ldots,1)$, for $1 \leq i \leq n$. The product of two of these is

$$
C_i C_j = \mathrm{diag}(1,\ldots,1,\underbrace{-1}_{i},1,\ldots,1,\underbrace{-1}_{j},1,\ldots,1), \quad \text{for every } i < j. \quad (9.2)
$$

The elements in (9.2) generate the subgroup $\mathrm{D}^+(n)$ of diagonal matrices in $SL(n,\mathbb{Z})$, of cardinality 2^{n-1}, and of index two in the group $\mathrm{D}(n)$ of diagonal $n \times n$ matrices with entries ± 1 on the diagonal. In the oriented case with holonomy group $\mathbb{Z}_2^{n-1}$, the rationalization of the integral holonomy representation h_Γ, (i.e. to consider $\mathbb{Z}^n$ as a $\mathbb{Q}$-module instead of a $\mathbb{Z}$-module), gives a unique representation $(h_\Gamma)_\mathbb{Q}$ whose image is $D^+(n)$. Hence, when n is even there are no oriented GHW manifolds, since $D^+(n)$ contains $-Id$, which always corresponds to an element of torsion.

Let us present an important structural property of the class of GHW manifolds.

Theorem 9.5. *The fundamental group Γ of a generalized Hantzsche-Wendt manifold is diagonal.*

We shall first prove two Lemmas.

Lemma 9.6. *Let $\rho\colon \mathbb{Z}_2^{n-1} \to GL(n, \mathbb{Z})$ be a diagonal faithful integral representation with $-Id \notin \mathrm{Im}(\rho)$. Then there is $g \in \mathbb{Z}_2^{n-1}$ such that $\rho(g) = \mathrm{diag}(-1, \ldots, -1, 1, -1, \ldots, -1)$. Moreover, if furthermore $\mathrm{Im}(\rho) \not\subseteq SL(n, \mathbb{Z})$ then there is $g \in \mathbb{Z}_2^{n-1}$ such that $\rho(g) = \mathrm{diag}(1, \ldots, 1, -1, 1, \ldots, 1)$.*

Proof of Lemma 9.6. We have $D(n) = \mathrm{Im}(\rho) \cup (-Id)\,\mathrm{Im}(\rho)$. Since the C_i's, for $1 \leq i \leq n$, are linearly independent (over $\mathbb{Z}_2$), they cannot lie all in $\mathrm{Im}(\rho)$ simultaneously, thus there is at least one of them in $(-Id)\,\mathrm{Im}(\rho)$, or equivalently, there is $-C_i$ in $\mathrm{Im}(\rho)$, as claimed. The last assertion in the statement follows from a similar argument. $\square$

Lemma 9.7. *Let Γ be a Bieberbach group of dimension n with the translation lattice Λ. Suppose that $(B, b) \in \Gamma$ and B has eigenvalues $1, -1$, with corresponding eigenspaces V^+ and V^- of dimensions 1 and $n - 1$, respectively. Then $\Lambda = (\Lambda \cap V^+) \oplus (\Lambda \cap V^-)$, and the orthogonal projection of b onto V^+ lies in $\frac{1}{2}(\Lambda \cap V^+) \smallsetminus (\Lambda \cap V^+)$.*

Proof of Lemma 9.7. We have

$$\Gamma \ni (B, b + \lambda)^2 = (Id, (B + Id)(b + \lambda)), \text{ for every } \lambda \in \Lambda. \qquad (9.3)$$

The torsion-free condition implies that $0 \neq (B + Id)(b + \lambda) \in \Lambda$, for every $\lambda \in \Lambda$. Now, $\frac{1}{2}(B \pm Id)$ is the orthogonal projection onto $V^\pm$. If we suppose that for some $\lambda \in \Lambda$, $\frac{1}{2}(B + Id)(b + \lambda) = \mu \in \Lambda$, then $(B + Id)(b + \lambda - \mu) = (B + Id)(b + \lambda) - (B + Id)(\mu) = 2\mu - 2\mu = 0$, a contradiction. Hence, the orthogonal projection of b (and also, $(b + \lambda)$, for every $\lambda \in \Lambda$) onto V^+ lies in $\frac{1}{2}\Lambda \setminus \Lambda$. The fact that V^+ is one-dimensional now implies that the orthogonal projection of Λ onto V^+ is exactly $\Lambda \cap V^+$. Hence Λ splits into the direct sum of two lattices, and the lemma follows. $\square$

Proof of the Theorem. We proceed by induction on the dimension n. If n is small, it is known that the theorem holds. Let h_Γ be the holonomy representation of Γ and let $\Lambda \simeq \mathbb{Z}^n$ be the lattice of translations of Γ. The nature of the holonomy group $G = \mathbb{Z}_2^{n-1}$ implies that each $(h_\Gamma)_\mathbb{Q}(g)$ diagonalizes, and since they commute, they diagonalize simultaneously, i.e., there is a $\mathbb{Q}$-basis $\mathcal{B}$ of Λ such that $[(h_\Gamma)_\mathbb{Q}(g)]_\mathcal{B}$ is diagonal for every $g \in G$ (and the elements on the diagonal are ± 1's). Then, Lemma 9.6 applied to these integral matrices, implies that $\exists\, g_0 \in \mathbb{Z}_2^{n-1}$ such that $[(h_\Gamma)_\mathbb{Q}(g_0)]_\mathcal{B} = \mathrm{diag}(1, -1, \ldots, -1)$ (after reordering $\mathcal{B}$). Hence, $h_\Gamma(g_0)$ has eigenvalues $+1$ and -1 with multiplicities 1 and $n-1$ respectively. Now, we can apply Lemma 9.7, and thus Λ splits into a direct sum $\Lambda = \Lambda_1 \oplus \Lambda_2$ where $\dim \Lambda_1 = 1$ and $\dim \Lambda_2 = n-1$. Also $\mathrm{span}_\mathbb{R} \Lambda_1$ (resp. $\mathrm{span}_\mathbb{R} \Lambda_2$) is the eigenspace of $h_\Gamma(g_0)$ with eigenvalue 1 (resp. -1) and the orthogonal projection of the translational part corresponding to $h_\Gamma(g_0)$ lies in $\frac{1}{2}\Lambda_1 \smallsetminus \Lambda_1$. Since $h_\Gamma(g_0)h_\Gamma(g) = h_\Gamma(g)h_\Gamma(g_0)$, for every $g \in \mathbb{Z}_2^{n-1}$, we have that Λ_1 and Λ_2 are stable under the action of $h_\Gamma(g)$, thus $\Lambda = \Lambda_1 \oplus \Lambda_2$ is a $\mathbb{Z}_2^{n-1}$-module decomposition. Hence we have that $h_\Gamma = \chi \oplus \rho'$, where χ is a character and ρ' has rank $n-1$, both defined on the same domain as h_Γ. Thus, $\forall g \in \mathbb{Z}_2^{n-1}$

$$
h_\Gamma(g) = \begin{array}{|c|c|} \hline \chi(g) & \\ \hline & \rho'(g) \\ \hline \end{array}
\qquad \text{and} \qquad
h_\Gamma(g_0) = \begin{array}{|c|c|} \hline 1 & \\ \hline & -Id \\ \hline \end{array}
$$

We may conjugate Γ by (Id, μ) with $\mu \in \frac{1}{4}\Lambda_1$ and obtain that $p_1(b) \in \frac{1}{2}\Lambda_1$ for every $(B, b) \in \Gamma$, where p_1 is the orthogonal projection onto $\mathrm{span}_\mathbb{R} \Lambda_1$ (cf. [103, Lemma 1.4]).

We define a character $\psi \colon \Gamma/\mathbb{Z}^n \simeq \mathbb{Z}_2^{n-1} \longrightarrow \mathbb{Z}_2 = \{0, 1\}$ by

$$
\psi((B, b)) = \begin{cases} 0 & \text{if } p_1(b) \in \Lambda_1; \\ 1 & \text{if } p_1(b) \in \frac{1}{2}\Lambda_1 \smallsetminus \Lambda_1. \end{cases}
$$

It is well defined since all the elements (B, b) in the same class in Γ/Λ have the same $p_1(b) \bmod \Lambda_1$. Now we consider

$$
\Gamma \xrightarrow{\ p\ } \Gamma/\Lambda \xrightarrow{\ \psi\ } \mathbb{Z}_2,
$$

where p is the canonical projection. Set $\Gamma' := \ker(\psi \circ p)$. It is a subgroup of Γ of index 2 and its elements (B, b) have the property that $p_1(b) \equiv 0 \pmod{\Lambda_1}$. The last fact shows that if we take $\Gamma'' := \Gamma'|_{\Lambda_2}$ (i.e. the restriction to the last $(n - 1)$ coordinates of B's and b's), it is still a Bieberbach group, it has rank $n - 1$ and holonomy group isomorphic to $\mathbb{Z}_2^{n-2}$. (Alternatively, we could have considered the subgroup $\Gamma_0 := \{(B, b) \in \Gamma \mid b_1 = 0\}$, where $b = (b_1, b_2) \in \Lambda_1 \oplus \Lambda_2$, and observed that the restriction of Γ_0 to $\mathrm{span}_{\mathbb{R}}\Lambda_2$ gives an isomorphism between Γ_0 and Γ''.) Then we can apply induction to Γ'', hence it is diagonal. By the relation between Γ and Γ'' it is clear that therefore Γ is diagonal. $\qquad\square$

In the proof, the group denoted by Γ'' is a subgroup of Γ (in the abstract sense, i.e., there is a monomorphism from Γ'' into Γ). This we shall use later in Section 9.3. In other words we have:

Corollary. *Every GHW group of dimension $n \geq 3$ has a subgroup isomorphic to a GHW group of rank $n - 1$.*

$\qquad\square$

In the oriented case, the existence of these manifolds in each odd dimension was proved in [97] and one realization in each odd dimension was shown in [131]. In [101] the oriented flat manifolds with (orthonormal) [1] diagonal holonomy and holonomy group $\mathbb{Z}_2^{n-1}$ were studied and they were parametrized by certain directed graphs; also lower bounds for their cardinality were shown, growing rapidly with n. It is an easy exercise (cf. Exercise 9.2) to show, that for every $k, 1 \leq k < n - 1$, there are flat manifolds with holonomy group $\mathbb{Z}_2^k$ not of diagonal type.

9.2 Non-oriented GHW groups

The class of non-oriented GHW manifolds includes two subclasses, depending on whether the first Betti number β_1 is 0 or 1, or equivalently (cf. [63, Proposition 1.4]), the classes of Bieberbach groups with trivial or non-trivial centre.

Let us consider first the case with non-trivial centre. From Theorem 9.5 and a similar observation as in the previous section for HW manifolds, this subclass has the integral holonomy representation uniquely determined. Let us present a simple example (cf. [85]). Let K_n be the subgroup of $\mathrm{E}(n)$ generated by the set $\{(C_i, s(C_i)) : 0 \leq i \leq n - 1\}$, where $C_0 = I$ and

---

[1] It is an exercise to show that here the words "orthonomal" and "orthogonal" are equivalent.

$C_i = \mathrm{diag}(1, \ldots, 1, \underbrace{-1}_{i}, 1, \ldots, 1)$ are the $n \times n$ orthogonal matrices defined in (9.1). Moreover $s(C_0) = e_1$ and $s(C_i) = e_{i+1}/2$ for other i. Here $e_1, e_2, \ldots, e_n$ denote the canonical basis of $\mathbb{Z}^n$. In [85, page 5] it was proved that K_n is a Bieberbach group and that there exists the following short exact sequence of groups

$$0 \longrightarrow K_{n-1} \longrightarrow K_n \longrightarrow \mathbb{Z} \longrightarrow 0, \qquad (9.4)$$

where the epimorphism $K_n \to \mathbb{Z}$ is given by the composition

$$K_n \to K_n/[K_n, K_n] \to \mathbb{Z}.$$

However in general we have:

Proposition 9.8. *Let Γ be a generalized Hantzsche-Wendt group of dimension n with a non-trivial centre. Then the kernel of the composition $f \colon \Gamma \to \Gamma/[\Gamma, \Gamma] \to \mathbb{Z}$ is either an oriented or a non-oriented generalized Hantzsche-Wendt group.*

Proof. First, it is not difficult to observe that $\ker f$ is a Bieberbach group (cf. Proposition 3.2). From the assumptions, the abelian group $\Gamma/[\Gamma, \Gamma] \simeq \mathbb{Z} \oplus F$ where F is a finite group (cf. [63]). Let $f((A, a)) \in \mathbb{Z}$ be a generator for some $(A, a) \in \Gamma$. We can assume, by using Lemma 9.6, that A has eigenvalues $1, -1$, with corresponding eigenspaces V^+ and V^- of dimensions 1 and $n-1$ respectively. Moreover, we have an isomorphism $\ker f/(\Lambda \cap \ker f) \simeq \ker f/(\Lambda \cap V^-) \simeq \mathbb{Z}_2^{n-2}$, where Λ is a translation lattice of Γ. We have to show that $\mathbb{Z}_2^{n-2}$ is the holonomy group of the Bieberbach group $\ker f$. But this follows from the assumptions and Lemma 9.7. To finish the proof, we observe that for odd n every $\ker f$ is non-oriented. On the other hand, using semi-direct product $\Gamma' \rtimes_\alpha \mathbb{Z}$ (see Theorem 3.3) with the integers, for any HW group Γ' of dimension n, we can define a $n + 1$-rank GHW group with non-trivial centre such that the kernel of the map onto $\mathbb{Z}$ is isomorphic to Γ'. Here $\alpha \colon \mathbb{Z} \to \mathrm{Out}(\Gamma')$ is a homomorphism which sends 1 to $-Id \in \mathrm{Out}(\Gamma')$. $\square$

As mentioned above, the holonomy representation of GHW manifolds with $\beta_1 = 1$ is always equivalent to that of the above example, hence all these manifolds have the same rational homology, namely

Proposition 9.9 ([124, Theorem 4.2]). *Let M be a generalized Hantzsche-Wendt flat manifold of dimension n with the first Betti number one. Then*

$$H_j(M) \simeq \begin{cases} \mathbb{Q} & \text{if } j = 0, 1; \\ 0 & \text{otherwise.} \end{cases}$$

Proof. From the definition and assumptions we have a short exact sequence

$$0 \longrightarrow \mathbb{Z}^n \longrightarrow \pi_1(M) \longrightarrow \mathbb{Z}_2^{n-1} \longrightarrow 0,$$

where the action of $\mathbb{Z}_2^{n-1}$ on $\mathbb{Z}^n$ is given by the matrices C_i, $1 \leq i \leq n-1$. This defines a representation $\rho \colon \mathbb{Z}_2^{n-1} \to GL(n, \mathbb{Z})$. Using an elementary representation theory, (see [131]) we have a character Φ of ρ given by the formula

$$\Phi = \sum_{i=1}^{n} \chi_i,$$

where $\chi_i = \chi_0' \cdots \chi_0' \chi_1' \chi_0' \cdots \chi_0'$, $1 \leq i \leq n-1$ with χ_1' the i-th of the $(n-1)$ factors in the product, and $\chi_n = (\chi_0')^{n-1}$. Here χ_0', χ_1' denote a trivial and a non-trivial character of the involution group $\mathbb{Z}_2$, respectively. From [131]

$$H_j(M, \mathbb{Q}) \simeq [\Lambda^j(\mathbb{Q}^n)]^{\mathbb{Z}_2^{n-1}}.$$

By definition, the representation of $\mathbb{Z}_2^{n-1}$ on $\Lambda^j(\mathbb{Q}^n)$ has the following character

$$\Lambda^j(\rho) = \sum_{1 \leq i_1 < i_2 < \cdots < i_j \leq n} \chi_{i_1} \chi_{i_2} \cdots \chi_{i_j}. \tag{9.5}$$

We have

$$\Lambda^j(\rho) = \chi_1 \sum_{2 \leq i_2 < \cdots < i_j \leq n} \chi_{i_2} \cdots \chi_{i_j} + \sum_{2 \leq i_1 < i_2 < \cdots < i_j \leq n} \chi_{i_1} \chi_{i_2} \cdots \chi_{i_j}.$$

Hence by induction and an immediate calculation the result follows. □

According to [109] and [24], the number of isomorphism classes of GHW groups and their holonomy representations in dimensions up to 6 are as in the following table.

dim	number of GHW groups					number of holon. repr.
	$\beta_1 = 0$	$\beta_1 = 1$	total	orient.	non-orient.	
2	0	1	1	0	1	1
3	1	2	3	1	2	2
4	2	10	12	0	12	2
5	23	100	123	2	121	3
6	352	2184	2536	0	2536	3

There are 62 oriented GHW groups in dimension 7 and more that 1000 in dimension 9 (see [102]).

If we consider the case of non-oriented generalized Hantzsche-Wendt Bieberbach groups with trivial centre then we have significantly more possibilities for the holonomy representations. Let us start establishing the following general result.

Proposition 9.10. *Let* $\rho\colon \mathbb{Z}_2^k \to GL(n,\mathbb{Z})$ *be a diagonal faithful integral representation with* $-Id \notin \mathrm{Im}(\rho)$ *and* $k < n$. *Then there is a Bieberbach group whose integral holonomy representation is* ρ.

Proof. It suffices to prove the case $k = n - 1$, since in the cases $k < n - 1$ we can use Lemma 9.11 below to induce from ρ a new representation defined on $\mathbb{Z}_2^{n-1}$; then apply the result for $n - 1$ to obtain a diagonal Bieberbach group Γ with the new integral holonomy representation; and finally, choose the appropriate proper subgroup of Γ of index 2^{n-1-k} corresponding to the original integral representation ρ. Note that this chosen subgroup has the same lattice of translations as Γ, and by construction, it is always a diagonal Bieberbach group.

Lemma 9.11. *If* $\sigma\colon \mathbb{Z}_2^k \to GL(n,\mathbb{Z})$ *is a diagonal faithful integral representation with* $-Id \notin \mathrm{Im}(\sigma)$ *and* $k < n - 1$, *then there exists an integral representation* $\tau\colon \mathbb{Z}_2^{k+1} \to GL(n,\mathbb{Z})$ *with the same properties as* σ *and containing properly* σ, *i.e.,* $\tau \restriction_{\mathbb{Z}_2^k \times \{0\}} = \sigma$.

Proof of Lemma 9.11. Denote $\sigma^- := -Id \circ \sigma$, i.e. $\sigma^-(g) := -(\sigma(g))$. Since $-Id \notin \mathrm{Im}(\sigma)$, we have $\mathrm{Im}(\sigma) \cap \mathrm{Im}(\sigma^-) = \emptyset$. Since $2 \times 2^k < 2^n$, there is $B \in \mathcal{D}(n) \setminus (\mathrm{Im}(\sigma) \cap \mathrm{Im}(\sigma^-))$. Now define $\tau\colon \mathbb{Z}_2^{k+1} \to \mathcal{D}(n)$ as the unique representation satisfying $\tau \restriction_{\mathbb{Z}_2^k \times \{0\}} = \sigma$ and $\tau \restriction_{\{0\}^k \times \mathbb{Z}_2} = \langle B \rangle$ (here we are identifying $\langle B \rangle$ with the representation of $\mathbb{Z}_2$ in $\mathbb{Z}^n$ whose non-trivial element acts by B). Since $B \notin \mathrm{Im}(\sigma)$, τ is a faithful representation; since $B \notin \mathrm{Im}(\sigma^-)$, $-Id \notin \mathrm{Im}(\tau)$, and therefore τ has the desired properties. $\square$

We continue the proof by induction on n. If $n = 2, 3$, it is easy to prove it. If $\mathrm{Im}(\rho) \subseteq SL(n,\mathbb{Z})$, then n must be odd and we know that there are manifolds with ρ as its integral representation, namely the Hantzsche-Wendt manifolds as in [97], [101], [131]. Otherwise $\mathrm{Im}(\rho)$ contains elements of determinant -1, then by Lemma 9.6, there is $C_i \in \mathrm{Im}(\rho)$, such that after reordering the coordinates we may assume $C_1 = \mathrm{diag}(-1, 1, \ldots, 1)$. Denote by p_1 the projection onto the first coordinate and p_r the projection onto the last $n - 1$ coordinates. Next, consider $\ker(p_1 \circ \rho)$, which is a subgroup of $\mathbb{Z}_2^{n-1}$ of index 2. Denote the inclusion $\ker(p_1 \circ \rho) \overset{\iota}{\hookrightarrow} \mathbb{Z}_2^{n-1}$, and define $\rho' := p_r \circ \rho \circ \iota$. Then ρ' is faithful, diagonal and $-Id \notin \text{Im}(\rho')$, however it is of rank $n - 1$ and defined from a group isomorphic to $\mathbb{Z}_2^{n-2}$. By induction, there is a Bieberbach group Γ' whose integral holonomy representation is ρ'. Moreover, we may assume that all the elements in Γ' are of the form (B', b') with $B' \in \text{Im}(\rho')$ and $b' \in \frac{1}{2}\mathbb{Z}^{n-1}$ (see [103, Lemma 1.4]).

To complete the proof we shall use the following method, which will be useful in the next section.

1	$\cdots$	1	-1	-1	$\cdots$	-1
			1			
	Γ'		$\vdots$		Γ'	
			1			

The previous picture gives an idea of the Bieberbach group we are constructing. C_1 is placed in the middle column. We need a vector $c \in \mathbb{R}^n$ to form a torsion-free element (C_1, c). We choose $c = e_2/2$ (any other choice $c = e_i/2$ for $i > 2$ would work well, too). Now we define new elements by using the elements in Γ' in the following way: if $(B', b') \in \Gamma'$, then we define (B, b) by $B := \begin{bmatrix} 1 & \\ & B' \end{bmatrix}$ and $b := (b_1', 0, b_2', \ldots, b_{n-1}')$, where b_i', $1 \le i \le n$, are the coordinates of b'. Then $\widetilde{\Gamma'} := \langle (B, b) : (B', b') \in \Gamma'; L_{e_i} : 1 \le i \le n \rangle$ is a n-rank Bieberbach group with holonomy group $\ker(p_1 \circ \rho)$ (it is torsion-free since Γ' is so). We claim that $\langle (B, b) \in \widetilde{\Gamma'}, (C_1, c) \rangle$ is a Bieberbach group with integral holonomy representation ρ. It is clear that its integral representation is ρ, so the only point to be explained is that it is torsion-free. This condition is clear for elements in $\widetilde{\Gamma'}$ and for (C_1, c). Now the products $(B, b)(C_1, c)$ are of the form (D, d) with $D = \begin{bmatrix} -1 & 0 \\ 0 & B' \end{bmatrix}$ and $d = (*, \frac{1}{2}, b_2', \ldots, b_{n-1}')$. Since Γ' is torsion-free, it follows that elements (D, d) are torsion-free, too. $\square$

Now we are interested in the holonomy representations of GHW groups.

Proposition 9.12. *The total number of inequivalent integral holonomy representations occurring in generalized Hantzsche-Wendt Bieberbach groups of dimension n equals $n/2$ for n even and $(n + 1)/2$ for n odd.*

Proof. By definition, each holonomy representation is defined by a hyperplane in $\mathbb{Z}_2^n$. From linear algebra, it is a kernel of the linear map with matrix $[x_1, x_2, \ldots, x_n]$, where $x_i \in \mathbb{Z}_2 = \{0, 1\}$. Hence, we have the following n classes of diagonal representations:

$$[1, 0, \ldots, 0], [1, 1, 0, \ldots, 0], \ldots, [1, 1, \ldots, 1, 0], [1, 1, \ldots, 1].$$

Since we are considering torsion-free groups, all hyperplanes, which correspond to holonomy representations, do not contain the vector $[1, 1, \ldots, 1]$, since $-Id$ corresponds to an element of torsion. Hence we have to consider only vectors with an odd number of non-zero elements. We claim that diagonal holonomy representations defined by such vectors are all inequivalent. In fact, let us consider a hyperplane (representation) as the set of $(n \times n)$ diagonal matrices. It suffices to count the number of matrices A such that the group of fix points $(\mathbb{Z}^n)^A$ of the action of A is isomorphic to integers $\mathbb{Z}$, i.e. matrices of the form $A = \text{diag}(-1, \ldots, -1, 1, -1, \ldots, -1)$. It is not difficult to see that for different hyperplanes the number of such A's is different. Summing up we have $n/2$ possibilities for n even, and $(n + 1)/2$ possibilities for n odd, which indeed occur by Proposition 9.10. □

As a consequence, the number of integral representations occurring in non-oriented GHW groups with trivial centre is $(n/2) - 1$ for n even and $((n + 1)/2) - 1$ for n odd.

Let us present an explicit example of GHW groups with trivial centre in each dimension n.

Example 9.13. In the previous notation, holonomy representation of GHW groups with trivial centre corresponds to hyperplanes $[1, 1, \ldots, 1, 1]$ in the odd case and $[1, 1, \ldots, 1, 1, 0]$ in the even case. Γ_2 is the fundamental group of the Klein bottle, cf. Example 2.3. For $n \geq 3$, let Γ_n be the subgroup of $E(n)$ generated by the set $\{(B_i, s(B_i) = x_i) : 1 \leq i \leq n - 1\}$. Here B_i's are the $n \times n$ orthogonal matrices:

$$B_i := -C_i = \text{diag}(-1, \ldots, -1, \underbrace{1}_{i}, -1, \ldots, -1) \qquad (9.6)$$

and

$$s(B_i) = e_i/2 + e_{i+1}/2 \text{ for } 1 \leq i \leq n - 1.$$

These groups have trivial centre for $n \geq 3$. In [131] (see also [101] and Example 5.22) it was proved that Γ_n are Bieberbach groups for odd n. For even n the groups Γ_n are also Bieberbach groups, in particular because we have:

Proposition 9.14. *For $n \geq 2$, there is a monomorphism $\phi_n \colon \Gamma_n \to \Gamma_{n+1}$ of groups. Moreover, $\phi_n(\Gamma_n)$ is not a normal subgroup of Γ_{n+1}.*

Proof. Let $\{(B_i, x_i) : 1 \leq i \leq n-1\}$ be a set of generators of Γ_n. We have an inclusion $k \colon \mathbb{Z}^n \to \mathbb{Z}^{n+1}$, where $b \in \mathbb{Z}^n$ is mapped to $k(b) = (b, 0) \in \mathbb{Z}^{n+1}$.

Next, we consider the $(n+1) \times (n+1)$ matrices $\bar{B}_i := \left[\begin{array}{c|c} B_i & * \\ \hline * & -1 \end{array}\right]$, $1 \leq i \leq n$, where $*$ denotes zeros. We define

$$\phi_n((B_i, x_i)) = (\bar{B}_i, k(x_i)).$$

Finally, it is enough to conjugate by the translation $(Id, (0, \ldots, 0, 1))$ an element $\phi_n((B_1, (1/2, 1/2, 0, \ldots, 0))$ to show that the image of Γ_n is not a normal subgroup. $\square$

By analogy with [131] we have:

Theorem 9.15. *For $n \geq 2$,*

$$H_j(\mathbb{R}^n/\Gamma_n, \mathbb{Q}) \simeq \begin{cases} \mathbb{Q} & \text{if } j = 0 \\ \mathbb{Q} & \text{if } n \text{ is even and } j = n-1 \\ \mathbb{Q} & \text{if } n \text{ is odd and } j = n \\ 0 & \text{otherwise} \end{cases}$$

Proof. For odd n it was proved in [131]. The proof for $n = 2k$ follows the same pattern as the proof of Proposition 9.9. We just have to replace n by $2k$, M by Γ_{n-1}, and the action of matrices C_i by the action of B_i as in (9.6), thus the characters change accordingly. Now, from the formula equivalent to (9.5) by an immediate calculation we get that no summand is trivial there, then $\dim_{\mathbb{Q}}[\Lambda^j(\mathbb{Q}^{2k})]^{\mathbb{Z}_2^{2k-1}} = 0$ for each j, $1 \leq j \leq 2k - 2$. Moreover,

$$\sum_{g \in \mathbb{Z}_2^{2k-1}} \frac{1}{2^{2k-1}} \sum_{1 \leq i_1 < i_2 < \cdots < i_{2k-1} \leq 2k} \chi_{i_1} \chi_{i_2} \cdots \chi_{ij}(g)$$

$$= \sum_{g \in \mathbb{Z}_2^{2k-1}} \frac{1}{2^{2k-1}} \sum_{i=1}^{2k} \chi_1 \cdots \hat{\chi}_i \cdots \chi_{2k}(g) = \frac{1}{2^{2k-1}} \sum_{i=0}^{2k-1} (2k - 2i) \binom{2k-1}{i} = 1.$$

Here $\chi_1 \ldots \hat{\chi}_i \ldots \chi_{2k}$ denotes the product of characters $\chi_1 \ldots \chi_i \ldots \chi_{2k}$ without the character χ_i. Hence $\dim_{\mathbb{Q}}[\Lambda^{2k-1}(\mathbb{Q}^{2k})]^{\mathbb{Z}_2^{2k-1}} = 1$. Finally, one can calculate that

$$\sum_{g \in \mathbb{Z}_2^{2k-1}} \chi_1 \chi_2 \cdots \chi_{2k}(g) = \sum_{i=0}^{2k-1} (-1)^i \binom{2k-1}{i} = 0.$$

In the formulas above we use the definition of characters on the elements of $\mathbb{Z}_2^{2k-1}$ considered as elements of $\mathbb{Z}_2$-vector space.   □

A small modification of the groups K_n defined at the beginning of this section produces examples as required at the end of section 9.1, (see also Exercise 9.2): we replace the $(1,1)$ entries in matrices C_i by 2×2 matrices as follows: $J := \begin{bmatrix} 0 & 1 \\ 1 & 0 \end{bmatrix}$ if the entry is a -1; and the 2×2 identity if the entry is a 1. Now the dimension has increased by one while the holonomy group remains the same, hence this is, as claimed, a non-diagonal example in the case that $k+2$ is the dimension. The verification that these are Bieberbach groups is analogous to that for Γ_n. The smaller the k (with fixed dimension), the easier the construction of these examples.

9.3   Graph connecting GHW manifolds

In this section we start the discussion of relations between the generalized Hantzsche-Wendt manifolds of different dimensions. For manifolds with the first Betti number one we have relations from E. Calabi's construction (see Proposition 9.8 and [148, page 125]). But we have also a lot of links between GHW groups of different dimensions in the case when one of them has trivial centre. We must say that this subject needs further research and we want to come back to it in the future. Most of the observations in this section are corollaries from the previous consideration, and we shall omit some proofs.

Let us start defining a graph $\mathbf{G}$ as follows:

- The vertices of $\mathbf{G}$ are the GHW groups (or manifolds).

- We say that a vertex has dimension n if it corresponds to a GHW manifold of dimension n.

- Edges are defined only between vertices in consecutive dimensions.

- There is an edge between two vertices (in consecutive dimensions) if one of the corresponding GHW groups is a subgroup of the other (in an abstract sense, i.e. there is a monomorphism from one into the other).

From the Corollary after Theorem 9.5 we see that in **G** there is always (at least) an edge 'down' from each vertex. As a direct consequence we have:

Corollary.

(i) The fundamental group of the Klein bottle is a subgroup of every GHW group.

*(ii) The graph **G** is connected.*

Moreover, from Figure 9.1 we see that **G** is not a tree.

On the other hand, from the method in the proof of Proposition 9.10 it is possible to deduce that any GHW group in dimension n is a subgroup of a GHW group in dimension $n + 1$. Moreover, we can prove that it is indeed a subgroup of at least two different GHW groups in dimension $n + 1$ (we omit the proof for brevity). In the graph **G** this means that from any vertex there are at least two edges 'up'.

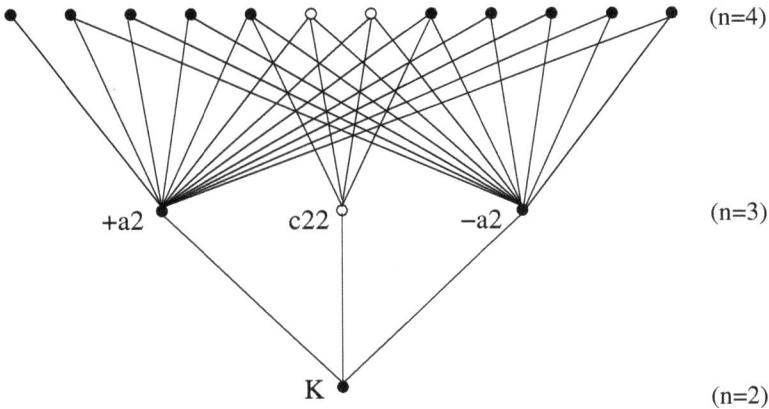

Figure 9.1: The graph of GHW manifolds up to $n = 4$

In Figure 9.1, a point ∘ (resp. a bullet •) denotes a GHW manifold with the first Betti number zero (resp. one); K is the Klein bottle; $+a2$ and $-a2$ are the *positive* and *negative amphidicosms*, respectively, they are the two non-oriented GHW manifolds of dimension 3; $c22$ is the *didicosm* (or the oriented Hantzsche-Wendt 3-dimensional flat manifold). The names as *didicosm, amphidicosm*, were defined in Section 3.3.

Other edges in Figure 9.1 are obtained by extensions to the methods of Theorem 9.5 and Proposition 9.10.

There are many other properties of **G** to be explored. For instance, a *distance* between vertices can be defined in the standard way for graphs.

In dimension 5 there are only two oriented GHW manifolds (see [101]). With the method of Theorem 9.5, it is not difficult to prove that they have both Γ_4 (defined in Example 9.13) as a subgroup, and therefore the distance between them is just two. We also have:

Proposition 9.16. *The fundamental group of the didicosm (or Hantzsche-Wendt 3-manifold) is a subgroup of every GHW group with trivial centre.*

Proof. Denote by Γ the given GHW group. By Lemma 9.1 there is $(B, b) \in \Gamma$ such that (after permuting coordinates) $B = \mathrm{diag}(1, -1, \ldots, -1)$ and $b = (\frac{1}{2}, *, \ldots, *)$, where the $*$'s are in $\{0, \frac{1}{2}\}$. Since Γ has trivial centre there is another element $(C, c) \in \Gamma$ with $C = \mathrm{diag}(-1, 1, -1, \pm 1, \ldots, \pm 1)$ and the second coordinate of c equals $1/2$. Now, the integral representation corresponding to $\langle B, C, BC \rangle$ is the direct sum of the three non-trivial characters of $\mathbb{Z}_2 \oplus \mathbb{Z}_2$ (with certain multiplicities). Then it is not difficult to check that the subgroup of Γ generated by (B, b) and (C, c) is isomorphic to the fundamental group of the didicosm. $\qquad\square$

In [55], the analogous result was proved for Bieberbach groups with holonomy group $\mathbb{Z}_2 \oplus \mathbb{Z}_2$ (instead of GHW groups) and trivial center (see also [143]).

By (9.4) the groups K_n, $n \geq 2$, form an infinite chain in **G**. Also the groups Γ_n, $n \in N$, $n \geq 3$ from the previous Section form an infinite chain by Proposition 9.14.

Finally, we remark that the edges in **G** sometimes correspond to normal subgroups and sometimes do not (or they can correspond to both simultaneously). For instance, the chain obtained by Proposition 9.14 is not normal. Also, $\pi_1(K)$ can be injected as a normal and as a non-normal subgroup in $\pi_1(+a2)$ and in $\pi_1(-a2)$ but it never is a normal subgroup of $\pi_1(c22)$.

9.4 Abelianization of HW group

We shall start with the statement from [101, page 4], which is the main technical tool in consideration of oriented Hantzsche-Wendt groups.

Lemma 9.17 ([101, Proposition 1.2]). *Let n be an odd number and $\Gamma = \langle (B_i, b_i) = \beta_i \colon b_{ij} \in \{0, \frac{1}{2}\}, 1 \le i \le n \rangle$, be a subgroup of $E(n)$. Then Γ is torsion free with translation subgroup $\Lambda = \langle e_i \colon 1 \le i \le n \rangle$ if and only if for any $\emptyset \ne I \subsetneq \{1, 2, \ldots, n\}$, there exists $j \in I$ if $|I|$ is odd (respectively $j \notin I$ if $|I|$ is even) such that*

$$|\{i \in I \colon b_{ij} = \tfrac{1}{2}\}| \text{ is odd.} \tag{9.7}$$

In particular, for each fixed $1 \le j \le n$, we have that $b_{jj} = \frac{1}{2}$ and $|\{i \in \{1, \ldots, n\} \colon b_{ij} = \frac{1}{2}\}|$ is even.

Proof. Let us assume that Γ is torsion free with translation lattice Λ. Hence, using a formula (II) from [101, p. 3], we can formulate the following condition:

$$(B_{i_1} B_{i_2} \ldots B_{i_s} + I) b(i_1, i_2, \ldots, i_s) \in \Lambda \setminus (B_{i_1}, B_{i_2} \ldots B_{i_s} + I)\Lambda \tag{9.8}$$

for any $1 \le i_1 < i_2 < \cdots < i_s \le r$, where

$$b(i_1, i_2, \ldots, i_s) = B_{i_s} \ldots B_{i_2} b_{i_1} + B_{i_s} \ldots B_{i_3} b_{i_2} + \cdots + B_{i_s} b_{i_{s-1}} + b_{i_s}.$$

Moreover, if the above condition is valid for any

$$I = \{i_1, i_2, \ldots, i_s\} \subset \{1, 2, \ldots, n-1\}$$

it is also valid for any $I = \{i_1, i_2, \ldots, i_s\} \subset \{1, 2, \ldots, n\}$. Indeed, if $i_s = n$, since $B_n = B_1 \ldots B_{n-1}$, then $B_{i_1} B_{i_2} \ldots B_{i_s} = B_{i'_1} B_{i'_2} \ldots B_{i'_{n-s}}$, where $I' = \{1, 2, \ldots, n\} \setminus I = \{i'_1, i'_2, \ldots, i'_{n-s}\}$. Thus $b(i_1, \ldots, i_s) - b(i'_1, \ldots, i'_{n-s}) \in \Lambda$. This implies that

$$(B_{i_1} B_{i_2} \ldots B_{i_s} + I) b(I) \equiv (B_{i'_1} B_{i'_2} \ldots B_{i'_{n-s}} + I) b(I')$$

mod $(B_{i_1} B_{i2} \ldots B_{i_s} + I)\Lambda = (B_{i'_1} B_{i'2} \ldots B_{i'_{n-s}} + I)\Lambda$. Hence condition (9.8) is satisfied for I if and only if it is so for I'.

Now, given a choice of $I = \{i_1, i_2, \ldots i_s\} \subset \{1, 2, \ldots, n\}$ we have that

$$(B_{i_1} B_{i_2} \ldots B_{i_s} + I)e_k = \begin{cases} \begin{cases} 0, & k \notin I, \\ 2e_k, & k \in I, \end{cases} & \text{s odd;} \\ \begin{cases} 0, & k \in I, \\ 2e_k, & k \notin I, \end{cases} & \text{s even.} \end{cases} \tag{9.9}$$

This implies,

$$(B_{i_1} B_{i_2} \ldots B_{i_s} + I)b(i_1, i_2, \ldots, i_s) = \begin{cases} 2 \sum_{h \in I} \left(\sum_{k=1}^s (-1)^{c_{k,h}} b_{i_k, h} \right) e_h, & \text{s odd;} \\ 2 \sum_{h \notin I} \left(\sum_{k=1}^s (-1)^{c_{k,h}} b_{i_k, h} \right) e_h, & \text{s even;} \end{cases}$$

where $(-1)^{c_{k,h}}$ is the eigenvalue of $B_{i_s} \ldots B_{i_{k+1}}$ on e_h. Since $b_{i,h} \in \frac{1}{2}\mathbb{Z}$ for all i, h we have, modΛ,

$$(B_{i_1} B_{i_2} \ldots B_{i_s} + I)b(i_1, i_2, \ldots, i_s) \equiv \begin{cases} \sum_{h \in I} 2 \left(\sum_{k=1}^s b_{i_k, h} \right) e_h, & \text{s odd;} \\ \sum_{h \notin I} 2 \left(\sum_{k=1}^s b_{i_k, h} \right) e_h, & \text{s even.} \end{cases}$$

On the other hand, by (9.9), $(B_{i_1} B_{i_2} \ldots B_{i_s} + I)\Lambda$ equals $\sum_{h \in I} \mathbb{Z} 2e_h$(s odd) and $\sum_{h \notin I} \mathbb{Z} 2e_h$(s even).

Hence we get that condition (9.8) is equivalent to condition (9.7) of the Lemma.

Concerning the second assertion, we note that if we choose $I_j = \{j\}$ we get from (9.8) that $b_{jj} = \frac{1}{2}$, for each j. Now, if we take $I'_j = \{1, 2, \ldots, \hat{j}, \ldots, n\}$, (9.8) implies that $|\{i \neq j : b_{ij} = \frac{1}{2}\}|$ is odd. Since $b_{jj} = \frac{1}{2}$, the last assertion follows.

For the proof of the sufficiency of the Lemma, we send the reader to [101]. $\square$

As an application of the above Lemma we present one of the most remarkable results about structure of GHW groups.

Theorem 9.18 ([119, Theorem 3.1]). *Let $\Gamma \subset E(n)$ be an oriented GHW group with $n > 3$. Then the commutator subgroup $[\Gamma, \Gamma]$ of Γ is equal to the translation subgroup of Γ.*

Proof. We can assume that $\Gamma = \langle (B_i, b_i) = \beta_i : b_{ij} \in \{0, \frac{1}{2}\}, 1 \leq i \leq n \rangle$. Put $\Lambda = \langle e_i : 1 \leq i \leq n \rangle$. Since the holonomy group of Γ is abelian, $[\Gamma, \Gamma] \subseteq \Lambda$. Moreover, $2\Lambda \subseteq [\Gamma, \Gamma]$. Indeed, for distinct $1 \leq i, j \leq n$, one can check that $\beta_j e_i^{-1} = e_i \beta_j$, which can be transformed into $[\beta_j, e_i^{-1}] = 2e_i$. Hence $2e_i \in [\Gamma, \Gamma]$ for all $i \in \{1, \ldots, n\}$.

In the rest of the proof we show that

$$e_i \in [\Gamma, \Gamma] \text{ for all } i \in \{1, \ldots, n\}. \tag{9.10}$$

For any $a \in \mathbb{R}^n$ we shall denote by $(a)_i$ the i-coordinate of a. Set $c(p, q, r) := [\beta_p, \beta_q][\beta_q, \beta_r][\beta_r, \beta_p]$, where $1 \leq p, q, r \leq n$ are pairwise distinct. By using

$$\beta_p^{-1} = (B_p, (b_{p1}, b_{p2}, \ldots, -b_{pp}, \ldots, b_{pn}))$$

one can check that

$$([\beta_p, \beta_q])_i = \begin{cases} 2(b_{pi} - b_{qi}) & \text{if } i \notin \{p, q\}, \\ 1 & \text{if } i = p, \\ -1 & \text{if } i = q. \end{cases} \tag{9.11}$$

The other two commutators are similar and we get

$$(c(p, q, r))_i = \begin{cases} 0 & \text{if } i \notin \{p, q, r\}, \\ 2(b_{qp} - b_{rp}) & \text{if } i = p, \\ 2(b_{rq} - b_{pq}) & \text{if } i = q, \\ 2(b_{pr} - b_{qr}) & \text{if } i = r. \end{cases} \tag{9.12}$$

We note that such elements of the commutator subgroup have at most three non-zero coordinates; the values at these coordinates can be -1 or 1. Each -1 can be changed to a 1 by just adding the square of a translation, and the resulting element is still in $[\Gamma, \Gamma]$. We also note that permuting p, q, r fixes or inverts $c(p, q, r)$.

We claim, that for each triple of pairwise distinct indices $1 \leq p, q, r \leq n$ at least one coordinate p, q or r of $c(p, q, r)$ must be 0. Set $I = \{p, q, r\}$. By Lemma 9.17, there is $j \in I$ such that $|\{i \in I : b_{ij} = \frac{1}{2}\}|$ is odd, thus equals 1 or 3. Without loss of generality we assume $j = p$. Since $b_{pp} = \frac{1}{2}$, the other two indices b_{qp} and b_{rp} must be equal to each other. Thus, from (9.12), we have $(c(p, q, r))_p = 0$. The cases $j = q$ and $j = r$ are similar. Hence $c(p, q, r)$ has at most two non-zero coordinates.

We separate the proof (9.10) into four steps.

Step 1. $[\Gamma, \Gamma]$ *contains some element of* $\{e_1, \ldots, e_n\}$. We fix an index p, and for all q, r such that p, q, r are distinct we compute the parity of the number of non-zero p-coordinates of $c(p, q, r)$. By (9.12), we have $(c(p, q, r))_p \neq 0$ if and only if $b_{qp} \neq b_{rp}$.

We know from Lemma 9.17 that

$$|\{i \in \{1, \ldots, n\} \colon b_{ip} = \tfrac{1}{2}\}| \text{ is even.}$$

Hence, since $b_{pp} = \tfrac{1}{2}$, we have $b_{ip} = \tfrac{1}{2}$ for an odd number of indices $i \neq p$ and $b_{ip} = 0$ for an odd number of indices i. Thus there are an odd number of choices of q, r such that $c(p, q, r)$ has non-zero p-coordinate. Because there is an odd number of coordinates, there are an odd number of non-zero coordinates in all elements $c(p, q, r)$. Hence not all elements $c(p, q, r)$ can have 0 or 2 non-zero coordinates, and thus one must have exactly one non-zero coordinate. Therefore there exist p, q, r for which $e_p = c(p, q, r)^{\pm 1} \in [\Gamma, \Gamma]$.

Step 2. $[\Gamma, \Gamma]$ *contains two distinct elements of* $\{e_1, \ldots, e_n\}$.

From Step 1, $e_p \in [\Gamma, \Gamma]$ for some index p. Consider the subset of indices

$$I = \{p\} \cup \{i \in \{1, \ldots, n\} \colon b_{ip} = 0\}.$$

Note that $I \neq \{1, \ldots, n\}$ and $|I|$ is even. By Lemma 9.17 there is some $j \notin I$ such that $|\{i \in I \colon b_{ij} = \tfrac{1}{2}\}|$ is odd. Hence the numbers b_{ij} for $i \in I$ cannot all be equal, and thus there exists an index $k \in I, k \neq p$, for which $b_{kj} \neq b_{pj}$. The three indices p, j, k are pairwise distinct, hence

$$(c(p, j, k))_j = 2(b_{kj} - b_{pj}) \neq 0$$

and, since $k \in I, j \notin I$,

$$(c(p, j, k))_p = 2(b_{jp} - b_{kp}) \neq 0.$$

This implies that $(c(p, j, k))_k = 0$. Finally, $c(p, j, k) = e_p^{\pm 1} e_j^{\pm 1} \in [\Gamma, \Gamma]$, which, since $e_p \in [\Gamma, \Gamma]$, shows that $e_j \in [\Gamma, \Gamma]$.

Step 3. If m is odd and $3 \leq m \leq n$, and $m - 1$ translations $e_{i_1}, \ldots, e_{i_{m-1}}$ are in $[\Gamma, \Gamma]$, then there is another translation $e_i \in [\Gamma, \Gamma]$.

Set $I = \{i_1, \ldots, i_{m-1}\}$. By Lemma 9.17 we can find some $j \notin I$ such that $|\{i \in I \colon b_{ij} = \tfrac{1}{2}\}|$ is odd. Then, there exist two distinct indices $k, l \in I$ such that $b_{kj} \neq b_{lj}$, and thus $(c(j, k, l))_j \neq 0$. Now $c(j, k, l) \in [\Gamma, \Gamma]$, and since $e_k, e_l \in [\Gamma, \Gamma]$ this shows that $e_j \in [\Gamma, \Gamma]$.

Step 4. If m is even and $4 \leq m \leq n$, and $m - 1$ translations $e_{i_1}, \ldots, e_{i_{m-1}}$ are in $[\Gamma, \Gamma]$, then there is another $e_i \in [\Gamma, \Gamma]$.

Set $I_1 = \{i_1, \ldots, i_{m-1}\}$. By Lemma 9.17 there is an index $j_1 \in I_1$ such that

$$|\{i \in I_1 \colon b_{ij_1} = \tfrac{1}{2}\}| \text{ is odd.} \tag{9.13}$$

Set $I_2 = I_1 \setminus \{j_1\}$, $|I_2|$ is even. By Lemma 9.17, there is $j_2 \notin I_2$ such that

$$|\{i \in I_2 \colon b_{ij_2} = \tfrac{1}{2}\}| \text{ is odd.} \tag{9.14}$$

We claim that $j_1 \neq j_2$. Suppose on the contrary that $j_1 = j_2$. As $b_{j_1 j_1} = \tfrac{1}{2}$ we get

$$\{i \in I_1 \colon b_{ij_1} = \tfrac{1}{2}\} = \{i \in I_2 \colon b_{ij_2} = \tfrac{1}{2}\} \cup \{j_1\},$$

which contradicts (9.13) and (9.14). Hence, $j_1 \neq j_2$ and $j_2 \notin I_1$. Now, by (9.14) there exist two distinct indices $k, l \in I_2 \subset I_1$ for which $b_{kj_2} \neq b_{lj_2}$. Thus $(c(j_2, k, l))_{j_2} \neq 0$, and $c(j_2, k, l)$, together with $e_k, e_l \in [\Gamma, \Gamma]$, gives $e_{j_2} \in [\Gamma, \Gamma]$. This finishes the proof. $\qquad\square$

9.5   Cohomological rigidity of oriented HW manifolds

Definition 9.19 (See [115]). Two flat manifolds M_1 and M_2 are cohomological rigid if and only if a homeomorphism between M_1 and M_2 is equivalent to an isomorphism of graded rings $H^*(M_1, \mathbb{F}_2)$ and $H^*(M_2, \mathbb{F}_2)$.

In this section we shall prove the following theorem.

Theorem 9.20 (Main Theorem). *Hantzsche-Wendt manifolds are cohomological rigid.*

The Theorem answers the question from [27, problem 4.3].

For the proof we introduce a new presentation of *HW-manifolds*. We consider these manifolds rather as a finite quotient of the torus than a quotient of the $\mathbb{R}^n$. Here, we use an obvious equivalence $\mathbb{R}^n/\Gamma = (\mathbb{R}^n/\mathbb{Z}^n)/G = T^n/G$, where Γ is a Bieberbach group from (2.9). According to the definition of n-dimensional *HW-manifold* we shall define a $(n \times n)$-*HW-matrix* A. The analysis of properties of the matrix A is used in the proof. Moreover, we apply the Lyndon-Hochschild-Serre spectral sequence $\{E_r^{p,q}, d_r\}$ of the covering $T^n \to T^n/G$ with $\mathbb{F}_2$ coefficients. Since a holonomy representation Φ_Γ is diagonal $E_2^{p,q} = H^p((\mathbb{Z}_2)^{n-1}) \otimes H^q(\mathbb{Z}^n)$. We shall only use the multiplicative structure of the first and second cohomology group. In particular, we shall consider the properties of the transgression homomorphism $d_2 \colon H^1(\mathbb{Z}^n) \to H^2((\mathbb{Z}_2)^{n-1})$. Finally, another important point of the proof is an isomorphism of cohomology groups $H^1((\mathbb{Z}_2)^{n-1})$ and $H^1(\Gamma)$, which was proved in [119, Theorem 3.1]. Hence, we can consider elements of the image of the transgression homomorphism d_2 as homogeneous polynomials of degree two which are equivalent to polynomial functions.

Let us present a structure of the section. In the next subsection, we give a "new-old" definition of HW-$manifold$ and we outline the proof of the theorem. In subsection 9.5.2 we define HW-$matrix$ and prove some of its properties. In the last section, we present the proof of the **Main Lemma**.

9.5.1 Proof of the Main Theorem

Let $\mathcal{D} = \{g_i \mid i = 0, 1, 2, 3\}$, where $g_i \colon S^1 \to S^1$, and $\forall z \in S^1 \subset \mathbb{C}$,

$$g_0(z) = z, g_1(z) = -z, g_2(z) = \bar{z}, g_3(z) = -\bar{z}. \tag{9.15}$$

Equivalently, if $S^1 = \mathbb{R}/\mathbb{Z}, \forall [t] \in \mathbb{R}/\mathbb{Z}$,

$$g_0([t]) = [t], g_1([t]) = [t + \frac{1}{2}], g_2([t]) = [-t], g_3([t]) = [-t + \frac{1}{2}]. \tag{9.16}$$

Let $(t_1, t_2, ..., t_n) \in \mathcal{D}^n$ and $(z_1, z_2, ..., z_n) \in T^n = \underbrace{S^1 \times S^1 \times ... \times S^1}_{n}$. It is easy to see that $\mathcal{D} = \mathbb{Z}_2 \times \mathbb{Z}_2$, and $g_3 = g_1 g_2$. For $k = 1, 2, 3$ we have different projections

$$p^{(k)} \colon \mathcal{D} \to \mathbb{F}_2 = \{0, 1\} \tag{9.17}$$

such that $p^{(k)}(g_k) = 1$ (cf. Table 9.1) and for $i = 1, 2, .., n$ we have homomorphisms

$$p^{(k)} \circ pr_i \colon \mathcal{D}^n \to \mathcal{D} \xrightarrow{p^{(k)}} \mathbb{F}_2 \tag{9.18}$$

given by the formula $p^{(k)} \circ pr_i(t_1, t_2, ..., t_i, ..., t_n) = p^{(k)}(t_i)$.

Table 9.1: Values of the projections $p^{(2)}$ and $p^{(3)}$

	g_0	g_1	g_2	g_3
$p^{(2)}$	0	1	1	0
$p^{(3)}$	0	1	0	1

The next, obvious formula

$$\forall x \in \mathcal{D} \quad x = p^{(2)}(x)2 + p^{(3)}(x)3 \tag{9.19}$$

will be useful later. We can define an action $\mathcal{D}^n$ on T^n as follows:

$$(t_1, t_2, ..., t_n)(z_1, z_2, ..., z_n) = (t_1 z_1, t_2 z_2, ..., t_n z_n). \tag{9.20}$$

We have:

Proposition 9.21. *Let M^n be a HW-manifold of dimension n. Then there exists a subgroup $(\mathbb{Z}_2)^{n-1} \subset \mathcal{D}^n$ such that $M^n = T^n/(\mathbb{Z}_2)^{n-1}$, where the action $(\mathbb{Z}_2)^{n-1}$ on T^n is defined by (9.6) and (9.20).*

Proof. Let $\pi_1(M^n) = \Gamma$ and $(B_l, b_l) \in \Gamma$ be the generators (9.6), $l = 1, 2, .., n$. On each coordinate, (9.16) defines $g_j \in \mathcal{D}, j = 0, 1, 2, 3$ which are determinated by projections $p^{(1)} \circ pr_i, p^{(2)} \circ pr_i, p^{(3)} \circ pr_i$. □

Let us start to prove that the graded ring $H^*(M^n, \mathbb{F}_2)$ defines a manifold M^n. We have an exact sequence

$$0 \to \mathbb{Z}^n \to \Gamma \xrightarrow{P} (\mathbb{Z}_2)^{n-1} \to 0, \tag{9.21}$$

where $\Gamma = \pi_1(M^n)$. As we mentioned already in the introduction the image of a holonomy representation $\Phi_\Gamma((\mathbb{Z}_2)^{n-1})$, is a subgroup of the group of all diagonal matrices of $GL(n, \mathbb{Z})$. Moreover (see [119]) $H^1(\Gamma, \mathbb{F}_2) = (\mathbb{F}_2)^{n-1}$ for any Hantzsche-Wendt group Γ of dimension n. That is an observation which we shall use during the proof.

Since $(\mathbb{Z}_2)^{n-1} \subset \mathcal{D}^n$ the above maps $p^{(k)} \circ pr_i, k = 1, 2, 3$ define homomorphisms from $(\mathbb{Z}_2)^{n-1} \to \mathbb{F}_2 \in \text{Hom}((\mathbb{Z}_2)^{n-1}, \mathbb{F}_2) = H^1((\mathbb{Z}_2)^{n-1}, \mathbb{F}_2) \overset{[6]}{=} H^1(M^n, \mathbb{F}_2)$. Hence we can define elements

$$T_i = (p^{(2)} \circ pr_i) \cup (p^{(3)} \circ pr_i) \in H^2((\mathbb{Z}_2)^{n-1}, \mathbb{F}_2),$$

where $\cup$ is a cup product. It is well known that $H^*((\mathbb{Z}_2)^{n-1}, \mathbb{F}_2)$ is isomorphic to $\mathbb{F}_2[x_1, x_2, ..., x_{n-1}]$. Hence the elements $p^{(k)} \circ pr_i = p_i^{(k)}$ correspond to

$$\sum_{j=1}^{n-1} p^{(k)}(pr_i(b_j))x_j = \sum_{j=1}^{n-1} p^{(k)}(A_{ji})x_j \in \mathbb{F}_2[x_1, x_2, ..., x_{n-1}], \tag{9.22}$$

where $b_1, b_2, ..., b_{n-1}$ is the basis of $(\mathbb{Z}_2)^{n-1}$ and $k = 2, 3; i = 1, 2, ..., n$. Here the matrix $A_{ij}, i = 1, 2, \ldots, n-1; j = 1, 2, \ldots, n$ is related to *HW-matrix* (Definition 9.26) from the next section.

We shall apply the Lyndon-Hochschild-Sere spectral sequence $\{E_r^{p,q}, d_r\}$ of (9.21). Since a holonomy representation Φ_Γ is diagonal $E_2^{p,q} = H^p((\mathbb{Z}_2)^{n-1}) \otimes H^q(\mathbb{Z}^n)$. Hence (see [40, Corollary 7.2.3 on p. 77]) we have an exact sequence (see [27, p. 770])

$$H^1(\mathbb{Z}^n, \mathbb{F}_2) \xrightarrow{d_2} H^2((\mathbb{Z}_2)^{n-1}, \mathbb{F}_2) \xrightarrow{p^*} H^2(\Gamma, \mathbb{F}_2), \tag{9.23}$$

where d_2 is a transgression and p^* is induced by the above homomorphism $p \colon \Gamma \to (\mathbb{Z}_2)^{n-1}$. In what follows we shall prove (see also [27, Theorem 2.7]) that a rank of

$$\mathrm{Im}(d_2) \subset H^2((\mathbb{Z}_2)^{n-1}, \mathbb{F}_2) \subset H^*((\mathbb{Z}_2)^{n-1}, \mathbb{F}_2) \simeq \mathbb{F}_2[x_1, x_2, ..., x_{n-1}]$$

is equal to n.

Let us define a basis $\hat{t}_i, i = 1, 2, \ldots, n$ of $H^1(\mathbb{Z}^n, \mathbb{F}_2) = \mathrm{Hom}(\mathbb{Z}^n, \mathbb{F}_2)$. For $k \in \mathbb{Z}$, we shall write $\bar{k} = 0$ if k is even and $\bar{k} = 1$ if k is odd. Let $(k_1, k_2, \ldots, k_n) \in \mathbb{Z}^n$ and let

$$\hat{t}_i(k_1, k_2, \ldots, k_n) = \bar{k}_i, i = 1, 2, \ldots, n.$$

We have:

Proposition 9.22. $d_2(\hat{t}_i) = T_i = (p^{(2)} \circ pr_i) \cup (p^{(3)} \circ pr_i)$. *Moreover elements* $T_i, i = 1, 2, \ldots, n$ *are a basis of* $\mathrm{Im}(d_2)$.

Proof. By Theorem 2.5 (ii) and Proposition 1.3 of [27] and using (9.22) it follows that

$$d_2(\hat{t}_i) = \sum_{A_{il}=1} x_i^2 + \sum_{i \neq j} x_i x_j,$$

where the second sum is taken for such i, j that

$$(A_{il}, A_{jl}) \in \{(1,2), (2,1), (1,3), (3,1), (3,1), (3,2), (2,3)\}.$$

On the other hand

$$T_l = p_l^{(2)} p_l^{(3)} = \sum_{i=1}^{n-1} p^{(2)}(A_{il}) p^{(3)}(A_{il}) x_i^2$$

$$+ \sum_{1 \leq i < j \leq n-1} p^{(2)}(A_{il}) p^{(3)}(A_{jl}) + p^{(2)}(A_{jl}) p^{(3)}(A_{il}) x_i x_j. \tag{9.24}$$

Comparing coefficients of the above two polynomials finishes the proof. □

The main idea of the proof of rigidity is an application of the above Proposition 9.22. It means, we show that any HW-*manifold* M, of dimension greater than three, define elements in the cohomology ring $H^*(M, \mathbb{F}_2)$ which determines M up to affine equivalence. In the **Main Lemma**, we shall prove an existence of n linear independent elements $T_1, T_2, \ldots, T_n \in \text{Im}(d_2)$ such that for any $i = 1, 2, \ldots, n$ $T_i = p_i q_i$. At the end of this section we give a method of a reconstruction of HW-*group* from the set $\{T_i\}_{i=1,2,\ldots,n}$.

Let us define

$$D = \{y \in \text{Im}(d_2) \mid y \text{ is a product of two polynomials of degree } 1\}. \quad (9.25)$$

We shall prove that D has less than $n + 2$ elements from which we can reconstruct the basis $T_1, T_2, \ldots, T_n$ of $\text{Im}(d_2)$.

Lemma 9.23 (Main Lemma). *Let $n > 3$, then there are the following possibilities for the structure of the set D:*

1. $D = \{T_1, T_2, \ldots, T_n\}$;
2. $D = \{T_1, T_2, \ldots, T_n, T_i + T_j\}$, *and we can find a polynomial* p *of degree one such that* p $\mid T_i$ *and* p $\mid T_j$ *for some* $1 \le i, j \le n$.

In the second case we can rediscover the set of generators $T_1, T_2, \ldots, T_n$.

Let M be HW-*manifold* of dimension n. From the **Main Lemma**, we know that there is a set $D = \{T_1, T_2, \ldots, T_n\} \subset \text{Im}(d_2)$ such that any T_i is a product of two polynomials p_i and $q_i, i = 1, 2, \ldots, n$ of a degree one. Let V be $(n-1)$-dimensional $\mathbb{F}_2$ vector space. We define a linear map $h: V^* \to \mathcal{D}^n$, which simple version is (9.19) such that

$$h_i(x) = p_i(x)2 + q_i(x)3, \text{ for } i = 1, 2, \ldots, n, \quad (9.26)$$

where $p_i, q_i \in V \simeq V^{**}$. Hence, through formulas (9.22), (9.24) and the Table 9.1, $\text{Im}(h)$, defines a Hantzsche-Wendt group.

Example 9.24. 1. Let $V = \text{gen}\{x_1, x_2, x_3\}$ and $D = \{x_1^2 + x_1 x_2, x_1 x_2 + x_1 x_3 + x_2^2 + x_2 x_3\}$. Put $p_1 = x_1, q_1 = x_1 + x_2, p_2 = x_1 + x_2, q_2 = x_2 + x_3$. Hence a homomorphism $h(x_1^*) = (1, 2), h(x_2^*) = (3, 1)$ and $h(x_3^*) = (0, 3)$. Here x_1^*, x_2^*, x_3^* is a dual basis of V^*. Finally we define a subgroup of $\mathcal{D}^2$

which generators are rows of the matrix

$$\begin{bmatrix} 1 & 2 \\ 3 & 1 \\ 0 & 3 \end{bmatrix}.$$

2. Let $\mathbb{Z}_2^{n-1} \subset \mathcal{D}^n$ be a HW-group, and D a set from the Proposition 9.22. Assume that $D = \{p_1 q_1, p_2 q_2, \ldots, p_n q_n\}$. Then

$$h_i(x) = p_i(x)2 + q_i(x)3 = p(x_i)2 + q(x_i)3 = x_i.$$

Hence for $x \in \mathbb{Z}_2^{n-1}, h(x) = x$ and $\text{Im}(h) = \mathbb{Z}_2^{n-1}$.

Let $\phi \colon H^*(M_1, \mathbb{F}_2) \to H^*(M_2, \mathbb{F}_2)$ be an isomorphism of cohomology rings of HW-manifolds M_1 and M_2. From the **Main Lemma** for the both manifolds we have the sets of elements D_1 and D_2 such that $\phi(D_1) = D_2$. Hence we obtain the affine equivalence manifolds M_1 and M_2.

$\square$

9.5.2   Properties of Hantzsche–Wendt matrices

Let us illustrate the Proposition 9.21 on two HW-manifolds of dimension 5, (see [124]). We shall denote by Γ_1 and Γ_2 its fundamental groups.

Example 9.25. A group $\Gamma_1 \subset E(5)$ is generated by

$$(B_1, (1/2, 1/2, 0, 0, 0)), (B_2, (0, 1/2, 1/2, 0, 0)),$$
$$(B_3, (0, 0, 1/2, 1/2, 0)), (B_4, (0, 0, 0, 1/2, 1/2)).$$

From above $\mathbb{R}^5/\Gamma_1 \simeq T^5/(\mathbb{Z}_2)^4$, where $(\mathbb{Z}_2)^4 \subset \mathcal{D}^5$ is defined by

$$(g_1, g_3, g_2, g_2, g_2), (g_2, g_1, g_3, g_2, g_2),$$
$$(g_2, g_2, g_1, g_3, g_2), (g_2, g_2, g_2, g_1, g_3).$$

Moreover a group $\Gamma_2 \subset E(5)$ is generated by

$$(B_1, (1/2, 0, 1/2, 1/2, 0)), (B_2, (0, 1/2, 1/2, 1/2, 1/2)),$$
$$(B_3, (1/2, 1/2, 1/2, 1/2, 1/2)), (B_4, (1/2, 0, 1/2, 1/2, 1/2)).$$

Hence, $\mathbb{R}^5/\Gamma_2 \simeq T^5/(\mathbb{Z}_2)^4$ where generators of a group $(\mathbb{Z}_2)^4 \subset \mathcal{D}^5$ are following

$$(g_1, g_2, g_3, g_3, g_2), (g_2, g_1, g_3, g_3, g_3),$$
$$(g_3, g_3, g_1, g_3, g_3), (g_3, g_2, g_3, g_1, g_3).$$

In what follows we shall write i for g_i, $i = 0, 1, 2, 3$. Let A be a $(n \times m)$ matrix with coefficients $A_{ij} \in \mathcal{D}$. For short $A \in \mathcal{D}^{n \times m}$. Let A_i (A^j) denote i-row (j-column) of a matrix A.

Definition 9.26. By *HW-matrix* we shall understand a matrix $A \in \mathcal{D}^{n \times n}$ such that

1. $A_{ii} = 1$, $A_{ij} \in \{2, 3\}$ for $i \neq j$, $1 \leq i, j \leq n$.

2. Sum of rows of A is equal to zero.

3. If $X \subset \{1, 2, ..., n\}$ and $1 \leq \#X \leq n - 1$ then the row $\sum_{i \in X} A_i$ has 1 on a some position.

Lemma 9.27. *Any HW-manifold of dimension n defines a $(n \times n)$ HW-matrix.*

Proof. Let (β_i, b_i), $1 \leq i \leq n - 1$ be generators of the fundamental group of some n-dimensional HW-manifold M. Then i-generator defines i-row of some $(n \times n)$ *HW-matrix*, cf. (9.6), (9.16). See also Example 9.25 and Proposition 9.21. The last row is defined by the product $\beta_1 \beta_2 \ldots \beta_{n-1}$ or equivalently is a sum of the first $(n - 1)$ rows. It is easy to see that the first property of the above matrix follows from a definition, see [101, p. 4]. Since a holonomy group $(\mathbb{Z}_2)^{n-1}$ acts freely on T^n (or equivalently $\pi_1(M)$ is a torsion free group) the last part of lemma follows. $\square$

We shall present some properties of *HW-matrices*.

Remark 9.28. Let $\sigma \in S_n$ and let P_σ be the corresponding permutation matrix. It is not difficult to see that if A is *HW-matrix* then $P_\sigma A P_\sigma^{-1}$ also satisfies conditions of the Definition 9.26. Moreover, if A' is a conjugation matrix of A, where conjugation means exchange at some column numbers 2 for 3, then A' is again a *HW-matrix*. The *HW-matrix* is related to the matrix defined on page 6 of *[101]*.

Remark 9.29. Let A be a $(n \times n)$ *HW-matrix*. Then

$$(p^{(2)} + p^{(3)})(A) = \begin{bmatrix} 0 & 1 & 1 & \cdots & 1 & 1 \\ 1 & 0 & 1 & \cdots & 1 & 1 \\ \vdots & \vdots & \vdots & \ddots & \vdots & \vdots \\ 1 & 1 & \cdots & 1 & 0 & 1 \\ 1 & 1 & \cdots & 1 & 1 & 0 \end{bmatrix}. \tag{9.27}$$

Let $A \in \mathcal{D}^{m \times n}$ be a $(m \times n)$ matrix with coefficients in $\mathcal{D}$ and $(\alpha_1, \alpha_2, ..., \alpha_n) \in \{2, 3\}^n$. By $p^{(\alpha)}(A)$ we shall understand a $(m \times n)$-matrix with coefficients in $\mathbb{F}_2$ which a i-column is equal to $p^{(\alpha_i)}(A^i)$.

Let M be a matrix. By defect of M we shall understand a number

$$d(M) = \{\text{number of columns of } M\} - \text{rk}(M).$$

Lemma 9.30.

1. Let M_1 be a matrix M from which we remove some columns. Then

$$d(M_1) \leq d(M),$$

2. If A is a HW-matrix of dimension n and $\alpha \in \{2, 3\}$, then

$$d(p^{(\alpha)}(A)) \leq 1.$$

Proof. The first statement is clear. For the proof of a second one, let us assume that $d(p^{(\alpha)}(A)) > 1$. Hence $rk(p^{(\alpha)}(A)) < n - 1$. By definition there exists a non-trivial $X \subset \{1, 2, ..., n-1\}$, such that $\sum_{i \in X} p^{(\alpha)}(A_i) = 0$. Finally $p^{(\alpha)}(\sum_{i \in X} A_i) = \sum_{i \in X} p^{(\alpha)}(A_i) = 0$. This contradicts the Definition 9.26. $\square$

Lemma 9.31. *Let $m < n$ and $W \in \mathcal{D}^{m \times n}$ is a sub-matrix of some $(n \times n)$ HW-matrix. Then $rk(p^{(\alpha)}(W)) = m$.*

Proof. Similar to the proof of the last Lemma. $\square$

A symmetric $(m \times m)$ matrix $A \in (\mathbb{F}_2)^{m \times m}$ defines a non-oriented graph, graph(A) with set of vertices $\{1, 2, ..., m\}$ and two different vertices i and j are connected if and only if $A_{ij} = 1$. We say that a matrix A is connected if a graph(A) is connected. Let $A \in \mathcal{D}^{m \times m}$ be a symmetric matrix, then $p^{(i)}(A)$ are symmetric with coefficients in $\mathbb{F}_2$, $i = 2, 3$. We shall write $i \sim_2 j$ if i, j are at the same connected component of a matrix $p^{(2)}(A)$. Similar definition is for a relation $i \sim_3 j$.

Lemma 9.32. *Let a HW-matrix M have the following decomposition on the blocks:*

$$M = \begin{bmatrix} * & 2 & * \\ C & A & D \\ * & 3 & * \end{bmatrix}, \tag{9.28}$$

where A is a symmetric matrix and $2, 3$ are block matrices with all rows and columns equal 2 and 3 correspondingly. Then

(I) *if $i \sim_2 j$* $\implies$ $D_i = D_j$;

(II) *if $i \sim_3 j$* $\implies$ $C_i = C_j$.

Proof. For the proof of (I) let us assume that i, j (where $i < j$) are connected by a 2-edge; i.e. $A_{i,j} = 2$. Let r be some column of a matrix D. Let us consider a diagonal submatrix of the matrix M related to (i, j, r). It looks like

$$\begin{bmatrix} 1 & 2 & a \\ 2 & 1 & b \\ 3 & 3 & 1 \end{bmatrix}. \tag{9.29}$$

The sums of the first two columns are zero. Since a Lemma 9.31 a sum of elements of the last one is not zero. Hence $a = b$. We have just proved that if $A_{i,j} = 2$ then $D_i = D_j$. It also means that if $i \sim_2 j$ then $D_i = D_j$. The proof of the second point of the lemma is similar. □

The next lemmas are about possibilities of complement of some matrices to a *HW-matrix*. We shall first consider an odd case.

Lemma 9.33. *Let $A \in \mathcal{D}^{m \times m}$ be a symmetric matrix with 1 on the diagonal and $\{2, 3\}$ off the diagonal with a column sums equal to 1. Assume that $m > 1$. Then a matrix*

$$K_A = \begin{bmatrix} 2 \\ A \\ 3 \end{bmatrix}, \tag{9.30}$$

cannot be complement to HW-matrix.

Proof. Let us assume that there exists a *HW-matrix*

$$\begin{bmatrix} * & 2 & * \\ C & A & D \\ * & 3 & * \end{bmatrix}. \tag{9.31}$$

From assumption m is an odd number and heights of the blocks 2 and 3 are also odd. We shall use induction. For $m = 3$

$$A = \begin{bmatrix} 1 & a & a \\ a & 1 & a \\ a & a & 1 \end{bmatrix}. \tag{9.32}$$

Here $a = 2$ or 3. If $a = 3$ then $\mathrm{rk}(p^{(2)}(A)) = 1$ and $d(p^{(2)}(A)) = 3 - 1 = 2 > 1$. From Lemma 9.30 it is impossible. For $a = 3$ the proof is the same. Let us assume that $m > 3$.

1. We shall consider a matrix $p^{(2)}(A)$. We claim that there is no such decomposition as

$$p^{(2)}(A) = B \oplus E,$$

such that a dimension of a matrix B is odd and > 1. In fact, in that case

$$A = \begin{bmatrix} \tilde{B} & 3 \\ 3 & \tilde{E} \end{bmatrix}. \tag{9.33}$$

Since a column sums of A are equal to 1 and height of a block 3 under $\tilde{B}$ is even, a column sums of $\tilde{B}$ are 1. If K_A has complement then $K_{\tilde{B}}$ has a complement (where a dimension of a block 3 is greater on a dimension of E). But by induction it is impossible, since $1 < \text{dimension}(\tilde{B}) < m$.

2. We claim that there is no such a non-trivial decomposition as

$$p^{(2)}(A) = B \oplus E \oplus F.$$

In fact since m is odd we have two possibilities:

(a) dimension of one component is odd and other components have dimension even

(b) dimension of all components are odd.

In the case (a) $\dim(B \oplus E) > 1$ and odd. Hence we consider decomposition $p^{(2)}(A) = (B \oplus E) \oplus F$. But it is a previous case 1.

In case (b), since $m > 3$ there exists a component (for example B) which dimension is > 1. In that case we have a decomposition $p^{(2)}(A) = B \oplus (E \oplus F)$ which was already considered in the point 1.

3. By definition we have a decomposition

$$p^{(2)}(A) = B_1 \oplus ... \oplus B_s,$$

where all components are connected matrices. From the above we can assume that $s = 2$ and odd component has a graph equal to one point or $s = 1$. Equivalently,

(a) $A = \begin{bmatrix} 1 & 3 \\ 3 & B \end{bmatrix}$ and $p^{(2)}(B)$ is connected or

(b) $p^{(2)}(A)$ is connected.

In the first case

$$p^{(3)}(A) = \begin{bmatrix} 1 & 1 \\ 1 & p^{(3)}(B) \end{bmatrix}. \tag{9.34}$$

Hence $p^{(3)}(A)$ is connected. Summing up, we have

(a) $A = \begin{bmatrix} 1 & 3 \\ 3 & B \end{bmatrix}$ and both $p^{(2)}(B)$ and $p^{(3)}(A)$ are connected or

(b) $p^{(2)}(A)$ is connected.

If we exchange $p^{(2)}$ for $p^{(3)}$ in the above points 1., 2. and 3. with the similar arguments, we obtain finally two cases:

(a) $A = \begin{bmatrix} 1 & 3 \\ 3 & B \end{bmatrix}$ and both $p^{(2)}(B)$ and $p^{(3)}(A)$ are connected or

(b)   both $p^{(2)}(A)$ and $p^{(3)}(A)$ are connected.

   We come back to the beginning of the proof. We shall try to figure out matrices C and D. From definition of $\sim_3$ and because $p^{(3)}(A)$ is connected we conclude that all rows of the matrix C are identical. By conjugation we can assume that $C = 2$. Using the same arguments and definition of $\sim_2$ together with a connectedness of $p^{(2)}(B)$ we conclude that with exception of the first row, all rows of the matrix D are the same. By conjugation and permutation we can assume that the first row of the matrix D is equal to $[2, ...2, 3, ..., 3]$. All other rows of a matrix D consist only 3. Summing up a matrix

$$W = [C \ A \ D]$$

is following

$$\begin{bmatrix} 2 & \begin{bmatrix} 1 & 3 \\ 3 & B \end{bmatrix} & 2 & 3 \\ 2 & & 3 & 3 \end{bmatrix}. \tag{9.35}$$

Apply homomorphisms: $p^{(3)}, [p^{(2)}, p^{(3)}], p^{(2)}, p^{(2)}$ to the corresponding columns we get a matrix

$$W' = \begin{bmatrix} 0 & \begin{bmatrix} 1 & 1 \\ 0 & p^{(3)}(B) \end{bmatrix} & 1 & 0 \\ 0 & & 0 & 0 \end{bmatrix}. \tag{9.36}$$

We have $\mathrm{rk}W' = 1 + \mathrm{rk}(p^{(3)}(B))$. From assumption a sums of columns of a matrix A are equal to 1. Hence a sums of columns of a matrix

$$(p^{(2)}, p^{(3)})A = \begin{bmatrix} 1 & 1 \\ 0 & p^{(3)}(B) \end{bmatrix} \tag{9.37}$$

are also equal to 1 and a sums of columns of a matrix $p^{(3)}(B)$ are equal to 0. It means $\mathrm{rk}(p^{(3)}(B)) < m - 1$ and also $\mathrm{rk}(W') < m$. From Lemma 9.31

$$\mathrm{rk}(W') = \mathrm{rk}(W) = \text{number of rows } (W) = m.$$

Hence a matrix W cannot be a matrix of some rows of HW-*matrix*.

We have to still consider a case when matrices $p^{(2)}(A)$ and $p^{(3)}(A)$ are connected. Similar to the above consideration, using relation $\sim_2$ and $\sim_3$ plus conjugation we can assume that

$$[C \ A \ D] = [2 \ A \ 3].$$

Hence all non-empty sums of rows of a matrix A include 1. For $m > 1$ it is impossible. □

The next lemma is an even version of the Lemma 9.33.

Lemma 9.34. *Let $A \in \mathcal{D}^{m \times m}$ be a symmetric matrix with 1 on the diagonal and $\{2, 3\}$ off the diagonal with a column sums equal to 3. Assume that $m > 2$. Then a matrix*

$$K_A = \begin{bmatrix} 2 \\ A \\ 3 \end{bmatrix}, \tag{9.38}$$

cannot be a complement to some HW-matrix.

Proof. As in the proof of the previous lemma let us assume that there exists a *HW-matrix*

$$\begin{bmatrix} * & 2 & * \\ C & A & D \\ * & 3 & * \end{bmatrix}. \tag{9.39}$$

From assumption and Definition 9.26 m is an even number and a height of the block 2 is even and 3 is odd. We shall use induction. For $m = 4$.

1. At the beginning let us consider the case, where $p^{(2)}(A)$ is not connected. We have two cases of matrices of dimension 4:

(a) $A = \begin{bmatrix} 1 & 3 \\ 3 & B \end{bmatrix}$, where B has a dimension 3 and

(b) $A = \begin{bmatrix} B & 3 \\ 3 & E \end{bmatrix}$, where matrices A, B have rank two.

The case (a) is impossible since $1 + 3 \neq 3$. In the case (b) matrices B and E are symmetric with columns sums equal to 3. Hence $B = E = \begin{bmatrix} 1 & 2 \\ 2 & 1 \end{bmatrix}$, and

$$p^{(2)}(A) = \begin{bmatrix} 1 & 1 & 0 & 0 \\ 1 & 1 & 0 & 0 \\ 0 & 0 & 1 & 1 \\ 0 & 0 & 1 & 1 \end{bmatrix}. \tag{9.40}$$

From the other side a matrix $p^{(2)}(K_A)$ has rows of 1 ($p^{(2)}(2) = 1$) and rows of 0 ($p^{(2)}(3) = 0$). These rows are linear combination of rows of $p^{(2)}(A)$ and

$$\mathrm{rk} p^{(2)}(K_A) = \mathrm{rk} p^{(2)}(A) = 2.$$

Finally $\mathrm{d}(K_A) = 4 - 2 = 2 > 1$ and from Lemma 9.30 we are done.

2. As the second step let us consider the case where $p^{(3)}(A)$ is not connected. We have to consider two cases of matrices of dimension 4:

(a) $A = \begin{bmatrix} 1 & 2 \\ 2 & B \end{bmatrix}$, and

(b) $A = \begin{bmatrix} B & 2 \\ 2 & E \end{bmatrix}$, and B and E have dimension 2.

In the case (a) a matrix B is symmetric of dimension 3 with sums of columns 1. If K_A has complement to HW-*matrix* then also a matrix K_B has this possibility. But it is impossible by Lemma 9.33. In case (b) matrices B, E are symmetric with sums of columns 3. Hence $B = E = \left[\begin{smallmatrix} 1 & 2 \\ 2 & 1 \end{smallmatrix}\right]$ and

$$p^{(2)}(A) = \begin{bmatrix} 1 & 1 & 1 & 1 \\ 1 & 1 & 1 & 1 \\ 1 & 1 & 1 & 1 \\ 1 & 1 & 1 & 1 \end{bmatrix}. \tag{9.41}$$

In the matrix $p^{(2)}(K_A)$ we have rows of 1 and 0. They are linearly dependent from the rows of $p^{(2)}(A)$. Hence

$$\mathrm{rk} p^{(2)}(K_A) = \mathrm{rk} p^{(2)}(A) = 1$$

and

$$d(K_A) = 4 - 1 = 3 > 1.$$

From Lemma 9.30 the matrix K_A has not complement to the HW-*matrix*.

3. By the above points 1. and 2. we have that $p^{(2)}(A)$ and $p^{(3)}(A)$ are connected matrices. As in the proof of Lemma 9.33 using relations $\sim_2, \sim_3$ and conjugations of matrices we can assume that

$$[C \ A \ D] = [2 \ A \ 3].$$

By assumption a sum of all rows of the above matrix has 1 on a some position. We can see easily that it is impossible at the first and the third block. For a matrix A it is also impossible, since its columns sums are equal to 3. This contradicts our assumption that $m < n$.

Let us assume that $m > 4$. We shall consider three steps.

1. Assume that $p^{(2)}(A)$ is not connected. We have to consider two cases:

(a) $p^{(2)}(A)$ is a direct sum of two odd blocks,

(b) $p^{(2)}(A)$ is a direct sum of two even blocks.

Hence $A = \left[\begin{smallmatrix} B & 3 \\ 3 & E \end{smallmatrix}\right]$. In the case (a) since dimensions of B, E are odd and sums of column of A are 3 we obtain that sums of column of B and E are 0. Moreover, if B is an odd diagonal submatrix of HW-*matrix* then a sum of rows of B should enclose 1 (see Proposition 9.37). But this is impossible and also case (a) is impossible.

In case (b) since dimensions of B, E are even and sums of column of A are 3 we obtain that sums of column of B and E are 3. Moreover either the matrix B or the matrix E has dimension > 2. Assume the matrix B has such a property. If a matrix K_A has complement, then a matrix K_B has complement to HW-*matrix*. But by induction it is impossible.

2. Assume that $p^{(3)}(A)$ is not connected. We have to consider two cases. The same as in the step 1.

(a) $p^{(3)}(A)$ is a direct sum of two odd blocks,

(b) $p^{(3)}(A)$ is a direct sum of two even blocks.

Hence $A = \begin{bmatrix} B & 2 \\ 2 & E \end{bmatrix}$. In the first case since dimensions of B, E are odd and sums of column of A are 3 we obtain that sums of column of B and E are 1. Moreover, either the matrix B or the matrix E has dimension > 2. Assume the matrix B has such a property. If a matrix K_A has complement then (after permutation of indexes) a matrix K_B has complement to HW-*matrix*. But by Lemma 9.33 it is impossible. In the second case, since dimensions of B, E are even and sums of column of A are 3 we obtain that sums of column of B and E are 3. Moreover, either the matrix B or E has dimension > 2. Assume the matrix B has such a property. If a matrix K_A has complement then a matrix K_B has complement to HW-*matrix*. But by induction it is impossible.

We can assume that matrices $p^{(2)}(A)$ and $p^{(3)}(A)$ are connected. As in the previous cases we can assume that

$$[C \ \ A \ \ D] = [2 \ \ A \ \ 3].$$

By definition 9.26 a sum of all rows should enclose 1. Since m is even and $m < n$ we have a contradiction. □

9.5.3 Proof of the Main Lemma

We keep the notation from previous sections, but we also need new definitions. Denote by $\mathcal{P}_n$ an algebra of all subsets of the set $\{1, 2, \ldots, n\}$. Let $|U|$ denote the number of elements of a set $U \in \mathcal{P}_n$ modulo two. We have an isomorphism of algebras $I \colon \mathbb{F}_2^n \to \mathcal{P}_n$, where

$$I(x) = \{i \mid x_i = 1\}, x \in \mathbb{F}_2^n \tag{9.42}$$

is an indicator.

Definition 9.35. Let A be a *HW*-matrix. The function $J \colon \mathcal{P}_n \to \mathcal{P}_n$ is defined by

$$J(U) = \{s \mid \sum_{i \in U} A_{is} = 1\},\tag{9.43}$$

where $U \in \mathcal{P}_n$.

Remark 9.36. In what follows we shall use a formula *(9.22)* with a basis $b_i, 1 \le i \le n - 1$. Let us consider a map $l \colon \mathcal{P}_n \to \mathbb{F}_2[x_1, \dots, x_{n-1}]$ given by a formula

$$l_Z := \sum_{i \in Z} x_i.\tag{9.44}$$

In this language the formula *(9.22)* for $k = 2, 3$ we can write as

$$\sum_{j=1}^{n-1} p^{(k)} A_{ji} x_j = l_S$$

where $S = \{j \mid p^{(k)}(A_{j,i}) = 1\}$.

Proposition 9.37. *The map J has the following properties:*
1. *$U \neq 0, 1$ then $J(U) \neq 0$, here $0, 1$ denote the trivial additive and multiplicative element of the algebra $\mathcal{P}_n$ respectively;*
2. *$J(U + 1) = J(U)$ where $U + 1 = U'$ denotes a complement of the subset U in the set $\{1, 2, \dots, n\}$;*
3. *$J(\{i\}) = \{i\}, i = 1, 2, \dots, n$;*
4. *if $|U| = 1$ then $J(U) \subset U$;*
5. *if $|U| = 0$ then $J(U) \subset U'$.*

Proof. Elementary calculations with support of the matrix (9.27). $\qquad\square$

Any polynomial of $\mathbb{F}_2[x_1, x_2, \dots, x_n]$ we shall identify with a polynomial map $\mathbb{F}_2^n \to \mathbb{F}_2$. Hence by indicator function (9.42) the formula (9.44) has the following presentation

$$l_Z(e_j) = \begin{cases} 1 & \text{if } j \in Z, \\ 0 & \text{if } j \notin Z, \end{cases}$$

where $Z \in \mathcal{P}_n$. Since the transgressive elements $T_i \in \mathbb{F}_2[x_1, \dots, x_{n-1}]$ we define a split monomorphism of rings $\mathbb{F}_2[x_1, \dots, x_{n-1}] \overset{\phi}{\to} \mathbb{F}_2[x_1, \dots, x_n]$ such that $\bar{T}_i = \phi(T_i) \in \mathbb{F}_2[x_1, \ldots, x_n], i = 1, \ldots, n$. Here, $\phi(x_i) = x_i + x_n, i = 1, 2, \ldots, n-1$. Obviously $\#D = \#\phi(D)$.

From definition, for polynomial functions $\bar{T}_i$ we have $\bar{T}_i(e_j) = \delta_{ij}$, where $1 \leq i, j \leq n$ and $e_i \in (\mathbb{F}_2)^n$ is the standard basis. Hence, by the isomorphism (9.42) a map J (see Definition 9.35) is equivalent to a function $T\colon \mathbb{F}_2^n \to \mathbb{F}_2^n, T(x) = (\bar{T}_1(x), \bar{T}_2(x), \ldots, \bar{T}_n(x))$, where $x \in \mathbb{F}_2^n$. Hence and from an equation (9.24) we have a commutative diagram

$$
\begin{array}{ccc}
\mathbb{F}_2^n & \xrightarrow{\;\;T\;\;} & \mathbb{F}_2^n \\
\Big\downarrow{\scriptstyle I} & & \Big\downarrow{\scriptstyle I} \\
\mathcal{P}_n & \xrightarrow{\;\;J\;\;} & \mathcal{P}_n
\end{array} \tag{9.45}
$$

We shall use these observations in the proof of the **Main Lemma**. Moreover, we shall apply a remark that homogeneous polynomials of degree 2 are recognized by their polynomial functions. Let $S, Z_1, Z_2 \in \mathcal{P}_n$. From definition if

$$\sum_{i \in S} \bar{T}_i = l_{Z_1} \cdot l_{Z_2}$$

then $S = Z_1 \cap Z_2$.

Proposition 9.38. *The following conditions are equivalent.*

(i) $\sum_{i \in S} \bar{T}_i = l_{Z_1} \cdot l_{Z_2}$

(ii) $\forall_{U \in \mathcal{P}_n} |J(U)S| = |UZ_1| \cdot |UZ_2|$.

Proof. We shall use (9.45) and an isomorphism I. Let $x \in \mathbb{F}_2^n, U = I(x)$. We have

$$\sum_{i \in S} \bar{T}_i(x) = \sum_{i \in S \cap I(T(x))} 1 = |I(T(x)) \cap S| = |J(I(x)) \cap S| = |J(U) \cap S|.$$

From the other side

$$l_{Z_1}(x) \cdot l_{Z_2}(x) = \sum_{i \in Z_1 \cap I(x)} 1 \cdot \sum_{i \in Z_2 \cap I(x)} 1 = |UZ_1| \cdot |UZ_2|.$$

This finishes a proof. $\qquad\qquad\qquad\qquad\qquad\qquad\qquad\qquad\qquad\qquad\qquad\square$

Corollary 9.39. *Let us assume the condition* (ii) *of Proposition 9.38, then*

1. $|Z_1|$ or $|Z_2|$ is even,
2. if $S \neq 0$ then $|Z_1|$ and $|Z_2|$ are even
3. if $S \neq Z_1$ and $S \neq Z_2$ then $Z_1 \cup Z_2 = 1$.

Proof. 1. Since $J(1) = J(\{1, 2, \ldots, n\}) = 0$ the condition is true.
2. Since $J(U) = J(U') = J(U+1)$ we have

$$|UZ_1||UZ_2| = |(1+U)Z_1||(1+U)Z_2|.$$

Hence

$$|Z_1||Z_2| + |Z_1||UZ_2| + |Z_2||UZ_1| = 0.$$

From a point 1. we can assume that $|Z_1| = 0$ (or $|Z_2| = 0$) and therefore $|Z_2||UZ_1| = 0$. If $|Z_2| = 1$ then $\forall_{U \in \mathcal{P}_n}|UZ_1| = 0$ and $Z_1 = 0$. Since $S = Z_1 \cdot Z_2 \neq 0$ we have a contradiction.
3. Let $a \in Z_1 \setminus S, b \in Z_2 \setminus S$ and $c \notin Z_1 \cup Z_2$. Put $U = \{a, b, c\}$. By Proposition 9.37 (4.) we have $J(U)S \subset US = 0$ and $UZ_1 = \{a\}, UZ_2 = \{b\}$. Hence

$$0 = |J(U)S| = |UZ_1||UZ_2| = 1 \cdot 1 = 1.$$

This is a contradiction. $\square$

Definition 9.40. Define

$$\sigma_a^S := \sum_{i \in S} A_{a,i},$$

where $a \in \{1, 2, \ldots, n\}, S \subset \{1, 2, \ldots, n\}$ and $A \in \mathcal{D}^{n \times n}$.

Let us present relations between the above definition and the function J.

Proposition 9.41. *Let A be $(n \times n)$ HW-matrix, $a, b \in \{1, 2, \ldots, n\}$ and $S \in \mathcal{P}_n$. Then*
1. $|J(\{a, b\})S| = \sigma_a^S + \sigma_b^S$, *where $a, b \notin S$;*
2. $|J(\{a, b\})S| = \sigma_a^S + \sigma_b^S + A_{a,b} + 1$, *where $a \notin S, b \in S$;*
3. $|J(\{a, b\})S| = \sigma_a^S + \sigma_b^S + A_{a,b} + A_{b,a}$, *where $a, b \in S$.*

Proof. 1. By a point 5. of Proposition 9.37 we know that $J(\{a, b\}) \subset \{a, b\}'$. If $J(\{a, b\})S = \emptyset$ we are done. On the contrary we shall consider the following cases.

(a) Assume $|S| = 1$ and $|J(\{a,b\})S| = 1$. We have two rows, which correspond to a and b,

$$\begin{matrix} 2 & 2 & \ldots & 2 & 2 \\ 2 & 3 & \ldots & 3 & 2 \end{matrix} \tag{9.46}$$

with a number of columns equal to $|S|$, and a number of columns with different coefficients equal to $J(\{a,b\})$. Hence a sum of the upper row is equal to 2 and a sum of the down row is equal to 3. This finishes a proof in this case.

(a') Assume $|S| = 1$ and $|J(\{a,b\})S| = 0$. We also have (9.46) and a sum of the upper row is equal to 2 and a sum of the down row is also equal to 2. This finishes a proof in this case.

(b) Assume $|S| = 0$. Then again we have two subcases $|J(\{a,b\})S| = 1$, then a sum of the upper row of (9.46) is equal to 0 and a sum of the down row is equal to 1. The proof of the case is complete. When $|J(\{a,b\})S| = 0$ a sum of the upper row of (9.46) is 0 and a sum of the down row is also 0. This finished a proof of point 1. The proofs of other cases are similar and we put it as an exercise. □

Using the above language we shall prove that for a *HW-manifold* there exists only a limited number of trangressive elements which are a product of degree one non-trivial polynomials.

Proposition 9.42. *Let A be a $(n \times n)$ HW-matrix, $(n > 3)$ then there does not exist not empty set $S \subset \{1, 2, \ldots, n\}$ such that*

$$\forall_{U \in \mathcal{P}_n} |J(U)S| = |US|. \tag{9.47}$$

Proof. It is the case $S = Z_1 = Z_2$. Let us assume (9.47). We are going to divide the proof into four steps.

Step 1. We claim that, if $a_1, a_2 \notin S$ and $b \in S$ then $A_{a_1,b} = A_{a_2,b}$. In fact, from (9.47) for $U = \{a_1, a_2\}$, $|J(\{a_1, a_2\})S| = |\{a_1, a_2\}S| = 0$. By Proposition 9.41 (1.), $\sigma := \sigma_{a_1}^S = \sigma_{a_2}^S$. If $a \notin S$ then from Proposition 9.41 (2.)

$$1 = |\{a,b\}S| = |J(\{a,b\})S| = \sigma_a^S + \sigma_b^S + A_{a,b} + 1 = \sigma + \sigma_b^S + A_{a,b} + 1.$$

Hence $\forall a \notin S, A_{a,b} = \sigma + \sigma_b^S$ and *Step 1.* is proved.

Step 2. We claim that, if $US = 0$ then $J(U)S = 0$. In fact from *Step 1.*

$$\forall_{a_1,a_2 \in U} \forall_{b \in S} A_{a_1,b} = A_{a_2,b},$$

hence

$$\forall_{b \in S} \sum_{a \in U} A_{a,b} \in \{0, 2, 3\}.$$

Then $J(U)S = 0$.

Step 3. We claim that, if $S \neq 0$ then $\#S = n - 1$. From *Step 2.* if $0 \neq U \subset S'$ then $J(U)S' \neq 0$. Let B be a diagonal submatrix of the matrix A related to the set S'. Then B is a square matrix with 1 on the diagonal and $2, 3$ elsewhere. Moreover all sums of rows of B have at some position an element 1. Hence, the only possible matrix B is (1×1) matrix.

Step 4. We claim that, if $S \neq 0$ then $n \leq 3$. For the proof, let us assume that $n > 3$. From the *Step 3.* we can assume that $S = \{2, 3, \ldots, n\}$. Let l_2 denote a number of 2 at the first column of A. We shall prove that $|l_2| = 0$. In fact, we can assume that $0 < l_2 < n - 1$ and at the first column, from the top we have first 2 then going down we have 3. On the contrary, suppose that l_2 is odd and let v be a sum of the first $l_2 + 1$ rows. Since $l_2 + 1$ is even v has not 1 on places $1, 2, \ldots, l_2 + 1$. Then it has 1 on the position $r > l_2 + 1$. Hence there exists $k \geq l_2 + 1$ such that $A_{1,r} \neq A_{k,r}$ or equivalently $A_{1,r} + A_{k,r} = 1$. Let us consider a diagonal submatrix

$$\begin{bmatrix} 1 & * & A_{1,r} \\ 2 & 1 & A_{k,r} \\ 3 & * & 1 \end{bmatrix}. \tag{9.48}$$

A sum of elements at the first column and at the third column is 0, then it at the second column has to be $\neq 0$. Let $U = \{1, k, r\}$. Since $j(U) \subset U$ and $n > 3$, $J(U) = \{k\}$. Finally

$$1 = |\{k\}S| = |J(U)S| = |US| = |\{k, r\}S| = 0.$$

That is a contradiction and l_2 is even. Moreover if l_3 is a number of 3 at the first column then $|l_3 = n - 1 - l_2| = 0$ and a sum $1 + l_2 * 2 + l_3 * 2 = 1$. But a sum of all rows is zero and we have a contradiction. This finishes a proof. □

Corollary 9.43. *In the space* $\mathrm{Im}(d_2)$ *we have not squares.*

Proof. If $l_Z^2 \in \mathrm{Im}(d_2)$, then we have $S = Z_1 = Z_2 = Z$ in Proposition 9.38. For $n > 3$ it is impossible. □

Proposition 9.44. *Let $S, Z \subset \{1, 2, \ldots, n\}$ such that $0 \neq S \neq Z$. Let A, J be as in Proposition 9.42. Assume that*

$$\forall_{U \in \mathcal{P}_n} |J(U)S| = |US| \cdot |UZ|$$

then $\#S = 2, |Z| = 0$ and $S \subset Z$.

Proof. At the beginning we claim that up to permutation and conjugation,

$$A = \begin{bmatrix} * & 2 & * \\ * & B & * \\ * & 3 & * \end{bmatrix}, \tag{9.49}$$

where B is a symmetric matrix with a column sums 3. Moreover a block 2 has rows indexed by numbers from the set $Z \setminus S$ and a block 3 has rows indexed by numbers from the set $1 + Z = Z'$. In fact, from Proposition 9.38, $S \subset Z$ and Corollary 9.39, $S \subset Z$ and $|S| = |Z| = 0$. Let us change the indexes of A such that

$$A = \begin{bmatrix} * & E & * \\ * & B & * \\ * & F & * \end{bmatrix}, \tag{9.50}$$

and E has rows indexed by numbers from the set $Z \setminus S$, B has rows indexed by numbers from S and F is indexed by $1 + Z = Z'$. From the point 1 of Proposition 9.41, for $a, b \notin S$

$$\sigma_a^S + \sigma_b^S = |J(\{a, b\})S| = |\{a, b\}S| \cdot |\{a, b\}Z| = 0.$$

Hence $\sigma_a^S = \sigma_b^S$. Let $\sigma := \sigma_a^S$, for $a \notin S$.

By the point 2 of Proposition 9.41 for $b \in S$ and $a \notin Z$,

$$A_{a,b} = \sigma + \sigma_b^S. \tag{9.51}$$

From the above all columns of the matrix F are constant. Again from the point 2 of Proposition 9.41 for $b \in S, a \in Z \setminus S$,

$$A_{a,b} = \sigma + \sigma_b^S + 1. \tag{9.52}$$

It follows that also columns of the matrix E are constant. Let us conjugate columns of the matrix A such that $E = 2$. In that case $\sigma = 0$ since for $a \in Z \setminus S$ we have $\sigma = \sigma_a^S = |S| \cdot 2 = 0$.

From (9.52), for $b \in S, 2 = 0 + \sigma_b^S + 1$. Hence $\sigma_b^S = 3$ and $F = 3$, because from the formula (9.51) $A_{a,b} = 0 + 3$, for $a \in Z'$ and $b \in S$. Finally, from Proposition 9.41 for $a, b \in S$ we have

$$A_{a,b} + A_{b,a} = 3 + 3 + A_{a,b} + A_{b,a} = \sigma_a^S + \sigma_b^S + A_{a,b} + A_{b,a}$$

$$= |J(\{a,b\})S| = |\{a,b\}S| \cdot |\{a,b\}Z| = 0. \tag{9.53}$$

To finish a proof it suffices to apply Lemma 9.34. $\qquad\square$

Proposition 9.45. *We keep the notation from the previous propositions. Let us assume* $S, Z_1, Z_2 \in \mathcal{P}_n$ *such that* $0 \neq S, S \neq Z_1, S \neq Z_2$ *and*

$$\forall_{U \in \mathcal{P}_n} |J(U)S| = |UZ_1| \cdot |UZ_2|$$

then $\#S = 1, |Z_1| = |Z_2| = 0$ *and* $Z_1 + Z_2 = 1$.

Proof. A proof is similar to the proof of Proposition 9.44. At the beginning we show that (up to permutation and conjugation)

$$A = \begin{bmatrix} * & 2 & * \\ * & B & * \\ * & 3 & * \end{bmatrix}, \tag{9.54}$$

where B is a symmetric matrix of odd dimension with sums of columns 1, a block 2 is indexed by the set $Z_1 \setminus S$ and a block 3 is indexed by the set $Z_2 \setminus S$. In fact, from assumption and Corollary 9.39, $S = Z_1 Z_2, |Z_1| = |Z_2| = 0$ and $Z_1 + Z_2 = 1$. Hence $|S| = 1$. Let us change the order of rows in the matrix A such that

$$A = \begin{bmatrix} * & E & * \\ * & B & * \\ * & F & * \end{bmatrix} \tag{9.55}$$

and E is indexed by $Z_1 \setminus S$, B by S and F by $Z_2 \setminus S$. From Proposition 9.41 we have $\forall a, b \in Z_1 \setminus S, \sigma_a^S = \sigma_b^S := \sigma_E$. With similar consideration we have $\forall a, b \in Z_2 \setminus S, \sigma_a^S = \sigma_b^S := \sigma_F$. Moreover, by Proposition 9.41 (2) for $b \in S$ and $a \in Z_1 \setminus S$,

$$A_{a,b} = \sigma_E + \sigma_b^S + 1. \tag{9.56}$$

From the above, all columns of the matrix E are the same. By analogy for $b \in S$ and $a \in Z_2 \setminus S$,

$$A_{a,b} = \sigma_F + \sigma_b^S + 1 \tag{9.57}$$

and columns of the matrix F are also constant. Let us conjugate columns of A such that $E = 2$. Then $\sigma_E = 2$, because for $a \in Z_1 \setminus S, \sigma_E = \sigma_a^S = |S| \cdot 2 = 2$ and for $b \in S, \sigma_b^S = 1$. The last equality follows from (9.56) because $2 = 2 + \sigma_b^S + 1$. Similarly, by (9.57) for $b \in S$ and $a \in Z_2 \setminus S$, we have $A_{a,b} = \sigma_F + 1 + 1 = \sigma_F$ and the matrix F is constant and equal to σ_F. Finally, a matrix B is symmetric since from Proposition 9.41

$$\sigma_a^S + \sigma_b^S + A_{a,b} + A_{b,a} = |J(\{a,b\})S|$$

which means,
$$A_{a,b} + A_{b,a} = |\{a,b\}Z_1| \cdot |\{a,b\}Z_2| = 0.$$
We have still to show that $\sigma_F = 3$. In fact from assumption columns' sums of B are 1. Since B is symmetric this same is true for rows. Let us calculate a sum of some column of A :
$$(|Z_1| - |S|)2 + 1 + (|Z_2| - |S|)\sigma_F = 2 + 1 + \sigma_F = 3 + \sigma_F = 0.$$
To finish a proof of Proposition we have to apply Lemma 9.33.   □

Summing up we have the following two possibilities:

I. $\#S = 1$ and $Z_1 + Z_2 = 1$;

II. $\#S = 2$ and $S = Z_1, S \neq Z_2$ or $S = Z_2, S \neq Z_1$.

Let us recall that $\mathrm{Im}(d_2)$ is a n-dimensional $\mathbb{Z}_2$-space generated by $T_i, i = 1, 2, 3, \ldots, n$. We are interested in description of the set D of elements in $\mathrm{Im}\, d_2$ which are a product of two non-trivial linear polynomials, see (9.25). We claim that $D \leq n + 1$. In what follows, if it does not give a contradiction we shall write T_i for $\bar{T}_i, i = 1, \ldots, n$.

Lemma 9.46. *Let $w \in D$, then $w = T_i$ or $w = T_j + T_k$ for some $1 \leq i, j, k \leq n$.*

Proof. In the beginning we shall prove that $T_i + T_j$ is a product of two non-trivial linear polynomials if and only if T_i, T_j have a common component. It means there exists p $\neq 0$ s.t. p$|T_i$ and p$|T_j$. If $T_i + T_j \in D$, then from the above case II and proof of Proposition 9.44 we can assume that $j = i + 1$ and the matrix A enclose:

$$\begin{bmatrix} \vdots & \vdots \\ 2 & 2 \\ 1 & 2 \\ 2 & 1 \\ 3 & 3 \\ \vdots & \vdots \end{bmatrix}.$$

By definition
$$T_i = (x_1 + \cdots + x_i + x_{i+1})(x_i + x_{i+2} + \cdots + x_n)$$
$$T_{i+1} = (x_1 + \cdots + x_i + x_{i+1})(x_{i+1} + \cdots + x_n).$$
For the proof of the opposite conclusion we shall need:

Definition 9.47. Let X be a subset of some monoid. By Γ_X we define a graph with the vertex set X and two vertices a, b are connected by an edge $a \xrightarrow{\ f\ } b$ if and only if $f|a$ and $f|b$. Put $\Gamma := \Gamma_{T_1, T_2, \ldots, T_n}$.

We claim that for $n > 3$ the graph

$$i \xrightarrow{\ f\ } j \xrightarrow{\ g\ } k \tag{9.58}$$

is not a subgraph of Γ, where $i := T_i, i = 1, 2, \ldots, n$. In fact we have two possibilities:

1. $f = g$. Let $i = 1, j = 2, k = 3$ and let $\mathfrak{J}$ be an ideal generated by $(f, T_4, \ldots, T_n)$ in the polynomial ring. Since there exist a non-trivial solution of system of $(n-2)$ linear equation in $(n-1)$ linear space, an algebraic set $V(\mathfrak{J})$ is not trivial. It means $0 \neq x \in V(\mathfrak{J})$. From definition $x \in V(\mathfrak{J}')$, where $\mathfrak{J}'$ is an ideal generated by $(T_1, T_2, \ldots, T_n)$. Hence, up to conjugation and permutation, A contains a submatrix of the form

$$\begin{bmatrix} \vdots & \vdots & \vdots \\ 2 & 2 & 2 \\ 1 & 2 & 2 \\ 2 & 1 & 2 \\ 2 & 2 & 1 \\ 3 & 3 & 3 \\ \vdots & \vdots & \vdots \end{bmatrix}.$$

By Lemma 9.33 this is impossible.

2. $f \neq g$. Using permutation of indexes and conjugation we can assume that in HW-*matrix* $A, j = i+1, k = i+2$. Recall that $S = \{i, i+1\}$ and A is as in Proposition 9.44. Hence it has a diagonal block related to rows (columns) $\{i, i+1, i+2\}$

$$\begin{bmatrix} 1 & 2 & b \\ 2 & 1 & a \\ 3 & 3 & 1 \end{bmatrix}, \tag{9.59}$$

and a matrix A has upper two first columns of (9.59) only elements 2, but lower only elements 3. Let us consider polynomials T_i, T_{i+1} and T_{i+2} for $x_s = 0, s \notin \{i, i+1, i+2\}$ and denote it by $\hat{T}_i$ respectively. We have

$$\hat{T}_i = (x_i + x_{i+1})(x_i + x_{i+2})$$

and

$$\hat{T}_{i+1} = (x_i + x_{i+1})(x_{i+1} + x_{i+2}).$$

The both polynomials are divisible by $(x_i + x_{i+1})$. Hence $\hat{T}_{i+1}$ and $\hat{T}_{i+2}$ are divided by $(x_{i+1} + x_{i+2})$. From the above we can observe that

$$\hat{T}_{i+2} = (x_{i+1} + x_{i+2})(x_{i+2} + x_i). \tag{9.60}$$

By (9.60) and definition we get $a \neq b$. Hence a sum of all columns of the matrix (9.59) are equal to 0. By Proposition 9.37 and definition of HW-matrices, it is impossible. This finishes a proof of our claim.   □

We have:

Corollary 9.48. *For $n > 3$ all connected components of a graph Γ are points or edges $i \underline{\quad f \quad} j$.*

□

Corollary 9.49. *For $n > 3, D = \{T_1, T_2, ..., T_n\}$ or $\mathrm{D} = \{T_1, T_2, ..., T_n, T_i + T_j\}$ for some $1 \le i, j \le n$.*

Proof. Conversely, suppose that edges

$$1 \underline{\quad f \quad} 2 \text{ and } 3 \underline{\quad g \quad} 4$$

are components of the graph Γ. Let us consider an ideal $\mathfrak{J} = (f, g, T_5, \ldots, T_n)$ in polynomial ring. Since there exist a non-trivial solution of system of $(n-2)$ linear equation in $(n-1)$ linear space an algebraic set $\mathrm{V}(\mathfrak{J})$ is not trivial. It means $0 \neq x \in \mathrm{V}(\mathfrak{J})$. But from definition $x \in \mathrm{V}(\mathfrak{J}')$, where $\mathfrak{J}'$ is an ideal generated by $(T_1, T_2, \ldots, T_n)$. But it is impossible. This finishes a proof.   □

Let us prove the **Main Lemma**.

Main Lemma *Let $n > 3$, then there are the following possibilities for the structure of the set D:*

1. $D = \{T_1, T_2, \ldots, T_n\}$;

2. $D = \{T_1, T_2, \ldots, T_n, T_i + T_j\}$, and we can find a polynomial p of degree one such that $\mathrm{p} \mid T_i$ and $\mathrm{p} \mid T_j$ for some $1 \le i, j \le n$. In the second case we can rediscover the set of generators $T_1, T_2, \ldots, T_n$.

Proof. We start from the simple observation. If $i \neq j$ and $T_i = \mathrm{p} \cdot \mathrm{q}, T_j = \mathrm{p} \cdot \mathrm{r}$ then $\forall k = 1, 2, \ldots, n$ $\mathrm{q} + \mathrm{r}$ does not divide T_k. In fact $\mathrm{q} + \mathrm{r} \neq \mathrm{p}$ since in other case $T_i + T_j = \mathrm{p}(\mathrm{q} + \mathrm{r}) = \mathrm{p}^2$. By Corollary 9.43 it is impossible. Hence T_i and T_j are also not divisible by $\mathrm{q} + \mathrm{r}$. Moreover, if $T_k = (\mathrm{q} + \mathrm{r})\mathrm{s}$ then $T_i + T_j + T_k = (\mathrm{q} + \mathrm{r})(\mathrm{p} + \mathrm{s})$. By Proposition 9.44, a decomposition for $\#S = 3$ is impossible. Let us prove the second point of the above lemma. From definition the graph Γ_D has connected components which are vertices T_k for $k \notin \{i, j\}$, of the triangle with vertices $T_i, T_j, T_i + T_j$ and a constant label which is a component of T_i and T_j. Let $T_i = \mathrm{p} \cdot \mathrm{q}$ and $T_j = \mathrm{p} \cdot \mathrm{r}$ then $T_i + T_j = \mathrm{p}(\mathrm{q} + \mathrm{r})$. The triangle is a connected component of a graph, by (9.58) for $k \notin \{i, j\}$ elements $\mathrm{p}, \mathrm{q}, \mathrm{r}$ do not divide T_k. Also from the above simple observation, the element $(\mathrm{q} + \mathrm{r})$ does not divide T_k.

We continue the proof of the **Main Lemma**. Let $w = \xi\eta$ where ξ and η are linear polynomials. Let us define $s(w) := \xi + \eta$. Since HW-*manifolds* are oriented $\sum_i s(T_i) = 0$. We claim that if $T_i + T_j \in D$, then $s(\xi) + s(\eta)$ recognizes subsets of order two of the set $\{T_i, T_j, T_i + T_j\}$. In fact, let $T_i = \mathrm{p} \cdot \mathrm{q}, T_j = \mathrm{p} \cdot \mathrm{r}$, then $T_i + T_j = \mathrm{p}(\mathrm{q} + \mathrm{r})$ and $s(T_i) + s(T_j) = \mathrm{q} + \mathrm{r}, s(T_i) + s(T_i + T_j) = \mathrm{r}, s(T_j) + s(T_i + T_j) = \mathrm{q}$.

Let $n > 3$, then there are the following possibilities for the structure of the set D:

1. $D = \{T_1, T_2, \ldots, T_n\}$;

2. $D = \{T_1, T_2, \ldots, T_n, T_i + T_j\}$, for some $1 \leq i, j \leq n$.

Let $n > 3$ if D has n elements we are done. If it has $(n + 1)$ elements then the graph Γ_D has $(n - 2)$ discrete connected components D^c and a triangle. We proceed in two steps:

1. Put $s_{D^c} := \sum_{a \in D^c} s(a)$

2. From the triangle we take a unique pair ξ, η such that

$$s(\xi) + s(\eta) + s_{D^c} = 0.$$

Hence $\{T_1, T_2, \ldots, T_n\} = \{\xi, \eta\} \cup D$. This finishes a proof of the **Main Lemma**. $\qquad\square$

For illustration of possibilities of the structure of the set D we present two examples.

Example 9.50. Let $G \subset \mathcal{D}^5$ correspond to HW-matrix $\begin{bmatrix} 1 & 2 & 2 & 2 & 2 \\ 2 & 1 & 3 & 2 & 2 \\ 3 & 2 & 1 & 3 & 2 \\ 3 & 2 & 3 & 1 & 3 \\ 3 & 3 & 3 & 2 & 1 \end{bmatrix}$.

The set

$$D = \{T_1 = (x_1 + x_2)(x_1 + x_3 + x_4), T_2 = (x_1 + x_2 + x_3 + x_4)x_2,$$
$$T_3 = (x_1 + x_3)(x_2 + x_3 + x_4), T_4 = (x_1 + x_2 + x_4)(x_3 + x_4), \quad (9.61)$$
$$T_5 = (x_1 + x_2 + x_3)x_4\}.$$

From Remark 9.28 the above group is isomorphic to the group Γ_1 of the Example 9.25. The next example illustrates the second case of the **Main Lemma**.

Example 9.51. Let a matrix $\begin{bmatrix} 1 & 2 & 2 & 2 & 2 \\ 2 & 1 & 3 & 2 & 2 \\ 2 & 2 & 1 & 3 & 2 \\ 2 & 2 & 2 & 1 & 3 \\ 3 & 3 & 2 & 2 & 1 \end{bmatrix} \in \mathcal{D}^{5 \times 5}$ be the second HW-matrix of dimension 5.

In this cases we have

$$D = \{T_1 = (x_1 + x_2 + x_3 + x_4)x_1, T_2 = (x_1 + x_2 + x_3 + x_4)x_2,$$
$$T_3 = (x_1 + x_3 + x_4)(x_2 + x_3), T_4 = (x_1 + x_2 + x_4)(x_3 + x_4), \quad (9.62)$$
$$T_5 = (x_1 + x_2 + x_3)x_4, T_1 + T_2\}.$$

9.6   Relation with Fibonacci groups

Let $F(r, n)$ be the group defined by the presentation

$$\langle a_1, a_2, ..., a_n \mid a_1 a_2 ... a_r = a_{r+1}, a_2 a_3 ... a_{r+1} = a_{r+2}, ...,$$
$$a_{n-1} a_n a_1 ... a_{r-2} = a_{r-1}, a_n a_1 a_2 ... a_{r-1} = a_r \rangle, \quad (9.63)$$

where $r > 0, n > 0$, and all subscripts are assumed to be reduced modulo n. This group is named a Fibonacci group. They were introduced in special case, by J. Conway. For more about it see [74] and [75].

Let $n \geq 3$ be an odd natural number and let $\Gamma_n \subset E(n)$ be a group generated by

$$\{(B_i, s(B_i) = x_i) : 1 \leq i \leq n - 1\},$$

where B_i are diagonal matrices

$$B_i := \mathrm{diag}(-1, \ldots, -1, \underbrace{1}_{i}, -1, \ldots, -1),$$

and $x_i = s(B_i) = e_i/2 + e_{i+1}/2, 1 \le i \le n-1$. They were considered in Examples 5.22 and 9.13.

Now, we are ready to formulate a result about relations between GHW groups and Fibonacci groups.

Theorem 9.52 ([141]). *For all odd natural numbers $n = 2k+1, k \in \mathbb{N}$, there are surjective group homomorphisms*

$$\Phi_n \colon F(n-1, 2n) \to \Gamma_n.$$

For $n = 3, \Phi_3$ is an isomorphism but for $n > 3$ the kernels of Φ_n are non-trivial.

Proof. We shall prove that there are non-trivial $(2n) = (2(2k+1))$-elements $x_1, x_2, ..., x_{4k}, x_{4k+1}, x_{4k+2}$ in the group Γ_n which are generators and satisfy the relations of the Fibonacci group $F(n-1, 2n) = F(2k, 4k+2)$. Let

$$x_i = (B_i, (0, ..., 0, 1/2, 1/2, 0, ..., 0)), 1 \le i \le 2k,$$

where $1/2$ is on the i and $(i+1)$ site. We define:

$$x_{j+1} = x_{(j+1)-2n} x_{((j+1)-2n)+1} ... x_{j-1} x_j, \qquad 2k \le j \le 2n-1.$$

We start with:

Lemma 9.53. *Let $i \ge 1$, and $i + l \le 2k$ then*

$$x_i x_{i+1} ... x_{i+l} = (B_i B_{i+1} ... B_{i+l}, (0, ..., 0, 1/2, 0, ..., 0, (-1)^l 1/2, 0, ..., 0)),$$

where the first $1/2$ is on the i site and second with given sign on the $(i+l+1)$ site.

Proof of Lemma 9.53. Assume $i = 1$ and use induction on l. For other i the lemma follows from the case $i = 1$ and property of the elements $x_i, (i \le 2k)$. □

Since the above lemma and immediate calculation we have:

$$x_{2k+2} = x_{(2k+1)+1} = (B_1, (-1/2, 1/2, 0, ..., 0, 1)).$$

The next lemma gives us the method of calculation of the products of x_i for $2k+1 \le i \le 4k+1$.

Lemma 9.54. *For* $2 \leq j \leq 2k$

$$x_{(2k+1)+j} = (B_j, (0, ..., 0, 1, -1/2, 1/2, 0, ..., 0)), \qquad (9.64)$$

where 1 *is on the* $(j-1)$ *site,*
and for given $2 \leq i \leq 2k - 1$ *and* $1 \leq l \leq 2k - i$

$$x_{(2k+1)+i}x_{(2k+1)+i+1}...x_{(2k+1)+i+l} =$$

$$(B_i B_{i+1}...B_{i+l}, (0, ..., 0, 1, 1/2, 0, ..., 0, (-1)^{l+1}1, (-1)^l 1/2, 0, ..., 0)),$$

where the first 1 *is on the* $(i-1)$ *site, and second with given sign on the* $(i+l)$ *site.*

Proof of Lemma 9.54. First we prove the lemma for $i = 2$. By straightforward calculation and Lemma 9.53 we can see that formula (9.64) is true for $j = 2, 3$. We shall prove by induction on l that the following formulas are satisfied for $1 \leq l \leq 2k - 2$:

(1) $\qquad x_{(2k+1)+2+l} = (B_{2+l}, (0, ..., 0, 1, -1/2, 1/2, 0, ..., 0))$

where 1 is on the $(l+1)$ site, and

(2) $\qquad x_{(2k+1)+2}x_{(2k+1)+2+1}...x_{(2k+1)+2+l} =$

$$(B_2...B_{2+l}, (1, 1/2, 0, ..., 0, (-1)^{l+1}1, (-1)^l 1/2, 0, ..., 0)),$$

where $(-1)^{l+1}1$ is on $(l+2)$ site. It is easy to see that the above formulas are true for $l = 1$. Let us assume that they are true for $l = t$. We must prove that the above formulas are true for $l = t + 1$. We have:

$$x_{(2k+1)+t+3} = x_{t+4}x_{t+5}...x_{2k}x_{2k+1}x_{(2k+1)+1}x_{(2k+1)+2}...x_{(2k+1)+t+2}.$$

We shall consider two cases: $t + 4 < 2k + 1$ and $t + 4 = 2k + 1$. In the first case we have from Lemma 9.53:

$$x_{t+4}...x_{2k} = (B_{t+4}...B_{2k}, (0, ..., 0, 1/2, 0, ..., 0, (-1)^t 1/2)),$$

where $1/2$ is on the $(t+4)$ site.
Moreover,

$$x_{2k+1}x_{2k+2} = (B_{2k+1}B_1, (1, -1/2, 0, ..., 0, 1/2)),$$

and by the induction assumption

$$x_{(2k+1)+2}\cdots x_{(2k+1)+t+2} =$$

$$(B_2...B_{t+2}, (1, 1/2, 0, ..., 0, (-1)^{t+1}1, (-1)^t 1/2, 0, ..., 0)),$$

where $(-1)^{t+1}1$ is on $(t+2)$ site. Hence

$$x_{(2k+1)+t+3} = (B_{t+3}, (0, ..., 0, 1, -1/2, 1/2, 0, ..., 0)),$$

where 1 is on the $(t+2)$ site.

In the second case $(t + 4 = 2k + 1)$ we have from the above, and by the induction assumption

$$x_{(2k+1)+2k} = (B_{2k}, (0, ..., 0, 1, -1/2, 1/2)).$$

This proves formula (1) and (9.64). Equation (2) follows now by induction and from (1).

For $i > 2$ the second formula follows from the formula $(**)$ and the definition of x_i, $(1 \le i \le 4k + 1)$.   □

Now we have to prove that

$$x_{4k+2} = (B_{4k+1}, (1/2, 0, ..., 0, 1, 1/2)).$$

In fact,

$$x_{4k+2} = x_{2k+2}x_{2k+3}...x_{4k+1} =$$

$$(B_1, (-1/2, 1/2, 0, ..., 0, 1))(B_2 B_3...B_{2k}, (1/2, 0, ..., 0, 1, 1/2)) =$$

$$(B_{2k+1}, (1/2, 0, ..., 0, 1, 1/2)).$$

We now go on to prove that for $2 \le j \le 2k + 1 = n$

(3) $$x_{(2k+1)+j}x_{(2k+1)+j+1}...x_{4k+2}x_1x_2...x_{j-2} = x_{j-1}.$$

Here x_0 denotes identity.

Let $j = 2$. In this case the above equation is the consequence of Lemma 9.54. Let us now consider the case $2 < j \le 2k$. By straightforward calculation, with support of Lemmas 9.53 and 9.54, one can see that formula (3) is satisfied. Finally, if $j = 2k + 1$, equation (3) follows from the formula for x_{4k+2} and Lemma 9.53.

Let us define

$$\Phi_n(a_i) = x_i,$$

where a_i, $(1 \leq i \leq 4k + 2)$ are the generators of the Fibonacci group. From the definition we have that, for all odd natural numbers $n \in \mathbb{N}$

$$\Phi_n \colon F(n - 1, 2n) \to \Gamma_n$$

are surjective homomorphisms. For the proof that Φ_{2n} are not isomorphisms for $(n > 1)$ we shall use Theorem 9.18 from the previous section. Hence, it is sufficient to show that abelianization of $F(2n, 2(2n + 1))$ has exponent > 2. But it is clear from the definition of Fibonacci groups which gives us for any n the following epimorphism (see [74, page 191])

$$f \colon F(2n, 2(2n + 1)) \to F(2n, 1) \cong \mathbb{Z}_{2n-1}.$$

Now, we prove that Φ_3 is an isomorphism. In fact, it is known (see [57]) that $F(2, 6) \cong \Gamma_3$. Hence it is sufficient to observe that any epimorphism $f \colon \Gamma_3 \to \Gamma_3$ is an isomorphism.

This completes the proof of Theorem 9.52.   □

9.7   An invariant of GHW

We begin with an example.

Example 9.55 ([63, p. 7]). Let us consider a short exact sequence of groups:

$$0 \to D \to F(2, 6) = \Gamma_3 \overset{\nu_2}{\to} F(2, 3) = Q_8 \to 0.$$

Here the map $\nu_2 \colon F(2, 6) \to F(2, 3)$ is the well known map from the theory of the Fibonacci groups (see [75, page 581]):

$$\nu_{2n} \colon F(2n, 2(2n + 1)) \to F(2n, 2n + 1)$$

defined by the formula $\nu_{2n}(a_i) = \nu_{2n}(a_{i+2n+1}) = a_i$ for $1 \leq i \leq 2n + 1$.

Let $D \subset \mathbb{Z}^n$ be generated by the following elements:

$$2e_1, e_1 - e_2, e_2 - e_3, ..., e_{n-1} - e_n,$$

where $e_i, i = 1, 2, ..., n$ is a standard basis of $\mathbb{Z}^n$. It is easy to see that D is a subgroup of index two in $\mathbb{Z}^n$. Moreover, for $n = 3, D$ is exactly from the Example 9.55.

This suggests the following:

Proposition 9.56. *Let $A \subset \mathbb{Z}^n$ be a subgroup of index two of the maximal abelian subgroup of the GHW group Γ. Then A is a normal subgroup of Γ, and $F = \Gamma/A$ is the central extension of $\mathbb{Z}_2 = \mathbb{Z}^n/A$ by $\mathbb{Z}_2^{n-1}$.*

Proof. Assume $\Gamma \subset E(n) = O(n) \ltimes \mathbb{R}^n$. For any $a \in A$ and $(G, g) \in \Gamma$ we have

$$(G, g)(I, a)(G, g)^{-1} = (G, G(a) + g)(G^{-1}, -G^{-1}(g))$$
$$= (I, -g + G(a) + g) = (I, G(a)).$$

By definition $2(\mathbb{Z}^n) \subset A$. Moreover, since G is the diagonal matrix, then $(G - I)x \in 2(\mathbb{Z}^n)$, for any $x \in \mathbb{Z}^n$. Hence $G(a) \in A$ and the Proposition is proved. $\square$

Now, we show how to associate a bilinear quadratic form over $\mathbb{Z}_2$ with any GHW group.

Theorem 9.57. *To any generalized Hantzsche-Wendt group Γ of dimension $n \geq 3$ and any subgroup A of index two of the maximal abelian subgroup $\mathbb{Z}^n$ we can associate a quadratic function*

$$Q_A^{\Gamma}: (\mathbb{Z}_2)^{n-1} \to \mathbb{Z}_2$$

and its associated alternating, bilinear quadratic form

$$B_A^{\Gamma}: (\mathbb{Z}_2)^{n-1} \times (\mathbb{Z}_2)^{n-1} \to \mathbb{Z}_2.$$

Moreover, if $n > 3$, and Γ is oriented, then $B_A^{\Gamma} \not\equiv 0$ for any A.

Proof. We shall define the quadratic form Q_A^{Γ} and its associated alternating, bilinear quadratic form B_A^{Γ} with the help of a group $\Gamma \subset E(n) = O(n) \ltimes \mathbb{R}^n$. Let $(X, x), (Y, y) \in \Gamma$, be mapped by p to $X, Y \in V = (\mathbb{Z}_2)^{n-1}$. We have

$$(X, x)(Y, y)(X, x)^{-1}(Y, y)^{-1} = (I, (X - I)y - (Y - I)x).$$

Define $B_A^\Gamma(X, Y)$ to be the image of $(X - I)y - (Y - I)x$ in $\mathbb{Z}_2 = \mathbb{Z}^n/A$. We have $(X, x)^2 = (I, (X + I)x)$; let $Q_A^\Gamma(X)$ be the image of $(X + I)x$ in $\mathbb{Z}_2 = \mathbb{Z}^n/A$. It is easy to see that B_A^Γ and Q_A^Γ are well defined. It means that they do not depend on the choice of an element $(X, x) \in \Gamma$. The bilinear form B_A^Γ is alternating. In fact,

$$B_A^\Gamma(X, X) = (X - I)(x + a) - (X - I)x = (X - I)a \in A,$$

where $(X, x), (X, x + a) \in p^{-1}(X)$ and $a \in \mathbb{Z}^n$.[2] We still have to prove the last assertion. But this follows because the commutator subgroup of Γ is equal to the translation subgroup, see [119, Theorem 3.1]. $\square$

Remark 9.58.

1. For different translation subgroups A_1, A_2 it can happen that $B_{A_1}^\Gamma = B_{A_2}^\Gamma$. Moreover, for different Γ_1 and Γ_2, $B_A^{\Gamma_1}$ can be equal to $B_A^{\Gamma_2}$.

2. The function Q_A^Γ and the form B_A^Γ correspond to the following short exact sequence of finite groups

$$0 \to \mathbb{Z}_2 = \mathbb{Z}^n/A \to F \to (\mathbb{Z}_2)^{n-1} \to 0.$$

Let us recall the basic facts about the quadratic forms over $\mathbb{Z}_2$. We shall follow [15] and [80]. Let V be a finite dimensional vector space over a finite field $\mathbb{Z}_2$ and $Q: V \to \mathbb{Z}_2$ be a function. We shall call Q a quadratic form if $Q(0) = 0$ and

$$B(x, y) = Q(x + y) + Q(x) + Q(y) \tag{9.65}$$

is bilinear, where $x, y \in V$.

Example 9.59. Let V be the 2-dimensional vector space with basis a, b and $B(a, b) = 1, B(a, a) = B(b, b) = 0$. There are two quadratic forms $Q_1: V \to \mathbb{Z}_2, Q_2: V \to \mathbb{Z}_2$ compatible with B, with $Q_1(a) = Q_1(b) = 1$ and $Q_2(a) = Q_2(b) = 0$. Note that $Q_1(a + b) = Q_2(a + b)$.

For a bilinear form B, satisfying (9.65) we define the radical

$$R = \{x \in V \mid B(x, y) = 0 \ \forall y \in V\},$$

and its subspace $R_1 = \{x \in R \mid Q(x) = 0\}$. Observe that B induces a non-degenerate alternating bilinear form $\bar{B}$ on V/R, hence $\dim_{\mathbb{Z}_2} V/R = 2m$ (cf.

---

[2]$p_1: E(n) = O(n) \ltimes \mathbb{R}^n$ is a projection onto $O(n)$, see Lemma 2.13.

[83, Theorem 8.1, p. 586]). Let us assume that $R = 0$. In this case we may find a basis $a_i, b_i, i = 1, 2, \ldots, m$ for V such that $B(a_i, b_j) = \delta_{ij}, B(a_i, a_j) = B(B_i, b_j) = 0$. Then, with respect to the symplectic basis $\{a_i, b_i\}$, we define for Q the Arf invariant (see [15, p. 54]):

$$\mathbf{c}(\mathbf{Q}) = \sum_{i=1}^{m} Q(a_i)Q(b_i) \in \mathbb{Z}_2.$$

It can be proved that $\mathbf{c}$ is independent of the choice of the basis, [15]. In the case $R \not\equiv 0$, let us assume that $R = R_1$. Then, it is easy to see that Q defines Q' on V/R and the radical Q' is zero. In this case we define $\mathbf{c}(Q) = \mathbf{c}(Q')$. Let $p(Q)$ be the number of elements of $x \in V$ such that $Q(x) = 1$ and let $n(Q)$ be the number of $x \in V$ such that $Q(x) = 0$. Obviously $p(Q) + n(Q) = |V|$. Put $r(Q) = p(Q) - n(Q)$. We have the following equivalent results.

Theorem 9.60 ([15, Theorem III.1.14]). *Let $Q\colon V \to \mathbb{Z}_2$ be a quadratic form over $\mathbb{Z}_2$, with the radical of the associated bilinear form R. Then the Arf invariant $\mathbf{c}(Q)$ is defined if and only if $R = R_1$. In general, if $R = R_1, Q$ is determined up to isomorphism by rank V, rank R and $\mathbf{c}(Q)$, while if $R \neq R_1$, then Q is determined by rank V and rank R. Note that in the last case $r(Q) = 0$.*

□

Theorem 9.61 ([80, Theorem 2.15]). *Any quadratic form $Q\colon V \to \mathbb{Z}_2$ is equivalent to $Q_1^r \oplus Q_2^s \oplus E^t \oplus G^h$, where $s = 0$ or 1, E is the one-dimensional form and G is the zero form. Moreover t can be chosen to be 0 or 1, and if $t = 1$ then s can be chosen to be 0.*

□

Suppose that G_1 and G_2 are 2-groups and that G_i has a unique normal subgroup $\langle z_i \rangle$ of order two for $i = 1, 2$. The central product $G_1 * G_2$ is defined by

$$G_1 * G_2 = (G_1 \times G_2)/\langle (z_1, z_2) \rangle.$$

It is not difficult to check that $D_8 * D_8 \simeq Q_8 * Q_8$ and that $D_8 * \mathbb{Z}_4 \simeq Q_8 * \mathbb{Z}_4$. Finally, we have the following corollary of Theorems 9.57 and 9.61 which characterize GHW groups. Recall, that $F = \Gamma/A$.

Corollary. *Let Γ be a generalized Hantzsche-Wendt group of dimension $n \geq 3$ and A be a subgroup of index two of the maximal abelian subgroup $\mathbb{Z}^n$.*

Moreover let $Q_A^\Gamma \sim Q_1^r \oplus Q_2^s \oplus E^t \oplus G^h$, then F is isomorphic to one of the following: $\underbrace{D_8 * \cdots * D_8}_{r} \times \mathbb{Z}_2^h,$ $\underbrace{D_8 * \cdots * D_8}_{r} * \mathbb{Z}_4 \times \mathbb{Z}_2^h$ *or* $\underbrace{D_8 * \cdots * D_8}_{r} * Q_8 \times$ $\mathbb{Z}_2^h.$

$\square$

If the form B_A^Γ is non-degenerate, then $\Gamma/A = F$ is an extraspecial group. Moreover, in that case F is either $D_8 * \cdots * D_8 * Q_8$ or $D_8 * \cdots * D_8$, and its isomorphism class depends on the value of the Arf invariant. We can see it more precisely in the case of the following example.

Example 9.62. Let $\Gamma_n \subset E(n)$ be oriented GHW considered above.[3] According to the definition we have

$$B_D^{\Gamma_n}(B_i, Bj) = \begin{cases} 0 & \text{if } i = j \\ 1 & \text{if } i = j + 1 \\ 0 & \text{if } i \geq j + 2 \end{cases}$$

and a matrix

$$X = \begin{bmatrix} 0 & 1 & 0 & \ldots & 0 \\ 1 & 0 & 1 & 0\ldots & 0 \\ \ldots & \ldots & \ldots & \ldots & \\ 0 & \ldots 0 & 1 & 0 & 1 \\ 0 & \ldots & 0 & 1 & 0 \end{bmatrix}.$$

Moreover, for any $B_i \in (Z_2)^{n-1}$, where $1 \leq i \leq n-1$

$$Q_D^{\Gamma_n}(B_i) = (B_i + I)(e_i/2 + e_{i+1}/2) = e_i \notin D.$$

By definition it follows that $n = 2k + 1$ for some $k \in \mathbb{N}$. In order to calculate the Arf invariant $\mathbf{c}$ of $Q_D^{\Gamma_n}$, we have to transform X to a symplectic matrix. Let us introduce a new basis $f_1, f_2, \ldots, f_k, f_{k+1}, \ldots, f_{2k}$:

$$f_i = B_{2i-1}, 1 \leq i \leq k$$

and

$$f_{k+i} = B_{2i} + B_{2i+2} + B_{2i+4} + \cdots + B_{2k}, 1 \leq i \leq k.$$

It is easy to see that the matrix X with respect to the new basis is a symplectic one. Then from the above

$$\mathbf{c}(Q_D^{\Gamma_n}) = \sum_{i=1}^{k} Q_D^{\Gamma_n}(f_i) Q_D^{\Gamma_n}(f_{k+i}) \in \mathbb{Z}_2 = \mathbb{Z}^n/D.$$

---

[3]See Examples 5.22, 9.13 and the beginning of Section 9.6.

Hence we have

Proposition 9.63 ([139]). *For any odd number $n = 2k + 1$, the group Γ_n/D is a group of order 2^{2k+1}. In addition:*

If $k = 2l - 1$ or $2l$ and if l is odd, then $\Gamma_n/D = \underbrace{D_8 * \cdots * D_8}_{k-1} *Q_8.$

If $k = 2l - 1$ or $2l$ and if l is even, then $\Gamma_n/D = \underbrace{D_8 * \cdots * D_8}_{k}.$

If $k = 1$, then $\Gamma_3/D = Q_8.$

Proof. The above bilinear form $B_A^{\Gamma_n}$ is non-degenerate. From the definition $Q_D^{\Gamma_n}(f_{2k}) = Q_D^{\Gamma_n}(f_j) = 1$ for $j = 1, 2, \ldots k$ and $Q_D^{\Gamma_n}(f_{2k-1}) = 0$. We have

$$\mathbf{c}(Q_D^{\Gamma_n}) = \sum_{i=1}^{k} Q_D^{\Gamma_n}(f_i) Q_D^{\Gamma_n}(f_{k+i})$$

$$= \sum_{i=1}^{k} Q_D^{\Gamma_n}(f_{k+i}),$$

where

$$Q_D^{\Gamma_n}(f_{k+i}) = \begin{cases} 1 & \text{if } i \equiv k \text{ mod } 2 \\ 0 & \text{if } i \equiv (k-1) \text{ mod } 2. \end{cases}$$

Finally,

$$\mathbf{c}(Q_D^{\Gamma_n}) = \begin{cases} 1 & \text{if } k = 2l - 1, 2l \text{ for l odd} \\ 0 & \text{if } k = 2l - 1, 2l \text{ for l even.} \end{cases}$$

$\square$

If the form B_A^{Γ} is degenerate then we should calculate the rank of its Gram-Matrix X. Then, we have to find the radical R. When $R = R_1$, we have $\Gamma/A = E \times \mathbb{Z}_2^k$, where $k = \dim_{\mathbb{Z}_2} R$ and E is a group which can be recognized by the previous case. However, if $R \neq R_1$ we know that the parameter t from Theorem 9.61 is equal to 1, and by Theorem 9.60 in that case the number $r(Q_A^{\Gamma}) = 0$.

Let us see what happens in low dimensions. In fact, already in dimension 3 we have all the cases. From the last Proposition, we know that $\Gamma_3/D = Q_8$. It is easy to see, that there are seven possibilities for the subgroup $A \subset \mathbb{Z}^3$ of index two. For example, if

$$A_1 = \text{gen}\{2e_1, e_2, e_3\}$$

we get the quadratic form Q_2 and hence $\Gamma_3/A_1 = D_8$. But for the subgroup

$$A_2 = \text{gen}\{e_1 + e_2, 2e_2, e_3\}$$

we have the zero Gram-Matrix. It is easy to see, that $\Gamma_3/A_2 = \mathbb{Z}_4 \times \mathbb{Z}_2$ and $r(Q_{A_2}^{\Gamma_3}) = 0$.

For the 12 GHW groups of dimension 4 (cf. [109]), and subgroup D of index two we get the following groups Γ/A: $\mathbb{Z}_4 \times (\mathbb{Z}_2)^2$, $Q_8 \times \mathbb{Z}_2$ and $Q_8 * \mathbb{Z}_4$.

9.8 Complex Hantzsche-Wendt manifolds

By Lemma 7.5 we cannot expect to find a Kähler flat manifold with holonomy group $\mathbb{Z}_2^{n-1}$ below the dimension n. For this reason we introduce analogously as in the real case a concept of complex (generalized) Hantzsche-Wendt manifold.

Definition 9.64. A flat Kähler n-manifold with a holonomy $\mathbb{Z}_2^{n-1}$ is called a complex generalized Hantzsche-Wendt manifold (abbreviated as complex GHW). It will be called complex Hantzsche-Wendt manifold if, in addition the holonomy representation $\mathbb{Z}_2^{n-1} \to U(n)$ has its image contained in $SU(n)$ (complex HW in short).

Lemma 9.65. *For each $n \geq 2$, there exists a complex GHW of complex dimension n. Complex HW only exists in odd dimensions and for each odd $n \geq 3$, there exists a complex HW of dimension n.*

Proof. First we show that complex HW exists in odd dimensions only. Let M be a complex HW of dimension n and with holonomy representation $\varphi\colon \mathbb{Z}_2^{n-1} \to SU(n)$. After conjugation inside $GL(n, \mathbb{C})$ we may assume that the image of φ consists of diagonal $n \times n$ matrices with ± 1's on the diagonal. As the total subgroup of $SU(n)$ consisting of diagonal matrices with ± 1's on the diagonal is isomorphic to $\mathbb{Z}_2^{n-1}$ and φ is faithful, the image of φ is completely determined. Now, if n is even $-I_n$, minus the $n \times n$-identity matrix belongs to $SU(n)$. However, this would imply that $-I_{2n}$ belongs to the image of the real holonomy representation $\mathbb{Z}_2^{n-1} \to GL(2n, \mathbb{R})$, which is a contradiction. Therefore n has to be odd.

Now, given n we show that there exists a complex (G)HW of complex dimension n. First of all, there exists a (real) GHW of real dimension n, which we take to be orientable (HW) when n is odd, and where the fundamental group $\pi_1(M)$ defines a short exact sequence

$$0 \to \mathbb{Z}^n \to \pi_1(M) \to \mathbb{Z}_2^{n-1} \to 0.$$

Hence $\pi_1(M)$ is given by a 2-cohomology class $\langle f \rangle \in H^2(\mathbb{Z}_2^{n-1}, \mathbb{Z}^n)$. Now, consider

$$\langle f \rangle \oplus \langle f \rangle \in H^2(\mathbb{Z}_2^{n-1}, \mathbb{Z}^n) \oplus H^2(\mathbb{Z}_2^{n-1}, \mathbb{Z}^n) \cong H^2(\mathbb{Z}_2^{n-1}, \mathbb{Z}^{2n}).$$

This 2-cohomology class determines a Bieberbach group π' and the direct sum of modules $\mathbb{Z}^n \oplus \mathbb{Z}^n$ automatically satisfies the criterion of Proposition 7.3. Therefore, π' is the fundamental group of a complex GHW, which is a complex HW in case n is odd. $\qquad\square$

As the image of the representation $\mathbb{Z}_2^{n-1} \to SU(n)$ is fixed for a complex HW, we are able to compute the Hodge diamonds for any complex HW and we prove:

Theorem 9.66. *Let $n \geq 3$ be an odd number and let M be a complex Hantzsche-Wendt flat manifold of complex dimension n. Then it is Calabi-Yau and has the following Betti numbers:*

$$\beta_1 = \beta_3 = \cdots = \beta_{n-2} = \beta_{n+2} = \beta_{2n-1} = 0 \text{ and } \beta_n = 2^n,$$

$$\beta_0 = \binom{n}{0}, \ \beta_2 = \binom{n}{1}, \ \ldots \beta_{2k} = \binom{n}{k}, \ \ldots \beta_{2n} = \binom{n}{n}.$$

Proof. We compute the Hodge diamond for these manifolds, from which the result follows easily. As in the proof of Lemma 9.17 we may assume that the representation $\varphi \colon \mathbb{Z}_2^{n-1} \to SU(n)$ is diagonal and that the image consists of all diagonal matrices with ± 1 on the diagonal (and of course with determinant 1).

Let us compute the upper left corner of the Hodge diamond, this is, the entries $h^{p,q}$ with $0 \leq q \leq p \leq n$. Then the other terms follow by symmetry. As the action of the holonomy group is diagonal, we have to look for those (p, q)-forms

$$dz_{i_1} \wedge dz_{i_2} \wedge \cdots \wedge dz_{i_p} \wedge d\bar{z}_{j_1} \wedge d\bar{z}_{j_2} \wedge d\bar{z}_{j_q} \tag{9.66}$$

which are fixed under the action of the holonomy group. As a conclusion, we have that $h^{p,q} = \binom{n}{p}$.

Summarizing all of the above situations, we find the following Hodge diamond for a complex HW of complex dimension n:

$$
\begin{array}{ccccccccccccc}
 & & & & & & \binom{n}{0} & & & & & & \\
 & & & & & 0 & & 0 & & & & & \\
 & & & & 0 & & \binom{n}{1} & & 0 & & & & \\
 & & & 0 & & 0 & & 0 & & 0 & & & \\
 & & 0 & & 0 & & \binom{n}{2} & & 0 & & 0 & & \\
\cdots & & & \cdots & & \cdots & & \cdots & & \cdots & & \cdots & \\
\binom{n}{0} & & \binom{n}{1} & & \binom{n}{2} & & \cdots & & \cdots & & \binom{n}{n-1} & & \binom{n}{n} \\
\cdots & & & \cdots & & \cdots & & \cdots & & \cdots & & \cdots & \\
 & & 0 & & 0 & & \binom{n}{n-2} & & 0 & & 0 & & \\
 & & & 0 & & 0 & & 0 & & 0 & & & \\
 & & & & 0 & & \binom{n}{n-1} & & 0 & & & & \\
 & & & & & 0 & & 0 & & & & & \\
 & & & & & & \binom{n}{n} & & & & & &
\end{array}
$$

The rest of the theorem now follows easily, because $\beta_i = \displaystyle\sum_{p+q=1} h^{p,q}$. $\square$

There are exactly four manifolds of this type in real dimension 6 (complex dimension 3), cf. [36]. Of all 174 six dimensional flat manifolds admitting a complex structure only five of them are having the first Betti number equal to zero and four of them are complex HW.

In the real case, all generalized HW manifolds are having a holonomy representation which is diagonalizable over $\mathbb{Z}$. This does no longer hold in the complex case.

Proposition 9.67. *Any complex Hantzsche-Wendt manifold has a spin structure.*

Proof (See also [99, Example 4.6]). Let M be a complex Hantzsche-Wendt manifold of complex dimension n. There is a short exact sequence

$$0 \to \mathbb{Z}^{2n} \to \pi_1(M) \to (\mathbb{Z}_2)^{n-1} \to 0,$$

inducing a holonomy representation $\varphi \colon \mathbb{Z}_2^{n-1} \to GL(2n, \mathbb{Z})$. When we consider $\mathbb{R}^{2n}$ as a $\mathbb{Z}_2^{n-1}$-module via φ, we have that $\mathbb{R}^{2n}$ is the direct sum $M \oplus M$ of two identical $\mathbb{Z}_2^{n-1}$-modules, where the action on M is given via matrices belonging to $SO(n)$. Hence it is enough to apply the definition of the spin structure for the "double" construction from the proof of Theorem 1 in [37]. $\qquad\square$

9.9 Exercises

Exercise 9.1. Give a proof of Theorem 9.15 for odd n.

Exercise 9.2. Show, that for every $k, 1 < k < n-1$, there are flat manifolds with holonomy group $\mathbb{Z}_2^k$ not of diagonal type.
Hint: Use Examples and results from Sections 6.3 and (6.3.1). See also comments after Theorem 9.15.

Exercise 9.3. Prove that the Fibonacci groups $F(2,3), F(2,4)$ and $F(2,5)$ are finite.

Exercise 9.4. Prove that abelianization of three dimensional orientable Hantzsche-Wendt group is equal to $\mathbb{Z}_4 \times \mathbb{Z}_4$.

Exercise 9.5. Give a proof of Lemma 9.17.
Hint: See [101, page 4].

Exercise 9.6. Prove that the subgroup $D \subset \mathbb{Z}^n$, from Example 9.55 is of index two.

Exercise 9.7. Prove that the groups $D_8 * \cdots * D_8$ and $D_8 * \cdots * D_8 * Q_8$ are extraspecial 2-groups, cf. Definition 5.26.

Exercise 9.8. Find all quadratic forms Q_A^Γ for the group Γ_3 and all 4-dimensional GHW groups.

Exercise 9.9. Let $\Gamma = \langle (B_i, b_i) \in E(5) : 1 \le i \le 4 \rangle$ be the 5-dimensional oriented Hantzsche-Wendt group, with $b_1 = (1/2, 0, 1/2, 0, 0), b_2 = (0, 1/2, 0, 0, 0), b_3 = (0, 0, 1/2, 1/2, 0), b_4 = (0, 0, 0, 1/2, 1/2)$. Using Proposition 9.63, prove that Γ is not isomorphic to Γ_5.

Exercise 9.10. Give an alternative proof of Proposition 9.63.
Hint: Try to find the number of involutions (elements of order two) in the extraspecial 2-group, cf. [38, Corollary 33.6].

10. Combinatorial Hantzsche-Wendt Groups

G is called a unique product group if given two nonempty finite subset X, Y of G, there exists at least one element $g \in G$ which has a unique representation $g = xy$ with $x \in X$ and $y \in Y$. A unique product group is torsion free, though the converse is not true in general. The original motivation for studying unique product groups was the Kaplansky zero divisor conjecture, namely that if k is a field and G is a torsion free group, then kG is a domain. It was proved in 1988 [118] that the group $G_2(\Gamma_3)$ is a nonunique product group. To prove it the author uses the combinatorial presentation ([111, Lemma 13.3.1, p. 606-607])

$$\Gamma_3 = \langle x, y \mid x^{-1}y^2xy^2, y^{-1}x^2yx^2 \rangle. \tag{10.1}$$

However the counterexample to the Kaplansky unit conjecture was given in 2021 by G. Gardam [50]. Again the counterexample was found in the group ring $\mathbb{F}_2[\Gamma_3]$. The Kaplansky unit conjecture states that every unit in $K[G]$ is of the form kg for $k \in K, k \neq 0$ and $g \in G$.

In [30] the following generalization of Γ_3 is proposed.

Definition 10.1. By a combinatorial Hantzsche-Wendt group we shall understand a finitely presented group

$$G_n = \langle x_1, ..., x_n | x_i^{-1}x_j^2 x_i x_j^2 \ \forall \ i \neq j \rangle.$$

It is easy to see that, $G_0 = 1$ and $G_1 = \mathbb{Z}$. Moreover G_2 is the Hantzsche-Wendt group of dimension 3.

Let

$$\mathbb{Z}^n \simeq \mathbb{A}_n \simeq \langle x_1^2, x_2^2, ..., x_n^2 \rangle, \tag{10.2}$$

be a free abelian subgroup of G_n. In [30, Lemma 3.1] is proved that $\mathbb{Z}^n \lhd G_n$. Later we shall denote $\mathbb{A}_n$ by $\mathbb{Z}^n$. Moreover, $W_n = G_n/\mathbb{Z}^n = \langle x_1, ..., x_n \mid x_1^2, x_2^2, ..., x_n^2 \rangle \simeq *^n \mathbb{Z}_2$. Finally in [30, Theorem 3.3] it is proved that G_n is torsion free for all $n \geq 1$. This is also the corollary from Theorem 10.13. For any $1 \leq m \leq n$, G_m embeds in G_n and for $n \geq 2, G_n$ is a nonunique product group [30, Corollary 3.5]. Another interesting result of [30, Theorem

3.6] is the following. There is for $n \geq 3$ and odd a surjective homomorphism $\Phi_n \colon G_{n-1} \to \Gamma_n$. It is easy to see that $\Phi_n(\mathbb{Z}^{n-1})$ is a free abelian subgroup of the translation subgroup of Γ_n of a rank $n-1$. Since $\Gamma_n/\Phi_n(\mathbb{Z}^{n-1})$ is an infinite group and $\ker(\Phi_n) \cap \mathbb{Z}^{n-1} = 1$ then $\ker(\Phi_n)$ is an infinitely generated free group. (See [30, Theorem 3.6] and [59, page 87].)

In the first part of a chapter we shall show two models of BG_n (or $K(G_n, 1)$). They are a topological realization of two algebraic representations of G_n. The first model is an appropriate gluing of n copies of generalized fat Klein bottles. It corresponds to an isomorphism of G_n with $*_{\mathbb{Z}^n}^n K_n$ where K_n is a generalized Klein bottle crystallographic group amalgamated over the translation lattices. The second model is some Borel construction. It corresponds to the representation of G_n as an extension:

$$1 \to \mathbb{Z}^n \to G_n \to W_n \to 1. \tag{10.3}$$

From the first model we obtain that the cohomological dimension of G_n is equal to $n+1$ for $n > 1$. In the second part we calculate the Hilbert-Poincaré series of $G_n, n \geq 1$ with $\mathbb{Q}$ and $\mathbb{F}_2$ coefficients and explain the algebra structure of the cohomology. Here our main tools will be the Lyndon-Hochschild-Serre (LHS) spectral sequence of the group extension (10.3). The case with $\mathbb{F}_2$ coefficients uses a multiplicative structure of LHS. In the $\mathbb{F}_2$ case it is enough to use $E_3^{\star,\star}$ groups, but for rational coefficients we only need the $E_2^{\star,\star}$-terms. (See formulas 10.11 and 10.15.)

In the last part we calculate some other invariants and properties of G_n. For example their abelianization and the Euler characteristic.

10.1   Two models of BG_n

10.1.1   Gluing fat Klein bottles

We start with an example.

Example 10.2. Let K_- be the fundamental group of the Klein bottle and $\mathbb{Z}^2$ its maximal abelian subgroup of index two. It is well known *(see [66, Chapter 8.2, p. 153])* that $\Gamma_3 \simeq K_- *_{\mathbb{Z}^2} K_-$.

A generalization of the above example gives us the following characterization of the combinatorial Hantzsche-Wendt group. Let $G_n^{(i)}$ denotes the subgroup of G_n generated by $\{x_i\}$ and the abelian subgroup (10.2) $\mathbb{Z}^n$. We shall call it a generalized Klein bottle.

Proposition 10.3. *The natural group homomorphism*

$$*_{\mathbb{Z}^n} G_n^{(i)} \to G_n \tag{10.4}$$

is an isomorphism.

Proof. This follows from the definition and the structure of the free product with amalgamation. $\qquad\square$

A topological interpretation of the above construction gives us a $n+1$ dimensional classification space BG_n as a gluing together n generalized Klein bottles after common the n dimensional torus $B\mathbb{Z}^n$. BG_n has dimension $n+1$ since a maps $B\mathbb{Z}^n \to BG_n^{(i)}$ we can consider as an inclusion.

Corollary 10.4. *A cohomological dimension of G_n is equal to $n+1$.*

$G_n^{(i)}$ is a torsion free crystallographic group of dimension n and acts freely on $\mathbb{R}^n$ (in a way analogous to K_-) so has a classifying space which is an n dimensional closed flat manifold K^i (the generalized Klein bottle). A topological interpretation of the isomorphism (10.4) gives us a $n+1$ dimensional classifying space BG_n as (homotopically) gluing together n generalized Klein bottles $K^{(1)}, K^{(2)}, ..., K^{(n)}$ along a common n dimensional torus $\mathbb{R}^n/\mathbb{Z}^n$. This space has dimension $n+1$ since we must convert maps $\mathbb{R}^n/\mathbb{Z}^n \to K^{(i)}$ to inclusions. More precisely it may be done as follows. Let us define an action of G_n on $\mathbb{R}^n$ by

$$x_i(v)_i = v_i + 1/2 \text{ and } x_i(v)_j = -v_j, j \neq i,$$

where $v = (v_1, v_2, ..., v_n)$ and an action on a segment $I = [-1, 1]$ by $x_i(t) = -t, t \in I$.

Definition 10.5. By a fat Klein bottle we shall understand the space $B_n^{(i)} := (\mathbb{R}^n \times I)/G_n^{(i)}$.

Let $S_n := \mathbb{R}^n/\mathbb{Z}^n$. Let us define maps $\alpha^{(i)}\colon S_n \to B_n^{(i)}$ by the formula $\alpha^{(i)}(v) = [(v, 1)]$.

Definition 10.6. By the space B_n we shall understand a *colim* of a diagram formed from maps $\alpha^{(i)}$, i.e.

$$B_n := colim_i \alpha^{(i)}.$$

Theorem 10.7 ([116, Theorem 1]). *The above space B_n is a classifying space for G_n.*

Proof. From the definition the action of the subgroup $G_n^{(i)}$ on $\mathbb{R}^n$ is free and the orbit space $K_n^{(i)}$ was called a generalized Klein bottle. Moreover, the fat Klein bottle is $(n+1)$ dimensional compact manifold with boundary and the projection on the first factor gives a bundle $B_n^{(i)} \to K_n^{(i)}$ with fibre I, hence in particular it is a homotopy equivalence. Finally, the map $\alpha^{(i)}$ is an embedding on the boundary of $B_n^{(i)}$. However, more geometrically we may write $B_n := \bigcup_i B_n^{(i)}$ treating the maps $\alpha^{(i)}$ as identifications (so $S_n \subset B_n^{(i)}$). In other words B_n is obtained from n copies of a fat Klein bottle by an appropriate identification of the boundaries of different copies. To finish our proof we observe that $\pi_1(B_n) \simeq G_n$ after van Kampen theorem. The space B_n is aspherical after JHC Whitehead's theorem in [146, Theorem 5]. $\square$

Corollary 10.8. *For $n > 1$ the cohomological dimension of G_n is equal to $n + 1$.*

Proof. For brevity we write $B = B_n$ and $S = S_n$. From the properties of B we have, that $\mathrm{cd}G_n \leq n+1$. Let H denote cohomology with $\mathbb{F}_2$ coefficients. We have an exact sequence

$$H^n(S) \to H^{n+1}(B,S) \to H^{n+1}(B).$$

Since $\dim H^n(S) = 1$ and $\dim H^{n+1}(B,S) = n$, then $\dim H^{n+1}(B) \geq n - 1$. Hence $\mathrm{cd}G_n \geq n+1$ for $n > 1$. $\square$

Remark 10.9. The space B_n is for $n = 1$ a Möbius band, for $n = 2$ a closed manifold (a classical 3-dimensional Hantzsche-Wendt manifold). However for $n > 2$ it is nonmanifold, since there is singularity along S_n.

10.1.2 Borel construction (homotopy quotient)

Let G be a discrete group and let $p_G\colon EG \to BG$ be the universal G bundle. The assignment $G \mapsto p_G$ may be done functorial in the group G and respecting products. If X is some G-space then the space

$$X_G := (EG \times X)/G$$

is called the Borel construction on X, [3, p. 10]. Here G acts on $EG \times X$ diagonally. Let $f_X\colon X_G \to BG$ be the quotient map. It is a fibration with fibre X. It is easy to see that, if X is aspherical then X_G is also aspherical.

Definition 10.10. (morphisms between maps) If $f\colon X_1 \to X_2$ and $g\colon Y_1 \to Y_2$ then a morphism from f to g is a pair (m_1, m_2) where $m_i\colon X_i \to Y_i$ and $gm_1 = m_2 f$.

The operation of taking pullback along ψ is denoted by $\psi^\star$. We shall write $f \simeq m_2^\star(g)$ if $(m_1, m_2)\colon f \to g$ and m_1 is an isomorphism on fibres.

Let $\xi, \eta \in W_2$ be generators of order 2 and $D := \mathbb{Z}_2 \oplus \mathbb{Z}_2$. The abelianization of W_2 defines

$$1 \to \mathbb{Z} \simeq \langle (\xi\eta)^2 \rangle \to W_2 \xrightarrow{\alpha} D \to 1. \tag{10.5}$$

Let Σ be the unit circle on the complex plane. Define an action of D on Σ by formulas

$$\xi(z) = \bar{z} \text{ and } \eta(z) = -\bar{z}.$$

Denote a resulting D-space by U. In the above language we have a map $f_U\colon U_D \to BD$ and we can observe that $\pi_1(f_U) \simeq \alpha$.

We define (for $i = 1, 2, ..., n$) homomorphisms $\phi_i\colon G_n \to W_2$

$$\phi_i(x_i) = \xi\eta \text{ and } \phi_i(x_j) = \xi \text{ for } j \neq i. \tag{10.6}$$

Then $\phi := (\phi_1, ..., \phi_n)$ gives a homomorphism from $G_n \to (W_2)^n$. Let $q_n\colon G_n \to W_n$ be the canonical surjection. The homomorphism ϕ factorizes and we obtain a map $(\phi, \psi)\colon q_n \to \alpha^n$ and ϕ is an isomorphism on fibres. We have:

Lemma 10.11. $q_n \simeq \psi^\star \alpha^n$.

Proof. With support of (10.5) we have the following commutative diagram

$$
\begin{array}{ccccccccc}
1 & \longrightarrow & \mathbb{Z}^n & \xrightarrow{\ i\ } & G_n & \xrightarrow{\ q_n\ } & W_n & \longrightarrow & 1 \\
& & \downarrow{\simeq} & & \downarrow{\phi} & & \downarrow{\psi} & & \\
1 & \longrightarrow & \mathbb{Z}^n & \xrightarrow{\ i_1\ } & (W_2)^n & \xrightarrow{\ \alpha^n\ } & (\mathbb{Z}_2 \oplus \mathbb{Z}_2)^n & \longrightarrow & 1
\end{array}
$$

Diagram 10.1

where i, i_1 are inclusions. $\square$

Define a W_n action on the space Σ^n

$$x_i(z)_i = -z_i \text{ and } x_i(z)_j = \bar{z}_j \text{ for } j \neq i, \tag{10.7}$$

where $z = (z_1, z_2, ..., z_n)$.

Denote the resulting W_n-space by T_n.

Proposition 10.12.

$$\pi_1(f_{T_n}) \simeq q_n$$

in particular $\pi_1((T_n)_{W_n}) \simeq G_n$.

Proof. The action on T_n is obtained by composing the product D^n action with the homomorphism ψ (i.e. $w(x) = \psi(w)(z)$ for $w \in W_n$). Hence, from naturality we have a map of fibrations

$$(\hat{\psi}_1, \hat{\psi}_2) = \hat{\psi} \colon f_{T_n} \to f_{U^n}.$$

Applying π_1 we get

$$\pi_1(\hat{\psi}) \colon \pi_1(f_{T_n}) \to \pi_1(f_{U^n}).$$

The map $\hat{\psi}$ gives an isomorphism (identity) on fibres so the map $\pi_1(\hat{\psi})$ also gives an isomorphism on fibres by an application of the long exact sequence for fibrations. We have (for codomain components) $(\hat{\psi})_2 = B\psi$ so $\pi(\hat{\psi})_2 = \pi(B\psi) = \psi$. Hence

$$\pi_1(f_{T_n}) \simeq \psi^\star \pi_1(f_{U^n}).$$

But

$$\psi^\star \pi_1(f_{U^n}) \simeq \psi^\star \pi_1((f_U))^n \simeq \psi^\star \alpha^n \simeq q_n.$$

$\square$

Theorem 10.13 ([116, Theorem 2]).
$$(T_n)_{W_n} = K(G_n, 1).$$

Proof. The space $(T_n)_{W_n}$ is aspherical because T_n is. And it has the appropriate fundamental group by Proposition 10.12. $\qquad\square$

Let $B = \bigvee_1^n \mathbb{R}P(\infty)$. The space $B = K(W_n, 1)$, cf. [146]. Let $E \to B$ be the universal-covering. Then:

Corollary 10.14. *We have the fibration*
$$T^n \to (T^n)_{W_n} \to E/W_n, \tag{10.8}$$
where a W_n action on E is by deck transformation.

$\qquad\square$

Remark 10.15. The W_n action on T^n is highly non-effective. The kernel of it is the commutator subgroup of W_n, which by the Kurosh subgroup theorem, is a free group of rank $1 + (n-2)2^{n-1}$.

See [28, Exercise 3, p. 212] and the proof of Proposition 10.36.

10.2 Cohomologies of G_n

In this part we shall calculate a cohomology of the group G_n with $\mathbb{Q}$ coefficients (Theorem 10.17) and $\mathbb{F}_2$ coefficients (Theorem 10.22). We shall apply the Leray-Serre spectral sequence of the fibration (10.8) and equivalently Lyndon-Hochschild-Serre spectral sequence for the short exact sequence of groups (10.3)
$$1 \to \mathbb{Z}^n \to G_n \xrightarrow{\nu} W_n \to 1.$$

10.2.1 Hilbert-Poincaré series

Definition 10.16 ([59, p. 230]). Let M be a topological space. For a fixed coefficient field k, define the Poincaré series of M the formal power series
$$P(x, k) = \sum_i a_i x^i$$
where a_i is the dimension of $H^i(M, k)$ as a vector space over k, assuming this dimension is finite for all i.

Theorem 10.17 ([116, Theorem 3]). *The rational Hilbert-Poincaré series of the space*

$$K(G_n, 1) = T^n \times_{W_n} E$$

is equal to

$$P_n(x, \mathbb{Q}) = ((1+x)(1 + \frac{(1-(-1)^n)}{2}x^n + x(\frac{n-2}{2}(1+x)^{n-1} - \frac{n}{2}(1-x)^{n-1})).$$

(10.9)

In particular, $P_0(x, \mathbb{Q}) = 1, P_1(x, \mathbb{Q}) = x + 1, P_2(x, \mathbb{Q}) = x^3 + 1.$

We start with Lemma.

Lemma 10.18. *For* $p > 1, H^p(W_n, \mathbb{Q}) = 0.$

Proof. We have a short exact sequence of groups related to the abelianization

$$1 \to \mathbb{F}_k \to W_n \to (\mathbb{Z}_2)^n \to 1,$$

(10.10)

where $\mathbb{F}_k$ is a non-abelian free group of a rank $k = 1 + (n-2)2^{n-1}$. Hence for $q > 1, H^q(\mathbb{F}_k, M) = 0$ for any $\mathbb{F}_k$-module M. Similar for any $p \geq 1, H^p((\mathbb{Z}_2)^n, N) = 0$ for any $(\mathbb{Z}_2)^n$-rational vector space N. Applying a Leray-Serre spectral sequence to (10.10) we have for $i \geq 2, H^i(W_n, S) = 0$. Where S is a W_n-rational vector space. □

Corollary 10.19. *For* $p > 1, q \geq 0, E_2^{p,q} = H^p(W_n, H^q(\mathbb{Z}^n, \mathbb{Q})) = 0$ *and the differentials* $d_i = 0$ *for* $i \geq 2$. *Moreover,* $E_2^{0,q}$ *and* $E_2^{1,q}, q \geq 0$ *are two non-trivial columns of the spectral sequence.*

The Hilbert-Poincaré polynomial (10.9) is the sum $f_0 + f_1$, where

$$f_p = x^p \sum_i \dim(E_2^{p,i})x^i.$$

(10.11)

Let us start to calculate dimensions of $E_2^{p,q} = H^p(W_n, H^q(\mathbb{Z}^n, \mathbb{Q}))$ for $p = 0, 1$ and $q \geq 0$. We shall use a W_n action on $H^q(\mathbb{Z}^n, \mathbb{Q}) = \Lambda^q(\mathbb{Q}^n)$, which follows from an action W_n on T^n, see (10.7).

We introduce sequences $\epsilon \in \{-1, 1\}^n$. Denote by

$$(-1)\epsilon = -\epsilon = (-\epsilon_1, -\epsilon_2, ..., -\epsilon_n).$$

Moreover, for $A \subset \{1, 2, ..., n\}$ the sequence e^A has -1 exactly on the positions from A. Finally let $1 = (1, 1, ..., 1) := e^{\varnothing}$ and $|\epsilon| := \sum_i \frac{1-\epsilon_i}{2}$ (the number of -1 in the sequence).

By $\mathbb{Q}_\epsilon$ we shall understand the rational numbers $\mathbb{Q}$ with the structure of a W_n-module such that the k-th generator of W_n acts as multiplication by $\epsilon_k, 1 \leq k \leq n$. In this language $H^1(\mathbb{Z}^n, \mathbb{Q}) \simeq \sum_i \mathbb{Q}_{-e^{\{i\}}}$ as W_n-module. Moreover, $H^*(\mathbb{Z}^n, \mathbb{Q}) \simeq \Lambda^*(\mathbb{Q}^n)$ is a sum of some $\mathbb{Q}_\epsilon$. Let

$$h^i(\epsilon) = \dim H^i(W_n, \mathbb{Q}_\epsilon).$$

From the definition

$$h^0(\epsilon) = \begin{cases} 1 & \text{if } \epsilon = 1 \\ 0 & \text{if } \epsilon \neq 1 \end{cases} \text{ and } h^1(\epsilon) = \begin{cases} 0 & \text{if } \epsilon = 1 \\ |\epsilon| - 1 & \text{if } \epsilon \neq 1 \end{cases}.$$

Using (10.11) and the above formulas gives us

$$f_i = x^i \sum_{A \subset \{1,2,...,n\}} x^{|A|} h^i((-1)^{|A|} e^A), i = 1, 2.$$

Hence for $n \geq 0$:

$$f_0 = 1 + \frac{1 - (-1)^n}{2} x^n$$

$$f_1 = x^{n+1}(n-1)\frac{1 + (-1)^n}{2} +$$

$$+x \sum_{0<k<n} x^k \binom{n}{k}\left((k-1)\frac{1 + (-1)^k}{2} + (n-k-1)\frac{1 - (-1)^k}{2}\right).$$

Lemma 10.20. *Formula (10.9) from Theorem 10.17 is equal to $f_0 + f_1$.*

Proof. We shall use two formulas:

$$\sum_{0<k<n} \binom{n}{k} x^k = (1+x)^n - 1 - x^n = g(x)$$

and

$$\sum_{0<k<n} k\binom{n}{k} x^k = nx(1+x)^{n-1} - nx^n = f(x).$$

On the beginning we shall prove that

$$S = \sum_{0<k<n} x^k \binom{n}{k}\left((k-1)\frac{1 + (-1)^k}{2} + (n-k-1)\frac{1 - (-1)^k}{2}\right) =^1$$

$^1 k(-1)^k + \frac{n-2}{2} - \frac{n}{2}(-1)^k = (k-1)\frac{1+(-1)^k}{2} + (n-k-1)\frac{1-(-1)^k}{2}.$

$$\sum_{0<k<n} x^k \binom{n}{k}\left(k(-1)^k + \frac{n-2}{2} - \frac{n}{2}(-1)^k\right) =$$

$$\sum_{0<k<n} x^k(-1)^k \binom{n}{k} + \frac{n-2}{2}\sum_{0<k<n} x^k \binom{n}{k} - \frac{n}{2}\sum_{0<k<n}(-1)^k x^k \binom{n}{k} =$$

$$f(-x) + \frac{n-2}{2} - \frac{n}{2}g(-x) =$$

$$\frac{n-2}{2}(1+x)^n - \frac{n}{2}(1+x)(1-x)^{n-1} + (1 - n\frac{1+(-1)^n}{2})x^n + 1.$$

Since

$$f_1 = x^{n+1}(n-1)\frac{1+(-1)^n}{2} + xS$$

then

$$f_1 = x + \frac{1-(-1)^n}{2}x^{n+1} + (1+x)x\left(\frac{n-2}{2}(1+x)^{n-1} - \frac{n}{2}(1-x)^{n-1}\right)$$

and

$$f_0 + f_1 =$$

$$(1+x)(1 + \frac{1-(-1)^n}{2}x^n + x(\frac{n-2}{2}(1+x)^{n-1} - \frac{n}{2}(1-x)^{n-1})).$$

$\square$

The above lemma finishes the proof of Theorem 10.17.

$\square$

As a complement to the above results we present an observation about the algebra structure of $H^*(G_n, \mathbb{Q})$.

Corollary 10.21. *If $x, y \in H^*(G_n, \mathbb{Q})$ are such that $\deg(x) > 0$ and $\deg(y) > 0$ then $xy = 0$.*

Proof. For $n = 1$ it is obvious. Let us assume $n \geq 2$. From the proof of Theorem 10.17 $H^1(G_n, \mathbb{Q}) = 0$ and $H^s(G_n, \mathbb{Q}) = 0$ for $s > n+1$. So, if $\deg(x) \geq n$ or $\deg(y) \geq n$ then $xy = 0$. Hence we can assume that $\deg(x) < n$ and $\deg(y) < n$. From the proof of Theorem 10.17 we know that the appropriate Serre spectral sequence converging to $H^*(G_n, \mathbb{Q})$ has the property:

$$(\star) \quad E_\infty^{p,q} \neq 0 \implies (p,q) \in \{(0,0), (0,n)\} \cup \{(1,i) : i \leq n\}.$$

Let (F_i) denote the filtration of $H^*(G_n, \mathbb{Q})$ associated with the spectral sequence. From $(\star)$ and $\deg(x) < n$ and $\deg(y) < n$ it follows that $x, y \in F_1$. Consequently $xy \in F_2$. But again from $(\star)$ $F_2 = 0$. $\square$

A calculation of the cohomology with $\mathbb{F}_2$ coefficients needs different tools. We shall also apply the Lyndon-Hochschild-Serre spectral sequence for the short exact sequence of groups (10.3)

$$1 \to \mathbb{Z}^n \to G_n \xrightarrow{\nu} W_n \to 1.$$

In this case we have $E_3^{*,*} \neq 0$ and we shall use multiplicative structures.

Theorem 10.22 ([116, Theorem 4]).

$$P_n(x, \mathbb{F}_2) = (1+x)(1+(n-1)x(1+x)^{n-1}). \tag{10.12}$$

In particular, $P_0(x, \mathbb{F}_2) = 1, P_1(x, \mathbb{F}_2) = x+1, P_2(x, \mathbb{F}_2) = 1 + 2x + 2x^2 + x^3$.

Proof. There are canonical isomorphisms over $\mathbb{F}_2$

$$E_2^{p,q} = H^p(W_n, H^q(\mathbb{Z}^n, \mathbb{F}_2)) \xleftarrow{\sim} H^p(W_n, \mathbb{F}_2) \otimes H^q(\mathbb{Z}^n, \mathbb{F}_2),$$

$$H^*(\mathbb{Z}_2^n, \mathbb{F}_2) = \mathbb{F}_2[z_1, z_2, ..., z_n],$$

where $z_i \in H^1(\mathbb{Z}_2^n, \mathbb{F}_2) = \mathrm{Hom}(\mathbb{Z}_2^n, \mathbb{F}_2)$ is the projection on the i-coordinate, $i = 1, 2, ..., n$ and

$$H^*(W_n, \mathbb{F}_2) = H^*(\mathbb{Z}_2^n, \mathbb{F}_2)/\{z_i z_j | i \neq j\}. \tag{10.13}$$

Let $g_1, ..., g_n \in H^1(\mathbb{Z}^n, \mathbb{F}_2)$ be a dual basis to $x_1^2, ..., x_n^2$. We shall denote by $\Lambda(g_1, ..., g_n)$ the exterior algebra over $\mathbb{F}_2$ generated by $g_1, ..., g_n$ ($H^*(\mathbb{Z}^n, \mathbb{F}_2)$). [2]

To begin we shall prove:

Proposition 10.23. *Let (E_r, d_r) be the above spectral sequence and on E_3 we use the total grading. $E_3^{0,0} = \langle 1 \rangle, E_3^{0,q} = 0, q > 0$ and $E_3^{p,q} = 0$ for $p > 2$. Moreover $d_2 \neq 0$ and $d_i = 0$ for $i \geq 3$.*

Proof of Proposition 10.23. To prove that $E_3^{0,q} = 0, q > 0$ it is enough to show that the kernel of $d_2^{0,q}$ is trivial. By contradiction assume that $\exists\ 0 \neq \omega \in E_2^{0,q}, q > 0$ and $d_2^{0,q}(\omega) = 0$. Then $\exists\ i$, s.t. $\omega = g_i \alpha + \beta$ and α, β do not depend on g_i. If $d_2^{0,q}(\omega) = 0$ then from the properties of the transgression $0 = d_2^{0,q}(\omega) = z_i \alpha + \gamma$, where γ is a linear combination of elements from the set $\{z_s^2 : s \neq i\}$ with coefficients in $\Lambda(g_1, ..., g_n)$. Hence $\alpha = 0$ a contradiction. From Lemmas 10.25 and 10.26 cycles of the differential d_2 are linear combinations of elements $z_i^k \omega$ (for some i, k) and $\omega \in \Lambda(g_1, g_2, ..., g_n)$ does not include g_i. Hence for $k \geq 3$ $z_i^k \omega = d_2^{k,s}(z_i^{k-2} g_i \omega)$ and for $p \geq 3, E_3^{p,q} = 0$. $\square$

---

[2] $\Lambda^*(g_1, ..., g_n) \simeq H^*(\mathbb{Z}^n, \mathbb{F}_2)$.

Lemma 10.24. *In the spectral sequence of the extension (10.5)*

$$0 \to \mathbb{Z} \to W_2 \to \mathbb{Z}_2 \oplus \mathbb{Z}_2 \to 0,$$

$\bar{d}_2(g) = z_1 z_2$, *where g is a generator of $H^1(\mathbb{Z}, \mathbb{F}_2)$.*

Proof of Lemma 10.24. From the above $z_1 z_2 \in H^*(W_2, \mathbb{F}_2)$ is equal to zero. Applying the five-term exact sequence (see [60, p. 16, 57]) we get the exact sequence

$$H^1(\mathbb{Z}, \mathbb{F}_2) \xrightarrow{\bar{d}_2} H^2(\mathbb{Z}_2 \oplus \mathbb{Z}_2, \mathbb{F}_2) \xrightarrow{f^*} H^2(W_2, \mathbb{F}_2)$$

and $\operatorname{Im} \bar{d}_2 = \ker f^*$. Hence $\bar{d}_2(g) = z_1 z_2$. $\qquad\square$

Lemma 10.25. *Let $g_1, ..., g_n \in H^1(\mathbb{Z}^n, \mathbb{F}_2)$ be a dual basis to $x_1^2, ..., x_n^2$. For the spectral sequence of the exact sequence of groups*

$$0 \to \mathbb{Z}^n \to G_n \to W_n \to 0,$$

using naturality and properties of homomorphisms (10.6) $\phi_i \colon G_n \to W_2$ we obtain $d_2(g_i) = z_i^2$.

Proof of Lemma 10.25. We have a commutative diagram

$$
\begin{array}{ccccccccc}
1 & \longrightarrow & \mathbb{Z}^n & \xrightarrow{\ i\ } & G_n & \xrightarrow{\ q_n\ } & W_n & \longrightarrow & 1 \\
 & & \downarrow{\scriptstyle \alpha_i} & & \downarrow{\scriptstyle \phi_i} & & \downarrow{\scriptstyle \gamma_i} & & \\
1 & \longrightarrow & \mathbb{Z} & \xrightarrow{\ i_1\ } & W_2 & \xrightarrow{\ \alpha\ } & \mathbb{Z}_2 \oplus \mathbb{Z}_2 & \longrightarrow & 1
\end{array}
$$

Diagram 10.2

Here, α_i and γ_i are defined by ϕ_i. From naturality $d_2 \circ \alpha_i^* = \gamma_i^* \circ \bar{d}_2$, (see Diagram 10.3) where α_i^*, γ_i^* are the induced maps on cohomology.

$$
\begin{array}{ccc}
\bar{E}_2^{0,1} \simeq H^1(\mathbb{Z}, \mathbb{F}_2) & \xrightarrow{\ \bar{d}_2\ } & \bar{E}_2^{2,0} \simeq H^2(\mathbb{Z}_2 \oplus \mathbb{Z}_2, \mathbb{F}_2) \\
\downarrow{\scriptstyle \alpha_i^*} & & \downarrow{\scriptstyle \gamma_i^*} \\
E_2^{0,1} \simeq H^1(\mathbb{Z}^n, \mathbb{F}_2) & \xrightarrow{\ d_2\ } & E_2^{2,0} \simeq H^2(W_n, \mathbb{F}_2)
\end{array}
$$

Diagram 10.3

We have
$$g_i = \alpha_i^*(g).$$

Hence

$$d_2(g_i) = d_2(\alpha_i^*(g)) = \gamma_i^*(\bar{d}_2(g)) \stackrel{\text{Lemma 10.24}}{=} \gamma_i^*(z_1 z_2) = \gamma_i^*(z_1)\gamma_i^*(z_2).$$

Moreover, let $\overline{\gamma_i} \colon W_n/[W_n, W_n] \to \mathbb{Z}_2 \oplus \mathbb{Z}_2$ be induced by γ_i, then

$$\overline{\gamma_i}(\lambda_1, \lambda_2, \cdots, \lambda_n) = (\textstyle\sum_s \lambda_s, \lambda_i).$$

Hence,

$$\gamma_i^*(z_1) = \textstyle\sum_s z_s \text{ and } \gamma_i^*(z_2) = z_i.$$

Finally, since in $H^*(W_n)$ $z_i z_j = 0$ for $i \neq j$, we get

$$d_2(g_i) = (\textstyle\sum_s z_s) z_i = z_i^2.$$

$\square$

The next observation is the following.

Lemma 10.26. *Let $k > 0$, then, $d_2(z_i^k \omega) = 0$ if and only if ω does not depend on g_i.*

Proof of Lemma 10.26. Let us write $\omega = g_i \alpha + \beta$ where $\alpha, \beta \in \Lambda$ do not depend on g_i. In fact, by definition $d_2(g_i \alpha + \beta) = z_i^2 \alpha + \gamma$, where γ is a linear combination of elements from the set $\{z_s^2 : s \neq i\}$ with coefficients from Λ. Hence, because for $i \neq s$, $z_i z_s = 0$ (10.13)

$$d_2(z_i^k(g_i \alpha + \beta)) = z_i^k d_2(g_i \alpha + \beta) = z_i^k(z_i^2 \alpha + \gamma) = z_i^{k+2}\alpha. \tag{10.14}$$

Assume $d_2(z_i^k \omega) = 0$. We can write $\omega = g_i \alpha + \beta$, where α and β are independent from g_i. From (10.14) $0 = d_2(z_i^k \omega) = z_i^k \alpha$. Hence $\alpha = 0$ and $\omega = \beta$ and so ω does not include g_i. Finally, if ω does not include g_i substituting $\alpha = 0$ and $\omega = \beta$ to the formula (10.14) we get $d_2(z_i^k \omega) = 0$. $\square$

Corollary 10.27. *Let $k > 0$ and $v = \sum_s z_s^k \omega_s \in E_2^{k,s}$, where $\forall s \ \omega_s \in \Lambda_s$ then*

$$d_2(v) = 0 \Longleftrightarrow \forall s \ \omega_s \text{ does not include } g_s.$$

Proof of Corollary 10.27. ($\Leftarrow$) Follows from the above Lemma 10.26. ($\Rightarrow$) For any i if $d_2(v) = 0$ then also $d_2(z_i v) = 0$. But $z_i v = z_i^{k+1} \omega_i$. Again, from Lemma 10.26 it follows that ω_i does not include g_i. $\square$

Remark 10.28. Let $Z_2^{i,j} = \ker d_2^{i,j}$ and $B_2^{i,j} = \operatorname{Im} d_2^{i,j}$. Let M be an $\mathbb{F}_2$-vector space and a trivial G-module, then

$$\dim H^i(G, M) = \dim H^i(G, \mathbb{F}_2) \dim M.$$

Moreover $\dim H^i(\mathbb{Z}^n, \mathbb{F}_2) = \binom{n}{i}$, $\dim H^i(W_n, \mathbb{F}_2) = n$ (for $i > 0$), $\dim E_2^{p,q} = n\binom{n}{q}$ (for $p > 0$).

Summing up the generating function for $H^*(G_n, \mathbb{F}_2)$ is a sum of three components: $f_0 + f_1 + f_2$ where

$$f_p = \sum_i \dim(E_3^{p,i}) x^{p+i}. \tag{10.15}$$

Lemma 10.29. *From the properties of the differentials d_2 we have:*

I. $f_0 = 1$;

II. $f_1 = nx(1+x)^{n-1}$;

III. $f_2 = nx^2(1+x)^{n-1} - x((1+x)^n - 1)$.

Proof of Lemma 10.29. By an application of the proof of Proposition 10.23 $f_0 = 1$.

From the above $d_2(z_i^k \omega) = 0$ if and only if ω does not include g_i. Hence $\dim Z_2^{k,s} = n\binom{n-1}{s}$ and $f_1 = nx(1+x)^{n-1}$, cf. Corollary 10.27.

For $p = 2$ we have $E_3^{2,i} = \ker d_2^{2,i} / \operatorname{Im} d_2^{0,i+1}$. Moreover, for $i > 0, d_2^{0,i}$ is a monomorphism. This follows from the proof of Proposition 10.23. Hence, $\dim \operatorname{Im} d_2^{0,i} = \dim E_2^{0,i} = \binom{n}{i}$. Summing up $f_2 = nx^2(1+x)^{n-1} - x((1+x)^n - 1)$ and the Lemma is proved. $\qquad\square$

Example 10.30. We have $\dim Z_3^{k,s} = n\binom{n-1}{s}$. In fact, a basis of $Z_2^{k,s}$ is the set $\{z_i^k \omega\}$, where $1 \le i \le n$ and ω is a Grassmann monomial of degree s on elements $\{g_1, ..., g_n\} \setminus \{g_i\}$.

For $n = 4$ the basis of $Z_2^{2,2}$ has 12 elements:

$$z_1^2 g_2 g_3, z_1^2 g_2 g_4, z_1^2 g_3 g_4, z_2^2 g_1 g_3, z_2^2 g_1 g_4, z_2^2 g_3 g_4,$$
$$z_3^2 g_1 g_2, z_3^2 g_2 g_4, z_3^2 g_2 g_4, z_4^2 g_1 g_2, z_4^2 g_1 g_3, z_4^2 g_2 g_3,$$

the basis of $E_2^{0,3}$ has the following 4 elements:

$$y_1 = g_1 g_2 g_3, y_2 = g_1 g_2 g_4, y_3 = g_1 g_3 g_4, y_4 = g_2 g_3 g_4.$$

Since $d_2^{0,3}$ is a monomorphism the basis of $B_2^{2,2}$ has four elements:

$$d_2^{0,3}(y_1) = z_1^2 g_2 g_3 + z_2^2 g_1 g_3 + z_3^2 g_1 g_2,$$
$$d_2^{0,3}(y_2) = z_1^2 g_2 g_4 + z_2^2 g_1 g_4 + z_4^2 g_1 g_2,$$
$$d_2^{0,3}(y_3) = z_1^2 g_3 g_4 + z_3^2 g_1 g_4 + z_4^2 g_1 g_3,$$
$$d_2^{0,3}(y_4) = z_2^2 g_3 g_4 + z_3^2 g_2 g_4 + z_4^2 g_2 g_3.$$

Hence dim $E_3^{2,2} = 12 - 4 = 8$.

Finally

$$f_0 + f_1 + f_2 = 1 + nxb + nx^2 b - xb(1 + x) + x$$
$$= x + 1 + nxb(1 + x) - xb(1 + x) = (x + 1)(1 + nxb - xb)$$
$$= (1 + x)(1 + (n - 1)x(1 + x)^{n-1}).$$

Here $b = (1 + x)^{n-1}$. $\qquad\square$

Finally, we would like to present some grading of $H^*(G_n, \mathbb{F}_2)$. We start with a definition.

Definition 10.31. Define the bigraded algebra $\mathcal{E}^{(n)}$ over $\mathbb{F}_2$ by (a direct sum of vector space):

$$\mathcal{E}^{(n)} = \mathcal{E}_0 + \mathcal{E}_1 + \mathcal{E}_2$$

where $\mathcal{E}_i$ are given by:
-   $\mathcal{E}_0 = \langle 1 \rangle$,
-   $\mathcal{E}_1$ is spanned (i.e. is a free vector space) by symbols: $z_i g_A$ where $1 \le i \le n$ and $A \subset \{1, 2, ..., n\}, i \notin A$,
-   $\mathcal{E}_2 = \mathcal{E}_2'/\mathcal{R}$ where $\mathcal{E}_2'$ is spanned by symbols $z_i^2 g_A$ with restrictions as above and $\mathcal{R} = \text{span}\{r_A : A \subset \{1, 2, ..., n\}, A \ne \emptyset\}$ where $r_A = \sum_{i \in A} z_i^2 g_{A \setminus \{i\}}$.

Bidegrees are given by:

$$\text{bideg}(1) = (0, 0), \text{bideg}(z_i g_A) = (1, |A|), \text{bideg}(z_i^2 g_A) = (2, |A|).$$

Multiplication is given by:
-   1 acts in obvious way;
-   $(z_i g_A)(z_i g_B) = z_i g_{A \cup B}$ if $A \cap B = \emptyset$;
-   All other products are zero.

The above definition summarizes explicitly the description of the bigraded algebra structure of E_3, namely:

Proposition 10.32. *If (E_r, d_r) is the spectral sequence of the short exact sequence (10.3) then*

$$E_3 \simeq \mathcal{E}^{(n)}$$

as bigraded algebras.

Example 10.33 (Bigraded algebra $H^*(G_2, \mathbb{F}_2)$). There are elements

$$a, b, A, B, w \in H^*(G_2, \mathbb{F}_2)$$

such that:

   - $a, b \in H^1(G_2, \mathbb{F}_2), A, B \in H^2(G_2, \mathbb{F}_2), w \in H^3(G_2, \mathbb{F}_2)$;

   - $aA = bB = w$ and all other products of elements from $\{a, b, A, B, w\}$ are zero;

   - $\{a, b, A, B, w\}$ is a basis of $H^*(G_2, \mathbb{F}_2)$.

Using Definition 10.31 we have:

$\mathcal{E}_0 = \langle 1 \rangle$,

$\mathcal{E}_1 = \langle z_1 g_\emptyset, z_2 g_\emptyset, z_1 g_{\{2\}}, z_2 g_{\{1\}} \rangle$,

$\mathcal{E}_2' = \langle z_1^2 g_\emptyset, z_2^2 g_\emptyset, z_1^2 g_{\{2\}}, z_2^2 g_{\{1\}} \rangle$,

$\mathcal{R} = \langle z_1^2 g_\emptyset, z_2^2 g_\emptyset, z_1^2 g_{\{2\}} + z_2^2 g_{\{1\}} \rangle$. So

$$(1, z_1 g_\emptyset, z_2 g_\emptyset, z_1 g_{\{2\}}, z_2 g_{\{1\}}, [z_1^2 g_{\{2\}}])$$

is a basis of $\mathcal{E}^{(2)}$, where $[\xi]$ denotes the class of ξ.

If $(1, a, b, A, B, w)$ are elements of $H^*(G_2, \mathbb{F}_2)$ which correspond (in this order) to elements of the above basis then they satisfy the conditions stated above.

Example 10.34. Let $X = P_2 \bigvee P_2 \bigvee S^3$ where S^3 is the 3-dimensional sphere and P_2 is the 2-dimensional real projective space. Let $Y = BG_2$. Then the Poincaré polynomials of X and Y over $\mathbb{F}_2$ and over $\mathbb{Q}$ are the same but $H^*(X, \mathbb{F}_2)$ and $H^*(Y, \mathbb{F}_2)$ are not isomorphic as algebras.

10.3   Additional observation

Proposition 10.35.

1. For $n > 1$, $(G_n)_{ab} \simeq \mathbb{Z}_4^n$;

2. For $n > 1$ the centre of G_n is trivial;

3. The Euler characteristic and the first Betti number of G_n are equal to zero.

Proof. 1. Follows from a direct calculation.

2. From (10.3) and the fact that the centre of W_n is trivial we have an inclusion $Z(G_n) \subset \mathbb{Z}^n$. Let $v \in Z(G_n)$. From the above $v = \Pi_i(x_i^2)^{\alpha_i}$ for some $\alpha_i \in \mathbb{Z}$. Using relations in G_n, we have

$$x_1 v x_1^{-1} = (x_1^2)^{\alpha_1} \Pi_{i \geq 2}(x_i^2)^{-\alpha_i}.$$

Since $v = x_1 v x_1^{-1}$, then $\alpha_i = 0$ for $i \geq 2$. Similar $x_2 v x_2^{-1} = v$, which gives us $\alpha_1 = 0$.

3. From the properties of the Euler characteristic of a fibration (10.8) (see [128, page 481]) $\chi(K(G_n, 1)) = \chi(T^n)\chi(E/W_n) = 0 \cdot \chi(E/W_n) = 0$. The conclusion about the Betti number follows directly from Theorem 1.   □

Proposition 10.36. *Let $n \geq 2$. The short exact sequence of groups (10.3) defines a representation*

$$h \colon W_n \to GL(n, \mathbb{Z}), \forall x \in W_n \ h(x)(e_i) = \bar{x} e_i \bar{x}^{-1},$$

where $e_i \in \mathbb{Z}^n$ is the standard basis $i = 1, 2, ..., n$ and $\nu(\bar{x}) = x$. However, $K = \ker h \neq 0$, because $[W_n, W_n] \subset \ker h$. In particular, K is a finitely generated free group of rank $1 + (n - 2)2^{2[n/2]-1}$.

Proof. From definition of h we have an extension $K \to W_n \to \mathbb{Z}_2^s$, where $s = 2[\frac{n}{2}]$ and $[x]$ is the largest integer not exceeding x. Again by the definition of h the commutator subgroup W_n' of W_n has index 1 for n even and index 2 for n odd in the group K. Hence $s = 2[\frac{n}{2}]$. Computing (fractional) Euler characteristics we get (see [15, Corollary 5.6, p. 245]) from the Euler characteristic formula $e(K) = e(W_n)/e(\mathbb{Z}_2^s) = e(W_n)2^s = (1 - \frac{n}{2})2^s$, which gives the announced rank.

Analogously $e(W_n') = (1 - \frac{n}{2})2^n$ and $e(K/W_n') = 2^{s-n}$.   □

Remark 10.37. Let n be an even number then

$$P_n(x, \mathbb{Q}) = P_n(x, \mathbb{F}_2) \bmod 2.$$

11. Open Problems

This part is related to the paper [138]. Let $\Gamma \subset E(n)$ be a crystallographic group which defines the exact sequence

$$0 \to \mathbb{Z}^n \to \Gamma \to H \to 0$$

and an element $\alpha \in H^2(H, \mathbb{Z}^n)$, see [92, Theorem 4.1, p. 112] and Theorem 3.1.

11.1　The classification problems

It is well known (cf. [23]) that any finite group G is the holonomy group of some flat manifold. Hence we have:

Problem 11.1. *Find the minimal dimension of a flat manifold with holonomy group G.*

Remark 11.2. The answer is known for cyclic groups ([23], [61]), elementary abelian p-groups, dihedral groups, semidihedral groups, generalized quaternion groups (see [64]) and simple groups $PSL(2, p)$ (p is a prime number) [113]. There also are some other classes of finite groups for which the minimal dimension is known. We do not pretend to give a complete list here, see also Section 4.2. Unfortunately, for most finite groups it seems to be a very difficult question. For example, almost nothing is known for the symmetric groups. The main obstruction is the calculation of the second cohomology of the finite group with special coefficients.

Problem 11.3. *Perform the calculations for those p-groups for which the cohomology is well-understood.*

Problem 11.4. *For a finite group G find a Bieberbach group (with trivial centre) of minimal dimension.*

Problem 11.5 (see Chapter III)**.** *Calculate the Vasquez invariant $n(G)$ for a finite group G.*

Remark 11.6. For some progress see [25] and [44].

Problem 11.7. *Classify finite groups which are holonomy groups of flat manifolds* $M^n = \mathbb{R}^n/\Gamma$ *with finite outer automorphism group* Out(Γ).

Remark 11.8. Some progress was made in [68], e.g. for finite abelian groups.

From the point of view of finite group representation theory, it might be interesting to consider a similar classification for Kähler flat manifolds (see [36] and [136]). Let us recall Theorem 5.9, that Out(Γ) is finite if and only if the holonomy representation is $\mathbb{Q}$-multiplicity free and any $\mathbb{R}$-irreducible component is $\mathbb{C}$-reducible.

The next conjecture is related to the previous problem.

Conjecture 11.9. *([135] and the end of Chapter IV) For any finite group* H, *there exists a Bieberbach group* Γ *with* $\mathbb{Q}$ *multiplicity free holonomy representation* ϕ_Γ.

Remark 11.10. If the first Betti number of a manifold M^n is zero then, from above, we have
$$\mathrm{Aff}(M^n) \simeq \mathrm{Out}(\Gamma).$$
This means that all information about symmetries of the manifold is built into the outer automorphism group.

It is interesting to compare Problem 5 with the work of M. Belolipetsky and A. Lubotzky [8] where they proved that for any finite group G there exist a fundamental group Γ of some compact hyperbolic manifold such that Out(Γ) = G. Hence we have, (cf. Chapter V).

Problem 11.11. *Which finite groups* G *occur as outer automorphism groups of Bieberbach groups with trivial centre.*

Moreover, in [91] there is proved (see also Proposition 5.30) that for any natural number n the symmetric group S_n and the group $\mathbb{Z}_3$ have a realization as an outer automorphism group of some Bieberbach group. In this connection we can ask if every finite subgroup of $GL(n, \mathbb{Z})$ has a realization as the centraliser of some finite subgroup of $GL(n, \mathbb{Z})$, for some n.

11.2 The Anosov relation for flat manifolds

Let $f: M \to M$ be a continuous map of a smooth manifold M. In fixed point theory, there exist two numbers associated to f which provide some information on the fixed point set of f: the Lefschetz number $L(f)$ and the Nielsen number $N(f)$. It is known that the Nielsen number provides more information about the fixed point set than $L(f)$ does, but $L(f)$ is easier to be calculated. For nilmanifolds however, D. Anosov [2] showed that $N(f) = |L(f)|$ for any self map f of the nilmanifold. On the other hand, he also observed that this result could not be extended to the class of all infra-nilmanifolds, since he was already able to construct a counter-example on the Klein bottle. It was recently shown that for large families of flat manifolds (e.g. all flat manifolds with an odd order holonomy group) the Anosov relation $N(f) = |L(f)|$ does hold for any self map (see [34] and [35]). It is therefore natural to ask:

Problem 11.12 (K. Dekimpe). *Describe the class of flat manifolds (or more generally the class of infra-nilmanifolds) on which the Anosov relation*

$$N(f) = |L(f)|$$

does hold for any map $f: M \to M$.

One can also consider this problem from a different point of view and fix the flat manifold M and try to find all self maps f for which the Anosov relation is valid.

Problem 11.13 (K. Dekimpe). *Let M be a flat manifold. Is it possible to determine classes of self maps $f: M \to M$ for which the Anosov relation*

$$N(f) = |L(f)|$$

does hold?

A first result in this direction can be found in [93].

11.3 Generalized Hantzsche-Wendt flat manifolds

Let us agree to call any fundamental group of the n-dimensional flat manifold with holonomy group $(\mathbb{Z}_2)^{n-1}$ a generalized Hantzsche-Wendt (GHW)

Bieberbach group (see [124] and Chapter IX). If the manifold is orientable, we call the GHW Bieberbach group orientable. For dimension 3 there is only one oriented flat manifold M^3 with holonomy group $\mathbb{Z}_2 \oplus \mathbb{Z}_2$. It was first considered by W. Hantzsche and H. Wendt, see [56] and recently, for example by J. Conway and J. P. Rossetti, see Section 3.3. On the other hand the fundamental group $\pi_1(M^3)$ is the group $F(2,6)$ (see Example 9.55), where $F(r,n)$ is the Fibonacci group (see (9.63) in Section 9.6).

Problem 11.14. *Explain the relation between orientable GHW Bieberbach groups and the Fibonacci groups.*

Remark 11.15. For any odd natural number n there exists an epimorphism

$$\Phi_n \colon F(n-1, 2n) \to \Gamma_n,$$

where $\Gamma_n \subset E(n)$ denotes an orientable GHW Bieberbach group introduced in Examples 5.22 and 9.13, see also [141]. Nothing is known about the kernel of the epimorphism Φ_n, for $n > 1$.

Problem 11.16. *Give a definition of any GHW Bieberbach group in terms of generators and relations.*

Let Γ be a GHW Bieberbach group of dimension n and with trivial centre. It is well known ([42, Theorem 1.1, page 183], [65]) that there is an epimorphism of Γ onto the infinite dihedral group $\mathbb{Z}_2 * \mathbb{Z}_2$. Hence we have a decomposition

$$\Gamma \simeq \Gamma_1 *_X \Gamma_2, \tag{11.1}$$

as the generalized free product of two Bieberbach groups of dimension $(n-1)$, amalgamated over a subgroup X of index two. Moreover any GHW Bieberbach group, with non-trivial centre, is the semi-direct product of $\mathbb{Z}$ with some lower dimensional GHW Bieberbach group. We have:

Conjecture 11.17. *For any GHW Bieberbach group Γ with trivial centre and dimension n, there exists a decomposition (11.1) such that Γ_1 and Γ_2 are GHW Bieberbach groups of dimension $n-1$.*

Remark 11.18. The conjecture is true for $n \leq 4$ (see [65, page 30 and 38]).

Let M^n be a flat oriented manifold with GHW fundamental Bieberbach group and dimension n. It is well known, see [124], that M^n is a rational homology sphere and $M^n, n \geq 5$ does not have a spin structure, cf. Exercise 6.8.

Problem 11.19. *What are the topological (geometrical) properties of M^n?*

11.4   Flat manifolds and other geometries

Let us recall the following well known question.

Question 11.20 (Farrell-Zdravkovska [41]). *Let M^n be a n-dimensional flat manifold. Is there a $(n+1)$-dimensional hyperbolic manifold W the boundary of which equals M^n?*

We would like to mention that the above question is very closely related to the following one. Let V^{n+1} be a hyperbolic Riemannian manifold (constant negative curvature) of finite volume. It is well known that V^{n+1} has finite number of cusps and each cusp topologically is equivalent to $M^n \times \mathbb{R}^{\geq 0}$, where M^n is n-dimensional flat manifold. We can formulate:

Problem 11.21. *Is there some V^{n+1} with one cusp homeomorphic to $M^n \times \mathbb{R}^{\geq 0}$?*

D. D. Long and A. Reid proved ([86]) that the answer to the above Problem 11.21 is negative if the η-invariant of the signature operator of M^n is not an integer. To prove it they used the Atiyah, Patodi, Singer formula, where signature operator D means some differential elliptic operator and its η-invariant measures the symmetry of the spectrum of D. See Section 6.4 and in particular the formula (6.13) together with Definition 6.28, where the η-invariant is defined for the Dirac operator. They also proved, that the η-invariant of the signature operator of a 3-dimensional flat manifold Γ_7 (see Section 3.3) is not in $\mathbb{Z}$. By assumptions this method works only in dimension $4k - 1$.

Problem 11.22. *Classify flat manifolds with non-integral signature η-invariant.*

They also proved, (see Chapter VIII and [87]) that any flat manifold has a "realization" as a cusp of some hyperbolic orbifold of finite volume. Finally, D. B. McReynolds in 2009 (see [98]) gave general, positive answer for the above Question 11.20, but problem 11.21 is still open.

Remark 11.23. We have the following flat manifolds of dimension two: the torus and the Klein bottle. The first one has a "realization" as a cusp of the eight knot complement and the second one as a cusp of the Gieseking manifold, (cf. [142]). In dimension three there are ten flat manifolds (see

Section 3.3) and each one has "realization" as a cusp of some four dimensional hyperbolic manifold W, which is not necessarily one-cusped, (see [106]). One major difficulty is the lack of examples of hyperbolic manifolds. In particular we have:

Problem 11.24 (See [87, p. 286]). *Give an example of an n-dimensional hyperbolic manifold of finite volume with only one cusp.*

An example of dimension 4 is given in [82].

The η-invariant of the signature operator for a flat manifold is an important quantity as we have observed above. Let us ask the same question about the Dirac operator, cf. Section 6.4. We shall need our orientable flat manifold M^n to have a spin structure. It turns out that this is equivalent (see Corollary before Proposition 6.15) to the existence of a homomorphism $\epsilon \colon \Gamma \to Spin(n)$ such that the following diagram is commutative

$$
\begin{array}{ccc}
 & & Spin(n) \\
 & \nearrow^{\epsilon} & \downarrow^{\lambda_n} \\
\Gamma & \xrightarrow{\ r\ } & SO(n)
\end{array}
$$

where λ_n is the universal covering and r is the projection onto the linear part, see Section 6.3 and [37].

Problem 11.25. *Classify the holonomy groups of flat manifolds which admit a spin structure.*

Remark 11.26. It is known [37, Proposition 1, Corollary 1] that any flat manifold with holonomy group of odd order or of order $2k$, k an odd number, has a spin structure. However, see [99, Theorem 3.2], for the holonomy group $\mathbb{Z}_2 \times \mathbb{Z}_2$, there exist orientable flat manifolds M_1, M_2 of dimension 6 with different holonomy representations $h_1, h_2 \colon \mathbb{Z}_2 \times \mathbb{Z}_2 \to GL(6, \mathbb{Z})$, only one of which has a spin structure. Similar examples are also given in Section 6.3 and (6.3.1). The complete classification is done for dimension four, [120].

Question 11.27. *Is there an example of an oriented flat spin-manifold for which the Dirac η-invariant is not equal modulo $\mathbb{Z}$ to the signature η-invariant?*

For the methods of calculation of the Dirac η-invariant (see Definition 6.28) we send to [125].

Let us recall Proposition 7.3, that a six dimensional flat manifold is complex Kähler if and only if its holonomy representation has the property that each $\mathbb{R}$-irreducible summand, which is also $\mathbb{C}$-irreducible occurs with even multiplicity. Hence the classification of such manifolds is possible, see Theorem 7.8. In Section 7.2 we introduced the n-dimensional Calabi-Yau flat manifold as Kähler flat one, whose holonomy representation is contained in $SU(n)$. Then in Section 9.8 examples of such objects are given. Let us propose our last question.

Problem 11.28. *What are the properties of the Calabi-Yau flat manifolds? Do they have some "mirror-symmetry"?*

11.5 The Auslander conjecture

Let us change the group $E(n)$ for $A(n)$. We can similarly define a discrete and cocompact subgroup of $A(n)$ as affine crystallographic group. In 1964 L. Auslander formulated the following conjecture.

Conjecture 11.29. *Let $\Gamma \subset A(n)$ be an affine crystallographic group. Is Γ a virtual solvable group?*

The above is true for $n = 2$. D. Fried and W. Goldman proved the above for $n = 3$. They classify all three dimensional affine crystallographic groups. H. Abels, G. A. Margulis and G. A. Soifer announced the proof of the above conjecture for $n \leq 6$ in 1997.

In 1997 J. Milnor asked a similar question without the assumption that Γ is cocompact. It is known that for $n = 3$ the answer is negative. For more information about the Auslander conjecture, we refer the reader to [33, pages 78-79].

Appendix I. Alternative Proof of Bieberbach Theorem

Below we give the second proof of the first Bieberbach Theorem (Theorem 2.4). It is based on the ideas of M. Gromov and we follow [20].

Theorem 11.30. *Let Γ be a crystallographic group of dimension n. Then its translation subgroup has n-linearly independent elements.*

Suppose $A \in O(n)$. We define

$$m(A) = max\{|Ax - x|/|x| | x \in \mathbb{R}^n \setminus \{0\}\}.$$

Let us see, that we always have $|Ax - x| \leq m(A)|x|$, for $x \in \mathbb{R}^n$. Moreover, the set

$$(i) \quad E_A = \{x \in \mathbb{R}^n \mid |Ax - x| = m(A)|x|\}$$

is a non trivial, A-invariant linear subspace. This follows from the so called *parallelogram condition*[1] and the sequences of equations

$$2m^2(A)(|x|^2 + |y|^2) = 2(|Ax - x|^2 + |Ay - y|^2) = |A(x + y) - (x + y)|^2$$

$$+|A(x - y) - (x - y)|^2 \leq m^2(A)(|x + y|^2 + |x - y|^2) = 2m^2(A)(|x|^2 + |y|^2),$$

where $x, y \in E_A$. Let $E_A^\perp$ be the A-orthogonal complement of E_A. We define

$$(ii) \quad m^\perp(A) = max\{|Ax - x|/|x| | x \in E_A^\perp \setminus \{0\}\},$$

when $E_A^\perp \neq 0$, and $m^\perp(A) = 0$ in the opposite case. Hence

$$(iii) \quad m^\perp(A) < m(A),$$

---

[1] $\|x + y\|^2 + \|x - y\|^2 = 2(\|x\|^2 + \|y\|^2)$

when $A \neq id$. Let $x = x^E + x^\perp \in E_A \oplus E_A^\perp$. Then

$$(iv) \quad |Ax^E - x^E| = m(A)|x^E|, |Ax^\perp - x^\perp| \leq m^\perp(A)|x^\perp|.$$

After these elementary observations, we see that for all $A, B \in O(n)$ we have

$$m([A, B]) \leq 2m(A)m(B).$$

In fact, we have

$$[A, B] - id = (A - id)(B - id) - (B - id)(A - id)A^{-1}B^{-1}.$$

Since $|A^{-1}B^{-1}x| = |x|$, it follows that

$$|[A, B]x - x| \leq m(A)m(B)|x| + m(B)m(A)|x|,$$

for all $x \in \mathbb{R}^n$.

Lemma A ("Mini Bieberbach"). *For each unit vector $u \in \mathbb{R}^n$ and for all $\epsilon, \delta > 0$ there exists $\beta = (B, b) \in \Gamma$, such that $b \neq 0$, $\angle(u, b) \leq \delta$, $m(B) \leq \epsilon$. $\angle(u, b)$ denotes the angle between the vectors u, b and $\cos(\angle(u, b)) = \frac{\langle u, b \rangle}{\|b\|}$.*

Proof. From the definition of Γ there exists d and elements $\beta_k \in \Gamma$ such that for any natural number k, we have

$$|b_k - ku| \leq d.$$

Moreover $|b_k| \to \infty$, $\angle(u, b_k) \to 0$ $(k \to \infty)$. Since $O(n)$ is compact, we find a subsequence such that for $i < j$ we have

$$m(B_j B_i^{-1}) \leq \epsilon, \quad \angle(u, b_j) \leq \delta/2, \quad |b_i| \leq \frac{\delta}{4}|b_j|.$$

Finally, the element $\beta_j \beta_i^{-1}$ has the required properties. $\quad\square$

Lemma B. *If $\alpha = (A, a) \in \Gamma$ and $m(A) \leq \frac{1}{2}$, then α is a translation.*

Proof. Suppose $\alpha = (A, a) \in \Gamma$ satisfies our assumptions and $m(A) \neq 0$. Since Γ is a discrete group, we can assume that the number $|a|$ is minimal. From Lemma A, for $u \in E_A$, there exists $\beta = (B, b)$, such that

$$b \neq 0, \quad |b^\perp| \leq |b^E|, \quad m(B) \leq \frac{1}{8}(m(A) - m^\perp(A)). \tag{11.2}$$

Among these we consider again the one for which $|b|$ is a non-zero minimum. Observe that $|a| \leq |b|$, when β is not a translation.[2] Let $\tilde{\beta} = [\alpha, \beta]$. From the considerations preceding Lemma A, we have

$$m(\tilde{B}) = m([A, B]) \leq 2m(A)m(B) \leq m(B)$$

and

$$\tilde{b} = (A - id)b^E + (A - id)b^\perp + r, \tag{11.3}$$

where

$$r = (id - \tilde{B})b + A(id - B)A^{-1}a.$$

If β is a translation then $B = id = \tilde{B}$ and $r = 0$. As we have already observed, from the choice of α, in the case when β is not a translation, we have an inequality

$$|a| \leq |b|.$$

Hence

$$|r| \leq (m(\tilde{B}) + m(B))|b| \leq 2m(B)|b| < 4m(B)|b^E|.$$

In each case we have

$$|r| < \frac{1}{2}(m(A) - m^\perp(A))|b^E|. \tag{11.4}$$

By definition and from (11.3) we have

$$b^{\tilde{E}} - (A - id)b^E - r^E = (A - id)b^\perp + r^\perp - b^{\tilde{\perp}} = 0.$$

Hence, using (11.2) and the orthogonality of r^E and $r^\perp$, we obtain

$$|b^{\tilde{\perp}}| \leq m^\perp(A)|b^\perp| + |r^\perp| < m^\perp(A)|b^E| + |r|.$$

Summing up, with support of (11.4), we have an inequality

$$|b^{\tilde{\perp}}| < \frac{1}{2}(m(A) + m^\perp(A))|b^E|.$$

On the other hand

$$|b^{\tilde{E}}| = |(A - id)b^E + r^E| \geq m(A)|b^E| - |r^E|$$

---

[2] $m(A) \neq 0$.

$$> m(A)|b^E| - \frac{1}{2}(m(A) - m^\perp(A))|b^E|$$

$$= \frac{1}{2}(m(A) + m^\perp(A))|b^E|.$$

Here we apply again (11.4) and

$$|x + y| \geq \big||x| - |y|\big|.$$

Finally, we can see that $\tilde{\beta}$ also satisfies the condition (11.2) and because

$$|\tilde{b}| \leq m(A)|b| + |r| < m(A) + \frac{1}{2}(m(A) - m^\perp(A))|b^E| < |b|$$

we have a contradiction. $\qquad\square$

By Lemma A it follows that there exist n elements of Γ, such that their translation parts are linearly independent, and their rotation parts have norm smaller than $\frac{1}{2}$. Now, by Lemma B we can define n linearly independent translations in Γ.

$\qquad\square$

For the proof of the first Bieberbach Theorem (Theorem 2.4) it is enough to observe that the image $p_1(\Gamma)$ of the homomorphism $p_1 \colon \Gamma \to O(n)$ (see Lemma 2.13) is a discrete subgroup of the compact group $O(n)$, cf. Exercise 2.3.

Appendix II. Burnside Transfer Theorem

This part is used in Chapter 4 and in particular in the proof of Lemma 4.6. We shall follow [38, §18]. Our goal is to present the Burnside transfer Theorem.

Theorem 11.31 (Burnside transfer Theorem). *Suppose that the Sylow p-subgroup S of G is in the center of $N_G(S)$. Then G has a normal subgroup N with $G = NS, S \cap N = 1, |N| = |G : S|$.*

Proof. We start with a lemma.

Lemma 11.32. *Let G be a finite group with subgroup H, and let T be a cross section of H in G, so $G = \bigcup_{t \in T} tH, t_1 H = t_2 H \Rightarrow t_1 = t_2$. For any $g \in G$, let $\bar{g}$ denote the unique element of $gH \cap T$. If $g \in G$ and $x \in G$, then $\overline{xg} = \overline{x\bar{g}}$ and $\bar{g}^{-1} g \in H$.*

Proof of Lemma 11.32. $gH = \bar{g}H$, so $xgH = x\bar{g}H$ and $\overline{xg} = \overline{x\bar{g}}$. The equation $gH = \bar{g}H$ implies $\bar{g}^{-1} g \in H$. $\qquad\square$

Definition 11.33. Let G be a finite group with subgroup H of finite index, and assume that K is a normal subgroup of H with H/K abelian. Let T be a cross section of H in G. We define the function $V = V_T \colon G \to H/K$ by

$$V(x) = \prod_{t \in T} \overline{xt}^{\,-1} xt K, \quad \text{for } x \in G.$$

Since H/K is abelian, the order of the factors in the product is non-essential. $V = V_T$ is the transfer of G into H/K.

Lemma 11.34. *$V = V_T \colon G \to H/K$ is a homomorphism which is independent of the choice of cross section T.*

Proof of Lemma 11.34. If $x \in G$, then $x: tH \to xtH$ permutes the cosets of H, so as t ranges over T so does $\overline{xt}$. Since H/K is abelian, we can write

$$V(xy) = \prod_{t \in T} \overline{xyt}^{\,-1} xytK = \prod_{t \in T} \overline{\overline{xyt}}^{\,-1} xytyt^{\,-1} ytK$$

$$= \prod_{t \in T} \overline{xyt}^{\,-1} xytK \cdot \prod_{t \in T} \overline{yt}^{\,-1} ytK = V(x)V(y),$$

by Lemma 11.32.

Let S be a second cross section, and for any $g \in G$ let $\tilde{g}$ be the unique element of $gH \cap S$, so $V_S(x) = \prod_{s \in S} \overline{xs}^{\,-1} xsK$, for any $x \in G$. Enumerate $T = \{t_1, t_2, \dots\}, S = \{s_1, s_2, \dots\}$, where $t_i H = s_i H$, say $s_i = t_i h_i$, for $h_i \in H$. For a fixed $x \in G$, let σ be the permutation of $\{1, 2, \dots\}$ defined by $\overline{xs_i} = s_{\sigma(i)}$. By definition we have

$$xs_i H = s_{\sigma(i)} H = t_{\sigma(i)} H = xT_i H,$$

and $\overline{xt_i} = t_{\sigma(i)}$. Hence

$$h_{\sigma(i)}^{-1} t_{\sigma(i)}^{-1} xt_i h_i = h_{\sigma(i)}^{-1} \overline{xt_i}^{\,-1} xt_i h_i,$$

and we see that

$$V_S(x) = \prod_i \overline{xs_i}^{\,-1} xs_i K = \prod_i s_{\sigma(i)}^{-1} xt_i h_i K$$

$$= \prod_i h_{\sigma(i)}^{-1} t_{\sigma(i)}^{-1} xt_i h_i K = \prod_i h_{\sigma(i)}^{-1} \overline{xt_i}^{\,-1} xt_i h_i K$$

$$= (\prod_i h_{\sigma(i)}^{-1})(\prod_i h_i) \prod_i \overline{xt_i}^{\,-1} xt_i K = \prod_i \overline{xt_i}^{\,-1} xt_i K = V_T(x).$$

$\square$

For any finite group G and prime number p, $G'(p)$ denotes the intersection of all normal subgroups N cf G with G/N an abelian p-group. Obviously, the commutator subgroup $G' \subset G'(p)$, so $G/G'(p)$ is an abelian p-group. Also $G' = \bigcap_p G'(p)$, the intersection is taken over all primes p. We want to determine $G/G'(p)$. If S is a Sylow p-subgroup of G, then $G = SG'(p)$, so $G/G'(p) \simeq S/S \cap G'(p)$.

Definition 11.35. If S is a Sylow p-subgroup of G, we define $S^* = S_G^*$, the focal subgroup of S in G, to be the subgroup of S generated by all $x^{-1}y$, where $x, y \in S$ are conjugate in G. Clearly $S' \subset S^*$, so S/S^* is abelian.

Lemma 11.36. $S^* = S \cap G'(p) = S \cap G'$, and $G/G'(p) \simeq S/S^*$.

Proof of Lemma 11.36. If $x, y \in S$ are conjugate in G, say $y = g^{-1}xg$, then $x^{-1}y = x^{-1}g^{-1}xg \in G'$, so $S^* \subset S \cap G' \subset S \cap G'(p)$. We must show that $S \cap G'(p) \subset S^*$. We consider the transfer $V \colon G \to S/S^*$, and evaluate it on any $x \in S$. x acts as a permutation on the cosets of S in G; let its orbit decomposition be

$$(S)(x_2 S x x_2 S \ldots x^{e_2 - 1} x_2 S)(x_3 S x x_3 S \ldots x^{e_3 - 1} x_3 S) \ldots.$$

For a cross section T of S in G we choose

$$x_1 = 1, x_2, xx_2, \ldots, x^{e_2 - 1} x_2, x_3, \ldots, x^{e_3 - 1} x_3, \ldots,$$

and denote $e_1 = 1$ so $|G : S| = \sum_i e_i$.

Of course, $V(x) = \prod_{y \in T} \overline{xy}^{\,-1} xy S^*$. We compute the terms $\overline{xy}^{\,-1} xy$. $\overline{xx_1}^{\,-1} xx_1 = \bar{x}^{\,-1} x = 1x = x$ since $x \in S$. $\overline{xx_2}^{\,-1} xx_2 = (xx_2)^{-1} xx_2 = 1$, if $e_2 \geq 2$. If $e_2 > 2$ we get

$$\overline{x^2 x_2}^{\,-1} x^2 x_2 = (x^2 x_2)^{-1} x^2 x_2 = 1.$$

We keep getting 1 until the term

$$\overline{x \cdot x^{e_2 - 1} x_2}^{\,-1} x \cdot x^{e_2 - 1} x_2 = \overline{x^{e_2} x_2}^{\,-1} \cdot x^{e_2} x_2 = x_2^{-1} x^{e_2} x_2.$$

We conclude that

$$V(x) = x \cdot x_2^{-1} x^{e_2} x_2 \cdot x_3^{-1} x^{e_3} x_3 \ldots S^*.$$

By definition of S^*, $x^{-e_i} x_i^{-1} x^{e_i} x_i \in S^*$, so $x_i^{-1} x^{e_i} x_i S^* = x^{e_i} S^*$, and

$$V(x) = xx^{e_2} x^{e_3} \ldots S^* = x^{\sum e_i} S^*,$$

proving

$$V(x) = x^{|G:S|} S^*.$$

$(p, |G : S|) = 1$ and S/S^* is a p-group, so the map $V \colon G \to S/S^*$ is onto. Hence $G/\ker V \simeq S/S^*$ is an abelian p-group. So $G'(p) \subset \ker V$, $|G : G'(p)| \geq |S : S^*|$. But $S^* \subset S \cap G'(p)$, so

$$|S : S^*| \geq |S : S \cap G'(p)| = |SG'(p) : G'(p)| = |G : G'(p)|.$$

Therefore equality holds everywhere, $S^* = S \cap G'(p)$, and the lemma is proved. $\qquad\square$

Moreover, we have:

Lemma 11.37. *Let H and K be two normal subsets of S which are conjugate in G. Then H and K are conjugate in $N_G(S)$.*

Proof of Lemma 11.37. Let $K = H^g, g \in G$. Then $N_G(K) = N_G(H^g) = N_G(H)^g$. $S^g \subset N_G(H)^g$, so S and S^g are Sylow p-subgroups of $N_G(K)$. Choose $n \in N_G(K)$ with $S = S^{gn}$. Then $gn \in N_G(S)$, and $H^{gn} = H^g = K$. $\qquad\square$

Summing up, we have that for the abelian Sylow p-subgroup S in G, the focal subgroup of S in G equals the focal subgroup of S in $H = N_G(S)$. Hence $G/G'(p) \simeq H/H'(p)$. For the proof of the Burnside theorem we have to observe that the focal subgroup of S in H is trivial, so $H/H'(p) \simeq S \simeq G/G'(p)$. Then it is enough to take $N = G'(p)$. This finishes the proof of the Burnside transfer Theorem. $\qquad\square$

Definition 11.38. A normal p-complement in a group G is a normal subgroup N such that, if S is a Sylow p-subgroup of G, then $G = SN$, $S \cap N = 1$, $|S| = |G : N|$, $|N| = |G : S|$.

The Burnside transfer Theorem says that if S is in the center of $N_G(S)$, then G has a normal p-complement.

Corollary. *If G has a cyclic Sylow 2-subgroup S, then G has a normal 2-complement.*

Proof. Let S has order 2^n. Then S has 2^{n-1} generators. Hence, the automorphism group of S has order 2^{n-1}. We have $S \subset C_G(S)$, so $N_G(S)/C_G(S)$ has odd order, and any $x \in N_G(S)$ performs an automorphism of odd order on S. From the above, this automorphism is trivial, so $x \in C_G(S)$, $N_G(S) = C_G(S)$, and Burnside transfer Theorem (Theorem 11.31) applies. $\qquad\square$

Appendix III. Example of a Flat Manifold without Symmetry

This Appendix should be considered as a complement to Example 5.29. There is presented an example of a Bieberbach group Γ of the rank 141 with a trivial center and a trivial outer automorphism group. Let $\delta \in H^1(M_{11}, \mathbb{Z}^{141})$ be a cocycle (see (5.4)) which corresponds to Γ. M_{11} is the Mathieu sporadic simple group. Let us recall, (see (5.3)) its presentation

$$\langle a, b \mid a^2 = b^4 = (ab)^{11} = (ab^2)^6 = ababab^{-1}abab^2ab^{-1}abab^{-1}ab^{-1} = 1 \rangle.$$

From (2.7) we have a short exact sequence of groups

$$0 \to \mathbb{Z}^{141} \to \Gamma \to M_{11} \to 0.$$

With support of a computer program CARAT [109] and properties of the representations of M_{11}, it was calculated by R. Waldmüller [147] that all elements of the sequence (5.2)

$$0 \to H^1(M_{11}, \mathbb{Z}^{141}) \to \mathrm{Out}(\Gamma) \to N_\delta/M_{11} \to 0$$

are trivial. Here, $\mathbb{Z}^{141} = \oplus_{i=1}^{4} L_i$, where L_i are M_{11}-lattices of rank $n_i, i = 1, 2, 3, 4$, defined in Example (5.29) and independently below. Moreover, by (5.1) $N_\delta = \{n \in N \mid n * \delta = \delta\}$, where N is the normalizer of a subgroup $h_\Gamma(M_{11}) \subset GL(141, \mathbb{Z})$ in the group $GL(141, \mathbb{Z})$.[3] Summing up, we obtain an example of a Bieberbach group $\Gamma \subset E(141)$ with trivial center, such that $\mathrm{Out}(\Gamma)$ is trivial. Hence we can say that a flat manifold $\mathbb{R}^{141}/\Gamma$ has no symmetries. We follow [147]. The M_{11}-lattices $L_i, i = 1, 2, 3, 4$ are given by the images of the generators a and b of the Mathieu group under the

---

[3] h_Γ is a holonomy representation, see (2.6).

corresponding integral representation of $G = M_{11}$. A cocycle δ (see (2.7)) is defined by

$$\delta = \delta_1 + \delta_2 + \delta_3 + \delta_4 \in \bigoplus_{i=1}^{4} H^1(G, \mathbb{Q}^{n_i}/L_i),$$

where δ_i are given by representative vectors $v_a^i, v_b^i \in \mathbb{Q}^{n_i}$ such that $\delta_i(a) = v_a^i + \mathbb{Z}^{n_i}$ and $\delta_i(b) = v_b^i + \mathbb{Z}^{n_i}$, $i = 1,2,3,4$.

$$a$$

```
⎡ -1  1  0  0  1 -1  1  0 -1  0  0 -1  0  0  0  0  0  0  0  0 ⎤
⎢  0  0  0  0  0  0  0  0  0  0  0  0 -1  0  0  0  0  0  0  0 ⎥
⎢  0 -2 -1  0 -1  0  0  0  0  0  1  1  1  1  0  0  0  0  0  0 ⎥
⎢  0  0  0  0  1  0  0 -1  1  0  0 -1  1 -1 -1 -2 -1  0  2  1 ⎥
⎢  0  1  1  0  1  0  1  0 -1  1 -1 -1  0  0  0  0  1  1  0 -1 ⎥
⎢  0  1  1 -1  1  1  0 -1  0  0 -1 -1  0 -1  0  0  0  0  0  0 ⎥
⎢  0  1  0 -1  0  1 -1 -1  0 -1  0 -1  0  0 -1 -1  0  1      ⎥
⎢  0  0  0 -1  0  0  0 -1  0 -1  0  0  0  1  2  1  0 -2 -1    ⎥
⎢  0  0  0 -1  1  0  0  0  0 -1  0  0  0  0 -1  0  0  1  0  0 ⎥
⎢  0 -2 -1  0 -1 -1  1  0  0  0  1  1  1  2  0  0  0  0  0  0 ⎥
⎢  0  0  1  0 -1  0  1  0 -1  0  1  0  0  1 -1  0  0  1  0  0 ⎥
⎢  0  0  0  0  0  0 -1  0  0 -1  0  0 -1  0  1  0  0 -1  0  0 ⎥
⎢  0 -1  0  0  0  0  0  0  0  0  0  0  0  0  0  0  0  0  0  0 ⎥
⎢  0  0  0  0  1  0  1  0  0  2 -1  0  0  0  0  0  1  1  0 -1 ⎥
⎢  0 -2 -1  0  0 -1  0  0  0  0  0  1  1  1  0  0  0 -1  0  0 ⎥
⎢  0  0  0  0 -1  1  1 -1  1  0  0  0  1  0  0  1  1 -1 -1    ⎥
⎢  0  0  1  0 -1  1  0  0  0  0  1  0  0 -1  0 -1  1  0  0    ⎥
⎢  0  0  0  1  1 -1  0  1  0  1 -1  0  0  0  0  0  1  0  0 -1 ⎥
⎢  0  0  0  0  0 -1  0  1 -1  0  0  1 -1  1  1  1  1  1  0 -2 -1⎥
⎣  0 -2 -1  0  0 -1  1  0  0  1  0  0  2  1  0  0  0  0  0 -1 ⎦
```

$$b$$

```
⎡ -1  0  0  0  0  0  0  0  0  0  0  0  0  0  0  0  0  0  0  0 ⎤
⎢ -1  1  0  0  1  0  1  0 -1  0  0 -1  0  0  0  0  0  0  0  0 ⎥
⎢  1  0  0  0 -1  0 -1  1  0  0  0  1  0  0  0  1  0  0 -1 -1 ⎥
⎢  0  0  0  0 -1  0 -1  0  0  0  0  0  0 -1  0  0  0  0  0  0 ⎥
⎢  0 -2 -1 -1  1 -1  1 -1  0  0 -1  0  1  1  1  0  1 -1  0  0 ⎥
⎢  0 -2 -1 -1 -1  0  0 -1  0 -1  1  0  1  1  0  0 -1 -1  0  1 ⎥
⎢  0  0  1  0 -1  1  0  0  0  0  1  0  0  0 -1  0 -1  1  0  0 ⎥
⎢  0 -2 -1  0 -1  0  0  0  0  0  1  1  1  1  0  0 -1  0  0  0 ⎥
⎢  0  0  0  0  1  0  0  0  0  0  0  0  0  0  0  0  0  0  0  0 ⎥
⎢ -1  1  0  1  0 -1  0  2 -1  0  0  1 -1  0  1  2  1  0 -2 -1 ⎥
⎢ -1  0 -1  0  1 -1  0  1  0 -1  0  0 -1  0  2  1  1 -1 -1  0 ⎥
⎢  0  0  0  0  0  0  1  0  0  0  0  0  0  0  0  0  0  0  0  0 ⎥
⎢  0  0 -1  0  0  0  0  0  0  0  0  0  0  0  0  0  0  0  0  0 ⎥
⎢  0  0  0 -1  1  0  0 -1 -1  0 -1 -1  1  0  0  0  0 -1  0  0 ⎥
⎢  0  1  0  1 -1  0 -1  1  0  0  0  1 -1  0  0  1  0  0 -1 -1 ⎥
⎢  0  0  0  0  1  0  0  0  1 -1  0  1  0  0  0  0  0  0  0  0 ⎥
⎢ -1 -1 -2  0  0  0  0  0  0 -1  1  0  0  0  1  1  0 -1  0  1 ⎥
⎢  1  0  0  0  1  0  0 -1  1  1 -2  0  0  0  0 -1  1  0  1  0 ⎥
⎢  0  0  0  0  0  1  0  0  0  1  0  0  1  0  0  0 -1  0  0 -1 ⎥
⎣ -1  0 -1  1  0 -1  0  1 -1  1 -1  1  0  0  0  2  1  0 -1 -1 ⎦
```

Lattice L_1.

a

```
⎡ 0  1 -1  0  0 -2 -1 -1  0 -2  0  1  0  0  1  0 -1  0  0 -1  1  0 -1  1  1 -1 -1  0 -1  2  0 -2⎤
⎢ 0 -1 -1  0  0 -1 -1 -1  0 -2 -1  0  0  0 -1  1  1  0  0  0 -1  1  0  2  0 -2  0  0  0  1  0 -2⎥
⎢ 0  0 -1  0  1 -1 -1 -1  0 -1  0  0  0  0  0  0  0  0  0  0  0  1  0 -1  0  0  0  1  0 -1⎥
⎢ 0  0 -1  0  0 -1 -1 -1  0 -1  0  0  0  0  0  0  1  0  0  0  0  0  1  0 -1  0  0  0  1  0 -1⎥
⎢ 0  0  0  0  0 -1 -1 -1  0 -1  0  0  0  0  0  0  0  0  0  0  0  1  0 -1  0  0  0  1  0 -1⎥
⎢-1  0 -1 -1 -1 -2 -1 -1 -1  1  1  0  0  0  0 -1  0  0  0  0 -1  0  1  0 -1  1  0  0  1  0 -3⎥
⎢ 0  0 -1  0  0 -1 -1 -1  1 -1  0  0 -1  0  0  0  0  1  0  0  1  0  0  2  1 -1  0  0  0  1  0  0⎥
⎢ 0  0 -2  0  0 -1 -2 -2  0 -1  0  0  1  0  0  0  0 -1 -1  1 -1  1  0  1 -1 -1  0 -1  0  1  0 -3⎥
⎢ 0 -1 -2  0  1 -1 -1 -1  1 -1 -1 -1  0 -1  1 -1  0  0  1  0  0  0  0  0  2  0 -1  0  0  1  1  0 -1⎥
⎢ 1  1 -1  1  0  0 -1 -1  0  1  0  0  1  0  0  0  1 -1  0  0  0  0  0 -1 -1  0  0  1  0  0⎥
⎢ 0  1 -1  0  0 -1 -1 -1  0 -1  1  1  0  0  1  0 -1  0  0  0  1 -1  0  1  0 -1  0  0  0  2  0  0⎥
⎢ 0  1 -1  1  0 -1 -1 -1 -1  0 -1  0  0  0  0  0 -1  0  0  0 -1  0  0  0  0 -1 -1  0  0  1  0 -2⎥
⎢ 0  0 -2  0  1 -1 -2 -2  0  0  0 -1  0  0  0 -1  0  0  0  0  0  1  1 -1 -1  0  0  1  1  0 -2⎥
⎢ 0  0 -1  0  0 -1 -1 -1  0 -1  0  0  0  0  0  0  0  0  0  0  0  0  1  1 -1  0  0  0  1  0 -1⎥
⎢ 0  0 -1  0  0 -1  0  0  0  0  0  1  0  0  0  1  0  0  0  0 -1  1  1  0  0  0  0  0  0⎥
⎢ 1  0  0  0  0  0  0  0  0 -1  0  0  0  0  1  1  0  1  0  0  0  0  1  0 -1  0  1  0  1  0  2⎥
⎢ 0  0 -1  1  0 -1 -1 -1 -1  0 -1  0  0  0  0  0  0  0  0  0  0  0  1  0 -1  0  0  0  1  0 -1⎥
⎢ 0  0  0  0  0 -1 -1 -1 -1 -1  1  0  0 -1  1  0  0 -1  1  0  0  0  1  0 -1 -1  0  1  0  1  0 -1⎥
⎢ 0  0 -1  0  0 -1 -1 -2  0 -1 -1 -1  0 -1 -1  0  1  0  0  0  0  1  1  0 -2  0  0  0  1 -1 -3⎥
⎢ 0 -1 -1  0  0 -1 -2 -2  0 -1  0 -1  0 -1  0  0 -1  1 -1  0  1  1  1  1 -1 -2  0  0  0  1  0 -3⎥
⎢ 0  0 -1  0  0 -1 -1 -2  0  0  0 -1  0 -1  0 -1  0  0  0  0  1  1  0 -1  0  0  0  1  0 -2⎥
⎢ 0  0 -1  0  0 -1 -1 -1  0 -1  0  0  0  0  0  0  0  0  0  0  0  1  1  0 -1  0  0  0  1  0 -1⎥
⎢ 0  0 -1  0  0 -1 -1 -1  0 -1  0  0  0  0  0  0  0  0  0  1  0  1  0 -1  0  0  0  1  0 -1⎥
⎢ 0  1 -2  0  0 -1 -1 -1  0 -1  0  0  1  0  0  0  0  0 -1  0  0  0 -1  1  0 -1  0 -1  0  1  0 -2⎥
⎢ 0  0 -1  0  0 -1 -1 -1  0 -1  0  0  0  1  0  0  0  0  0  0  0  1  0 -1  0  0  0  1  0 -1⎥
⎢ 0 -1 -1  0  0 -1 -1  0  0 -1  0  0  0  0  1  0  0  0  0  0  0  1 -1 -1  1  0  0  1  0 -1⎥
⎢-1 -1 -1  0  0 -2 -1 -1  0 -2 -1  0 -1  1 -1  0  0  0  1  0 -1  0  0  0  2  1 -1  0  0  0  1  0 -3⎥
⎢ 0 -1 -2  0  0 -1 -2 -1  0 -1  0  0  0  1  0  0  0  0  0  0 -1  0  0  1 -1 -1  0  0  1  1  1 -2⎥
⎢ 0  0 -2  0  1 -1 -1 -1 -1  1 -1  0  0  0  1  0  0  0  1  0  0  0  0  2  0 -1  0  0  0  1  0  0⎥
⎢ 0 -1 -1  0  0 -1 -2 -1  0 -1  1  1  0  0  0  0  1  0 -1  0  1  0  0  1  1 -1 -1  1  0  0  1  0 -1⎥
⎢ 0 -1  0  0  0 -1 -1 -2  0 -1 -1 -1 -1 -1 -1  0  1  0  0  0  0  1  1  2  0 -2  0  1  0  1  0 -2⎥
⎣ 0  0  3  0  0  3  3  3  0  3  0  0  0  0  0  0  0  0  0  0  0  0  0 -3  0  3  0  0  0 -3  0  4⎦
```

b

```
⎡ 1  0  2  0  0  1  0  1  1  0  0  0  0  0  0  1  1  0  0  0  1  0  0  0  0  0  0  0  0  0  0  3⎤
⎢ 1  1  1  0  0  1  0  1  1  0  0  0  0  0  0  1  1  0  0  0  2 -1  0  0  1  0  0  0  0  0 -1  3⎥
⎢ 0  0  1 -1 -1  0  0  1  0  0  1  0  0  0  0  1  0  0  0  0  1 -1  0  0  0  1  0  0  0  0  0  1⎥
⎢ 1  0  2  0 -1  2  1  2  1  0  0  0  0  0  0 -1  1  2  0  0  0  1  0  0  0  1  0  0  0  0 -1  0  4⎥
⎢ 0 -1  0  0  0  0 -1  0  1  0  0 -1  0  0 -1  1  1  0  0  0  0  1  1 -1 -1  1  0  0  0  0  0  0⎥
⎢ 2  0  2  1  0  2  0  1  1  0  0  0  0 -1  1  1  2 -1  0  0  1  1  0 -1  1  0 -1  1  0  0  1  4⎥
⎢ 1  0  1  0  0  2  0  2  2  0  0  1  0  1  0  2  1  0  0  0  1  0 -1  0  0  0 -1  0  0  0  0  4⎥
⎢ 1  0  1  0  0  1  0  0  1  0 -1 -1  0 -1 -1  1  2  0  0  0  1  0  1  0  0 -1  0  0  0  0 -1  1⎥
⎢ 1  0  0 -1  0  1  0  2  1  1  0  0  0  1  0  1  0  1  0  0  0  1 -1 -1  1  1  0  0  3⎥
⎢ 1  0  1  0  0  1  0  1  2  0  0  0  0  0  1  1  0  0  0  1  0  1  0  1  0  0  0  0  0  0  3⎥
⎢ 1  0  1  0  0  1  0  1  2  0  0  0  0  0  1  1  0  0  0  0  0  0  0  0  0  0  0  0  0  0  3⎥
⎢ 1  0  0  0  0  1 -1  0  1  0  0  0  1  0  0  1  1 -1 -1  1  0  1  0  0 -1  0  0 -1  0  0  0  1⎥
⎢ 1  0  1 -1 -1  1  0  1  1  1  0  0  0 -1  0  1  1  0 -1  1  1  0  0  0  0  0  1 -1  0  0  0  2⎥
⎢ 1  1  1  0  0  1  0  1  1  0  1  1  0  0  1  1  0  0  0  0  2 -1  0  0  0  0  0  0  0  1  0  4⎥
⎢ 1  0  1  0  0  2  0  1  1  0  0  0  0  0  1  1  0  0  0  1  0  0  0  0  0  0  0  0  0  0  3⎥
⎢ 1  0  1  0  0  1  0  1  1  0  0  0  0  0  0  1  1  0  0  0  1  0  0  0  1  0  0  0  0  0  0  3⎥
⎢ 1  0  0  0  1  1  0  1  2  0  0  0  0  0  1  0  1  1  0  0  0  1  0  0  0  0  0  0  0  0  0  4⎥
⎢ 1  0  1  0  0  1  0  0  1  0  0  0  0 -1  0  1  1 -1 -1  1  1  1  0  0  0  0  0  0  0  0  2⎥
⎢ 1 -1  1  0  0  1  0  2  1  0  0  0  0  1  0  1  1  0  0  0  1  0  0  0  0  1  0  0  0 -1  0  3⎥
⎢ 1  0  1  0  0  1  0  1  1  0  0  0  0  0  1  1  0  0  0  1  0  0  0  0  0  0  0  0  0  0  3⎥
⎢ 1  0  1  0  0  1  0  1  1  0  0  1  0  0  0  1  1  0  0  0  1  0  0  0  0  0  0  0  0  0  3⎥
⎢ 1  0  1  0  0  1  0  1  1  0  0  0  0  0 -1  0  1  1 -1  0  1  1  0  1  0  0  0  0  0 -1  2⎥
⎢ 1  1  1  0  0  2  0  0  1  1  0  0  1 -1  0  1  1 -1 -1  1  1  0  0 -1  0  0  0 -1  0 -1  2⎥
⎢ 1  0  2  0 -1  2  0  2  1  0  0  0  0  0  1  2  0  0  0  1  0  0 -1  0  0  0  0 -1  0  3⎥
⎢ 1  0  1  0  0  1  0  0  1  1  0 -1  0 -1  0  0  1  1  0  0  0  0  0  1  0  0  0  0  0  0  2⎥
⎢ 1  0  1  0  0  1  1  2  1  0  0  0  0  1  0  1  1  1  0  0  1  0 -1  0  1  1  0  0 -1  0  4⎥
⎢ 1  0  1  0  0  1  0  1  1  0  0  0  0  0  0  1  1  0  0  0  1  0  0  0  1  0  0  0  0  0  0  3⎥
⎢ 2  1  1  0  0  1  0  1  1  0  0  0  0 -1  1  1  1  0  0  0  2  0  0  0  0  0  0  0  1  0  4⎥
⎢ 0  0  1 -1 -1  1  0  1  1  0  0  0  0 -1  1  1  0  0  0  1 -1  0  0  1  0  1 -1  0  0 -1  1⎥
⎢ 1  0  1  0  0  1  1  1  1  0  0  0  0  0  1  1  0  0  0  1  0  0  0  0  0  0  0  0  0  0  3⎥
⎢ 1  0  1  0  0  1  0  1  1  1  0  0  0  0  0  1  1  0  0  0  1  0  0  0  0  0  0  0  0  0  0  3⎥
⎣-3  0 -3  0  0 -3  0 -3 -3  0  0  0  0  0  0 -3 -3  0  0  0 -3  0  0  0  0  0  0  0  0  0  0 -8⎦
```

Lattice L_2.

a

```
⎡ 0 1 0 0 0 0 0 0 0 0 0 0 0 0 0 0 0 0 0 0 0 0 0 0 0 0 0 0 0 0 0 0 0 0 0 0 0 0 0 0 0 0 0 0 0 ⎤
  1 0 0 0 0 0 0 0 0 0 0 0 0 0 0 0 0 0 0 0 0 0 0 0 0 0 0 0 0 0 0 0 0 0 0 0 0 0 0 0 0 0 0 0 0
  0 0 0 0 0 0 0 0 0 0 0 0 0 0 0 0 0 0 0 0 0 0 0 0 0 0 0 0 0 0 0 0 0 0 0 0 0 0 0 0 1 0 0 0 0
  0 0 0 0 0 1 0 0 0 0 0 0 0 0 0 0 0 0 0 0 0 0 0 0 0 0 0 0 0 0 0 0 0 0 0 0 0 0 0 0 0 0 0 0 0
  0 0 0 0 1 0 0 0 0 0 0 0 0 0 0 0 0 0 0 0 0 0 0 0 0 0 0 0 0 0 0 0 0 0 0 0 0 0 0 0 0 0 0 0 0
  0 0 0 0 0 0 0 1 0 0 0 0 0 0 0 0 0 0 0 0 0 0 0 0 0 0 0 0 0 0 0 0 0 0 0 0 0 0 0 0 0 0 0 0 0
  0 0 0 1 0 0 0 0 0 0 0 0 0 0 0 0 0 0 0 0 0 0 0 0 0 0 0 0 0 0 0 0 0 0 0 0 0 0 0 0 0 0 0 0 0
  0 0 0 0 0 0 0 0 0 0 0 0 1 0 0 0 0 0 0 0 0 0 0 0 0 0 0 0 0 0 0 0 0 0 0 0 0 0 0 0 0 0 0 0 0
  0 0 0 0 0 0 0 0 0 0 0 0 0 1 0 0 0 0 0 0 0 0 0 0 0 0 0 0 0 0 0 0 0 0 0 0 0 0 0 0 0 0 0 0 0
  0 0 0 0 0 1 0 0 0 0 0 0 0 0 0 0 0 0 0 0 0 0 0 0 0 0 0 0 0 0 0 0 0 0 0 0 0 0 0 0 0 0 0 0 0
  0 0 0 0 0 0 0 0 0 1 0 0 0 0 0 0 0 0 0 0 0 0 0 0 0 0 0 0 0 0 0 0 0 0 0 0 0 0 0 0 0 0 0 0 0
  0 0 0 0 0 0 0 0 0 0 0 0 0 0 0 0 0 1 0 0 0 0 0 0 0 0 0 0 0 0 0 0 0 0 0 0 0 0 0 0 0 0 0 0 0
  0 0 0 0 0 0 0 0 0 0 0 0 0 0 0 0 0 0 1 0 0 0 0 0 0 0 0 0 0 0 0 0 0 0 0 0 0 0 0 0 0 0 0 0 0
  0 0 0 0 0 0 0 0 0 0 0 0 0 0 0 0 0 0 0 1 0 0 0 0 0 0 0 0 0 0 0 0 0 0 0 0 0 0 0 0 0 0 0 0 0
  0 0 0 0 0 0 0 1 0 0 0 0 0 0 0 0 0 0 0 0 0 0 0 0 0 0 0 0 0 0 0 0 0 0 0 0 0 0 0 0 0 0 0 0 0
  0 0 0 0 0 0 0 0 1 0 0 0 0 0 0 0 0 0 0 0 0 0 0 0 0 0 0 0 0 0 0 0 0 0 0 0 0 0 0 0 0 0 0 0 0
  0 0 0 0 0 0 0 0 0 0 0 0 0 0 0 0 0 0 0 0 0 0 0 0 0 0 0 0 0 0 0 0 0 0 0 1 0 0 0 0 0 0 0 0 0
  0 0 0 0 0 0 0 0 0 0 0 0 0 0 0 0 0 0 0 0 0 0 0 0 0 0 0 0 0 0 0 0 0 0 0 0 1 0 0 0 0 0 0 0 0
  0 0 0 0 0 0 0 0 0 0 0 0 1 0 0 0 0 0 0 0 0 0 0 0 0 0 0 0 0 0 0 0 0 0 0 0 0 0 0 0 0 0 0 0 0
  0 0 0 0 0 0 0 0 0 0 0 1 0 0 0 0 0 0 0 0 0 0 0 0 0 0 0 0 0 0 0 0 0 0 0 0 0 0 0 0 0 0 0 0 0
  0 0 0 0 0 0 0 0 0 0 0 1 0 0 0 0 0 0 0 0 0 0 0 0 0 0 0 0 0 0 0 0 0 0 0 0 0 0 0 0 0 0 0 0 0
  0 0 0 0 0 0 0 0 0 0 0 0 0 0 0 0 0 0 0 0 0 0 0 0 0 0 0 0 0 0 0 0 0 0 0 0 0 1 0 0 0 0 0 0 0
  0 0 0 0 0 0 0 0 0 0 0 0 0 0 0 0 0 0 0 0 0 0 0 0 0 0 0 0 0 0 0 0 0 0 0 0 0 0 1 0 0 0 0 0 0
  0 0 0 0 0 0 0 0 0 0 0 0 0 0 0 0 0 0 0 0 0 0 0 0 0 0 0 0 0 0 0 0 0 0 0 0 0 0 0 1 0 0 0 0 0
  0 0 0 0 0 0 0 0 0 0 0 0 0 0 0 0 0 0 0 0 0 0 0 0 0 0 1 0 0 0 0 0 0 0 0 0 0 0 0 0 0 0 0 0 0
  0 0 0 0 0 0 0 0 0 0 0 0 0 0 0 0 0 0 0 0 0 0 0 0 0 0 0 0 0 0 0 0 0 0 0 0 0 0 0 0 0 1 0 0 0
  0 0 0 0 0 0 0 0 0 0 0 0 0 0 0 0 0 0 0 1 0 0 0 0 0 0 0 0 0 0 0 0 0 0 0 0 0 0 0 0 0 0 0 0 0
  0 0 0 0 0 0 0 0 0 0 0 0 0 0 0 0 0 0 0 0 0 0 0 0 0 0 0 0 0 0 0 0 0 0 0 0 0 0 0 0 0 0 0 1 0
  0 0 0 0 0 0 0 0 0 0 0 0 0 0 0 0 0 0 0 0 0 0 0 0 0 0 0 0 0 0 0 0 0 0 0 0 1 0 0 0 0 0 0 0 0
  0 0 0 0 0 0 0 0 0 0 0 0 0 0 0 0 0 1 0 0 0 0 0 0 0 0 0 0 0 0 0 0 0 0 0 0 0 0 0 0 0 0 0 0 0
  0 0 0 0 0 0 0 0 0 0 0 0 0 0 0 0 1 0 0 0 0 0 0 0 0 0 0 0 0 0 0 0 0 0 0 0 0 0 0 0 0 0 0 0 0
  0 0 0 0 0 0 0 0 0 0 0 0 0 0 0 0 0 0 0 0 0 0 0 0 0 0 0 0 0 0 0 0 0 0 0 0 0 0 0 0 0 0 0 0 1
  0 0 0 0 0 0 0 0 0 0 0 0 0 0 0 0 0 0 0 0 0 0 0 0 0 0 0 0 0 0 0 0 0 0 0 0 0 0 1 0 0 0 0 0 0
  0 0 0 0 0 0 0 0 0 0 0 0 0 0 0 0 0 0 0 1 0 0 0 0 0 0 0 0 0 0 0 0 0 0 0 0 0 0 0 0 0 0 0 0 0
  0 0 0 0 0 0 0 0 0 0 0 0 0 0 0 0 0 0 0 0 0 0 0 0 0 0 0 0 0 0 0 0 0 0 0 0 0 1 0 0 0 0 0 0 0
  0 0 0 0 0 0 0 0 0 0 0 0 0 0 0 0 0 0 0 0 0 0 0 0 1 0 0 0 0 0 0 0 0 0 0 0 0 0 0 0 0 0 0 0 0
  0 0 0 0 0 0 0 0 0 0 0 0 0 0 0 0 0 0 0 0 0 0 0 0 0 1 0 0 0 0 0 0 0 0 0 0 0 0 0 0 0 0 0 0 0
  0 0 0 0 0 0 0 0 0 0 0 0 0 0 0 0 0 0 0 0 0 0 0 1 0 0 0 0 0 0 0 0 0 0 0 0 0 0 0 0 0 0 0 0 0
  0 0 0 0 0 0 0 0 0 0 0 0 0 0 0 0 0 0 0 0 0 0 0 1 0 0 0 0 0 0 0 0 0 0 0 0 0 0 0 0 0 0 0 0 0
  0 0 0 0 0 0 0 0 0 0 0 0 0 0 0 0 0 0 0 0 0 0 0 0 0 0 0 0 0 0 0 0 0 0 0 0 0 0 0 1 0 0 0 0 0
  0 0 1 0 0 0 0 0 0 0 0 0 0 0 0 0 0 0 0 0 0 0 0 0 0 0 0 0 0 0 0 0 0 0 0 0 0 0 0 0 0 0 0 0 0
  0 0 0 0 0 0 0 0 0 0 0 0 0 0 0 0 0 0 0 0 0 0 0 0 0 0 0 0 0 0 0 0 0 0 0 1 0 0 0 0 0 0 0 0 0
  0 0 0 0 0 0 0 0 0 0 0 0 0 0 0 0 0 0 0 0 0 0 0 0 0 0 0 0 0 0 0 0 0 0 0 0 0 0 0 0 0 0 1 0 0
  0 0 0 0 0 0 0 0 0 0 0 0 0 0 0 0 0 0 0 0 0 0 0 1 0 0 0 0 0 0 0 0 0 0 0 0 0 0 0 0 0 0 0 0 0
⎣ 0 0 0 0 0 0 0 0 0 0 0 0 0 0 0 0 0 0 0 0 0 0 0 0 0 1 0 0 0 0 0 0 0 0 0 0 0 0 0 0 0 0 0 0 0 ⎦
```

Lattice L_3.

$$b$$

```
[ 1  0  0  0  0  0  0  0  0  0  0  0  0  0  0  0  0  0  0  0  0  0  0  0  0  0  0 -1  0  0  0  0  0  0  0  0  0  0  0  0  0  0  0 ]
[ 0  0  0  0  0  1  0  0  0  0  0  0  0  0  0  0  0  0  0  0  0  0  0  0  0 -1  0  0  0  0  0  0  0  0  0  0  0  0  0  0  0  0  0 ]
[ 0 -1  0  0  0  0  0  0  0  0  0  0  0  0  0  0  0  0  0  0  0  0  0  0  0 -1  0  0  0  0  0  0  0  0  1  0  0  0  0  0  0  0  1 ]
[ 0  0  0  0  0  0  0  0  1  0  0  0  0  0  0  0  0  0  0  0  0  0  0  0 -1  0  0  0  0  0  0  0  0  0  0  0  0  0  0  0  0  0  1 ]
[ 0  0  0  0  0  0  0  0  0  0  0  0  0  0  0  0  0  0  0  0 -1  0  0  0  0  0  0  0  0  0  0  0  0  0  0  0  0  0  0  0  0  0  0 ]
[ 0  0  0  0  1  0  0  0  0  0  0  0  0 -1  0  0  0  0  0  0  0  0  0  0  0  0  0  0  0  0  0  0  0  0  0  0  0  0  0  0  0  0  0 ]
[ 0  0  0  0  0  0  0  0  0  0  0  1  0  0  0  0  0  0  0  0  0  0  0  0  0  0  0  0  0  0  0  0  0  0  0  0  0  0  0  0  0  0 -1 ]
[ 0 -1  0  1  0  0  0  0  0  0  0  0  0  0  0  0  0  0  0  0  0  0  0  0  0  0  0  0  0  0  0  0  0  0  0  0  0  0  0  0  0  0  1 ]
[ 0  0  0  0  0  0  0  1  0  0  0  0  0  0  0  0  0  0  0 -1  0  0  0  0  0  0  0  0  0  0  0  0  0  0  0  0  0  0  0  0  0  0  1 ]
[ 0  0  0  0  0  0  0  1  0  0  0  0  0  0 -1  0  0  0  0  0  0  0  0  0  0  0  0  0  0  0  0  0  0  0  0  0  0  0  0  0  0  0  1 ]
[ 0 -1  0  0  0  0  0  0  0  0  0  0  0 -1  0  1  0  0  0  0  0  0  0  0  0  0  0  0  0  0  0  0  0  0  0  0  0  0  0  0  0  0  1 ]
[ 0  0  0  0  0  1  0  0  0  0  0  0  0  0  0  0  0  0  0  0  0  0  0  0  0  0  0  0  0  0  0  0  0  0  0  0  0  0  0  0  0  0 -1 ]
[ 0  0  0  0  0  0  0  0  0  0  1  0  0  0  0  0  0  0  0  0  0  0  0  0  0  0  0  0  0  0  0  0  0  0  0  0  0  0  0  0  0  0 -1 ]
[ 0  0  0  0  0  0  0  0  0  0  0  0  1  0  0  0  0  0  0  0  0  0  0  0  0  0  0  0  0  0  0  0  0  0  0  0  0  0  0  0  0  0 -1 ]
[ 0 -1  0  0  0  0  0  0  0  0  0  0  0  0  0  0  0  0  0  0  0  0  1  0  0  0  0  0  0  0  0  0  0  0  0  0  0  0  0  0  0  0  0 ]
[ 0  0  0  0  0  0  0  0  0  0  0  0  0  0  0  0  0  0  0  0  0  0  0  1  0  0  0  0  0  0  0  0  0  0  0  0  0  0  0  0  0  0 -1 ]
[ 0  0  0  0  0  0  0  0  0  0  0  0  1  0  0  0  0  0  0 -1  0  0  0  0  0  0  0  0  0  0  0  0  0  0  0  0  0  0  0  0  0  0  0 ]
[ 0  0  0  0  0  0  0  0  0  0  0  0  0  0  1  0  0  0  0  0  0  0  0  0  0  0  0  0  0  0  0  0  0  0  0  0  0  0  0  0  0  0 -1 ]
[ 0  0  0  0  0  0  0  0  0  1  0  0  0  0  0  0  0 -1  0  0  0 -1  0  0  0  0  0  0  0  0  0  0  0  0  0  0  0  0  0  0  0  0  1 ]
[ 0  0  0  0  0  0  0  0  0  0  0  0  0  0  0  0  0  0  0  0 -1  0  0 -1  0  0  1  0  0  0  0  0  0  0  0  0  0  0  0  0  0  0  1 ]
[ 0 -1  0  0  0  0  0  0  0  0  0  0  0  0  0  1  0  0  0  0  0  0 -1  0  0  0  0  0  0  0  0  0  0  0  0  0  0  0  0  0  0  0  1 ]
[ 0 -1  0  0  0  0  0  0  0  0  0  0  0  0  0  1  0  0  0  0  0  0  0  0  0  0  0  0  0  0  0  0  0  0  0  0  0  0  0  0  0  0  0 ]
[ 0  0  0  0  0  0  0  0  0  0  0  0  0  0 -1  0  0  0  0  0  0  0  0  0  0  0  0  0  0  0  0  0  0  0  0  0  1  0  0  0  0  0  0 ]
[ 0  0  0  0  0  0  0  0  0  0  0  0  0  0 -1  0  0  0  0  0 -1  0  0  0  0  0  0  0  0  0  0  0  0  0  1  0  0  0  0  0  0  0  1 ]
[ 0  0  0  0  0  0  0  0  0  0  0  0  0  0  0  0  0  0  0  0  0  0 -1  0  0  0  0  0  0  0  0  0  0  0  0  0  0  0  0  0  0  0  0 ]
[ 0  0  0  0  0  0  0  0  0  0  0  0  0  0 -1  0  0  0  0  0  0  0  0  0  0  0  0  0  1  0  0  0  0  0  0  0  0  0  0  0  0  0  0 ]
[ 0  0  0  0  0  0  0  0  0  0  0  0  0  0  0  0  1  0  0  0  0  0  0  0  0  0  0  0  0  0  0  0  0  0  0  0  0  0  0  0  0  0 -1 ]
[ 0  0  0  0  0  0  0  0  0  0  0  0  0  0  0  0  0  0  0  0  0  0  0  0  0  0  1  0  0  0  0  0  0  0  0  0  0  0  0  0  0  0 -1 ]
[ 0  0  0  0  0  0  0  0  0  0  0  0  0  0  0  0  0  0  0  0  0  0  0  0  0 -1  0  0  0  0  0  0  0  0  0  0  1  0  0  0  0  0  0 ]
[ 0  0  0  0  0  0  0  0  0  0  0  0  0  0  0  0  0  0  0  0  0  0  0  0  0 -1  1  0  0  0  0  0  0  0  0  0  0  0  0  0  0  0  1 ]
[ 0  0  0  0  0  0  0  0  0  0  0  0  0  0  0  0  0  0  1 -1  0  0  0  0  0  0 -1  0  0  0  0  0  0  0  0  0  0  0  0  0  0  0  1 ]
[ 0  0  0  0  0  0  0  0  0  0  0  0  0  0 -1  0  0  0  0  0  0  0  0  0  0  0 -1  0  1  0  0  0  0  0  0  0  0  0  0  0  0  0  1 ]
[ 0  0  0  0  0  0  0  0  0  0  0  0  0  0  0  0  0  0  0  0  0 -1  0  0  0  0  0  0  0  0  0  0  1  0  0  0  0  0  0  0  0  0  0 ]
[ 0  0  0  0  0  0  0  0  0  0  0  0  0  0 -1  0  0  0  0  0  0  0  0  0  0  0  0  0  0  0  0  0  0  0  0  0  0  0  0  0  0  0  0 ]
[ 0 -1  0  0  0  0  0  0  0  0  0  0  0  0  0  0  0  0  0  0  0  0  0  0  0  0  0  0  0  0  0  0  0  0  0  0  0  0  0  0  0  0  0 ]
[ 0 -1  1  0  0  0  0  0  0  0  0  0  0  0  0  0  0  0 -1  0  0  0  0  0  0  0  0  0  0  0  0  0  0  0  0  0  0  0  0  0  0  0  1 ]
[ 0  0  0  0  0  0  0  0  0  0  0  0  0  0 -1  0  0  0  0  0  0  0  1  0 -1  0  0  0  0  0  0  0  0  0  0  0  0  0  0  0  0  0  1 ]
[ 0  0  0  0  0  0  0  0  0  0  0  0  0  0  0  0  0  0  0  0  1  0  0 -1  0  0  0  0  0  0  0  0  0  0  0  0  0  0  0  0  0  0  0 ]
[ 0 -1  0  0  0  0  0  0  0  0  0  0  0  0  0  0  0  0  0  0  0  0  0  0  0  0  0  0  0  0  0  0  0  0  1  0  0  0  0  0  0  0  0 ]
[ 0  0  0  0  0  0  0  0  0  0  0  0  0  0  0  0  0  0  0 -1  0  0  0  0  0  0  0  0  0  0  0  0  0  1  0  0  0  0  0  0  0  0  0 ]
[ 0  0  0  0  0  0  0  0  0  0  0  0  0  0  0  0  0  0  0  0  0  0  0  0  0 -1  0  0  0  0  0  0  0  0  0  0  0  0  0  0  0  0  0 ]
[ 0  0  0  0  0  0  0  0  0  0  0  0  0  0  0  0  0  0  0  0  0  0  0  0  0  0 -1  0  0  0  0  0  0  0  0  0  1  0  0  0  0  0  0 ]
[ 0  0  0  0  0  0  0  0  0  0  0  0  0  0  0  0  0  0  0  0  0  0  0  0  0  0  0  0  0  0  0  0  0  0  0  0  0  0  0  0  0  0 -1 ]
```

Lattice L_3.

a

Lattice L_4.

$$b$$

```
⎡ 1  0 0 0 0 0 0  0 0 0  0 0 0 0 0 0 0 0 0 0 0 0 0  0  0 0 0 0 0 0 0 0 0 0 0 0 0 0 0 0 -1  0 -1 ⎤
⎢ 0  0 1 0 0 0 0  0 0 0  0 0 0 0 0 0 0 0 0 0 0 0 0  0  0 0 0 0 0 0 0 0 0 0 0 0 0 0 0 0  1  0  1 ⎥
⎢ 0  0 0 1 0 0 0  0 0 0  0 0 0 0 0 0 0 0 0 0 0 0 0  0  0 0 0 0 0 0 0 0 0 0 0 0 0 0 0 0  1  0  1 ⎥
⎢ 0  0 0 0 1 0 0  0 0 0  0 0 0 0 0 0 0 0 0 0 0 0 0  0  0 0 0 0 0 0 0 0 0 0 0 0 0 0 0 0  1  0  1 ⎥
⎢ 0  1 0 0 0 0 0  0 0 0  0 0 0 0 0 0 0 0 0 0 0 0 0  0  0 0 0 0 0 0 0 0 0 0 0 0 0 0 0 0  1  0  1 ⎥
⎢ 0  0 0 0 0 0 0  0 0 0  0 0 0 0 0 0 0 0 0 0 0 0 0  0  0 0 0 0 0 0 0 0 0 0 0 0 0 0 0 0  1  1  1 ⎥
⎢ 0  0 0 0 0 1 0  0 0 0  0 0 0 0 0 0 0 0 0 0 0 0 0  0  0 0 0 0 0 0 0 0 0 0 0 0 0 0 0 0  1  0  1 ⎥
⎢ 0  0 0 0 0 0 0  0 0 0  0 0 0 0 0 0 0 0 0 0 0 0 0  0  0 0 1 0 0 0 0 0 0 0 0 0 0 0 0 0  1  0  1 ⎥
⎢ 0  0 0 0 0 0 0 -1 0 0  0 0 0 0 0 0 0 0 0 0 0 0 0  0  0 0 0 0 0 0 0 0 0 0 0 0 0 0 0 0  1  0  0 ⎥
⎢ 0  0 0 0 0 0 0  0 0 0  0 0 1 0 0 0 0 0 0 0 0 0 0  0  0 0 0 0 0 0 0 0 0 0 0 0 0 0 0 0  1  0  1 ⎥
⎢ 0  0 0 0 0 0 0  0 0 0  0 0 0 0 1 0 0 0 0 0 0 0 0  0  0 0 0 0 0 0 0 0 0 0 0 0 0 0 0 0  1  0  1 ⎥
⎢ 0  0 0 0 0 0 0  0 0 0  0 0 0 0 0 0 1 0 0 0 0 0 0  0  0 0 0 0 0 0 0 0 0 0 0 0 0 0 0 0  1  0  1 ⎥
⎢ 0  0 0 0 0 0 0  0 0 0  0 0 0 1 0 0 0 0 0 0 0 0 0  0  0 0 0 0 0 0 0 0 0 0 0 0 0 0 0 0  1  0  1 ⎥
⎢ 0  0 0 0 0 0 0  0 0 0  0 0 0 0 0 1 0 0 0 0 0 0 0  0  0 0 0 0 0 0 0 0 0 0 0 0 0 0 0 0  1  0  1 ⎥
⎢ 0  0 0 0 0 0 0  0 0 1  0 0 0 0 0 0 0 0 0 0 0 0 0  0  0 0 0 0 0 0 0 0 0 0 0 0 0 0 0 0  1  0  1 ⎥
⎢ 0  0 0 0 0 0 0  0 0 0  0 0 0 0 0 0 0 1 0 0 0 0 0  0  0 0 0 0 0 0 0 0 0 0 0 0 0 0 0 0  1  0  1 ⎥
⎢ 0  0 0 0 0 0 0  0 0 0  0 0 0 0 0 0 0 0 1 0 0 0 0  0  0 0 0 0 0 0 0 0 0 0 0 0 0 0 0 0  1  0  1 ⎥
⎢ 0  0 0 0 0 0 0  0 0 0  0 1 0 0 0 0 0 0 0 0 0 0 0  0  0 0 0 0 0 0 0 0 0 0 0 0 0 0 0 0  1  0  1 ⎥
⎢ 0  0 0 0 0 0 0  0 0 0  0 0 0 0 0 0 0 0 0 1 0 0 0  0  0 0 0 0 0 0 0 0 0 0 0 0 0 0 0 0  1  0  1 ⎥
⎢ 0  0 0 0 0 0 0  0 0 0  0 0 0 0 0 0 0 0 0 0 1 0 0  0  0 0 0 0 0 0 0 0 0 0 0 0 0 0 0 0  1  0  1 ⎥
⎢ 0  0 0 0 0 0 0  0 0 0  0 0 1 0 0 0 0 0 0 0 0 0 0  0  0 0 0 0 0 0 0 0 0 0 0 0 0 0 0 0  1  0  1 ⎥
⎢ 0  0 0 0 0 0 0  0 0 0  0 0 0 0 0 0 0 0 0 0 0 1 0  0  0 0 0 0 0 0 0 0 0 0 0 0 0 0 0 0  1  0  1 ⎥
⎢ 0  0 0 0 0 0 0  0 0 0  0 0 0 0 0 0 0 0 0 0 0 0 0  0  0 0 0 1 0 0 0 0 0 0 0 0 0 0 0 0  1  0  1 ⎥
⎢ 0  0 0 0 0 0 0  0 0 0  0 0 0 0 0 0 0 0 0 0 0 0 0  0  0 0 0 0 0 1 0 0 0 0 0 0 0 0 0 0  1  0  1 ⎥
⎢ 0  0 0 0 0 0 0  0 0 0  0 0 0 0 0 0 0 0 0 0 0 0 0  0  0 0 0 0 0 0 0 1 0 0 0 0 0 0 0 0  1  0  1 ⎥
⎢ 0  0 0 0 0 0 0  0 0 0  0 0 0 0 0 0 0 0 0 0 0 0 0  0  0 0 0 0 0 0 0 0 0 1 0 0 0 0 0 0  1  0  1 ⎥
⎢ 0  0 0 0 0 0 0  0 0 0  0 0 0 0 0 0 0 0 0 0 0 0 0  0  0 0 1 0 0 0 0 0 0 0 0 0 0 0 0 0  1  0  1 ⎥
⎢ 0  0 0 0 0 0 0  0 0 0  0 0 0 0 0 0 0 0 0 0 0 0 0 -1  0 0 0 0 0 0 0 0 0 0 0 0 0 0 0 0  1  0  0 ⎥
⎢ 0  0 0 0 0 0 0  0 0 0  0 0 0 0 0 0 0 0 0 0 0 0 0  0  0 0 0 0 0 0 0 0 0 0 0 1 0 1 0 1  1  0  1 ⎥
⎢ 0  0 0 0 0 0 0  0 0 0  0 0 0 0 0 0 0 0 0 0 0 0 0  0  0 0 0 0 1 0 0 0 0 0 0 0 0 0 0 0  1  0  1 ⎥
⎢ 0  0 0 0 0 0 0  0 0 0  0 0 0 0 0 0 0 0 0 0 0 0 0  0  0 0 0 1 0 0 0 0 0 0 0 0 0 0 0 0  1  0  1 ⎥
⎢ 0  0 0 0 0 0 0  0 0 0  0 0 0 0 0 0 0 0 1 0 0 0 0  0  0 0 0 0 0 0 0 0 0 0 0 0 0 0 0 0  1  0  1 ⎥
⎢ 0  0 0 0 0 0 0  0 0 0  0 0 0 0 0 0 0 0 0 0 0 0 0  0  0 0 0 0 0 1 0 0 0 0 0 0 0 0 0 0  1  0  1 ⎥
⎢ 0  0 0 0 0 0 0  0 0 0  0 0 0 0 0 0 0 0 0 0 0 0 0  0  0 0 0 0 0 0 0 1 0 0 0 0 0 0 0 0  1  0  1 ⎥
⎢ 0  0 0 0 0 0 0  0 0 0  0 0 0 0 0 0 0 0 0 0 0 1 0  0  0 0 0 0 0 0 0 0 0 0 0 0 0 0 0 0  1  0  1 ⎥
⎢ 0  0 0 0 0 0 0  0 0 0  0 0 0 0 0 0 0 0 0 0 0 0 0 -1  0 0 0 0 0 0 0 0 0 0 0 0 0 0 0 0  1  0  0 ⎥
⎢ 0  0 0 0 0 0 0  0 0 0  0 0 0 0 0 0 0 0 0 0 0 0 0  0 -1 0 0 0 0 0 0 0 0 0 0 0 0 0 0 0  1  0  0 ⎥
⎢ 0  0 0 0 0 0 0  0 0 0  0 0 0 0 0 0 0 0 0 0 0 0 0  0  0 0 0 0 0 0 1 0 0 0 0 0 0 0 0 0  1  0  1 ⎥
⎢ 0  0 0 0 0 0 0  0 0 0  0 0 0 0 0 0 0 0 0 0 0 0 0  0  0 0 0 0 0 0 0 0 1 0 0 0 0 0 0 0  1  0  1 ⎥
⎢ 0  0 0 0 0 0 0  0 0 0  0 0 0 0 0 0 0 0 0 0 0 0 0  0  1 0 0 0 0 0 0 0 0 0 0 0 0 0 0 0  1  0  1 ⎥
⎢ 0  0 0 0 0 0 0  0 0 0  0 0 0 0 0 0 0 0 0 0 0 0 0  0  0 0 0 0 0 0 0 0 0 0 0 0 0 0 0 0  1  1  1 ⎥
⎢ 0  0 0 0 0 0 0  1 0 0  0 0 0 0 0 0 0 0 0 0 0 0 0  0  0 0 0 0 0 0 0 0 0 0 0 0 0 0 0 0  1  0  1 ⎥
⎢ 0  0 0 0 0 0 0  0 0 0  0 0 0 0 0 0 0 0 0 0 0 0 0  0  0 0 0 0 0 0 0 0 0 0 0 0 0 0 0 0  1  0  1 ⎥
⎢ 0  0 0 0 0 0 0 -1 0 0  0 0 0 0 0 0 0 0 0 0 0 0 0  0  0 0 0 0 0 0 0 0 0 0 0 0 0 0 0 0  1  0  0 ⎥
⎣ 0  0 0 0 0 0 0  0 0 0  0 0 0 0 0 0 0 0 0 0 0 0 0  0  0 0 0 0 0 0 0 0 0 0 0 0 0 0 0 0 -2  0 -1 ⎦
```

Lattice L_4.

$$
v_a^1 = \begin{bmatrix} \tfrac{1}{2} \\ \tfrac{1}{3} \\ -\tfrac{1}{6} \\ -\tfrac{1}{6} \\ \tfrac{1}{3} \\ 0 \\ \tfrac{1}{2} \\ 0 \\ -\tfrac{5}{6} \\ \tfrac{1}{3} \\ -\tfrac{5}{6} \\ \tfrac{1}{3} \\ -\tfrac{5}{6} \\ -\tfrac{5}{6} \\ -\tfrac{1}{3} \\ 0 \\ -\tfrac{2}{3} \\ \tfrac{1}{2} \\ -\tfrac{2}{3} \end{bmatrix}
\qquad
v_b^1 = \begin{bmatrix} 0 \\ 0 \\ -\tfrac{1}{3} \\ -\tfrac{2}{3} \\ 0 \\ 0 \\ 0 \\ 0 \\ \tfrac{1}{2} \\ 0 \\ 0 \\ 0 \\ \tfrac{5}{6} \\ 0 \\ 0 \\ -\tfrac{5}{6} \\ 0 \\ 0 \\ 0 \end{bmatrix}
$$

$$
v_a^2 = \begin{bmatrix} \tfrac{4}{5} \\ 0 \\ 0 \\ -\tfrac{1}{5} \\ \tfrac{2}{5} \\ \tfrac{2}{5} \\ 0 \\ \tfrac{2}{5} \\ 0 \\ 0 \\ -\tfrac{2}{5} \\ -\tfrac{2}{5} \\ -\tfrac{2}{5} \\ \tfrac{2}{5} \\ -\tfrac{3}{5} \\ -\tfrac{3}{5} \\ -\tfrac{1}{5} \\ -\tfrac{1}{5} \\ -\tfrac{4}{5} \\ -\tfrac{1}{5} \\ 0 \\ 0 \\ \tfrac{2}{5} \\ \tfrac{3}{5} \\ 0 \\ -\tfrac{3}{5} \\ -\tfrac{4}{5} \\ -\tfrac{1}{5} \\ \tfrac{2}{5} \end{bmatrix}
\qquad
v_b^2 = \begin{bmatrix} \tfrac{4}{5} \\ \tfrac{2}{5} \\ \tfrac{2}{5} \\ 0 \\ -\tfrac{3}{5} \\ \tfrac{1}{5} \\ -\tfrac{1}{5} \\ 0 \\ 0 \\ 0 \\ 0 \\ 0 \\ 0 \\ 0 \\ 0 \\ 0 \\ 0 \\ 0 \\ 0 \\ 0 \\ 0 \\ 0 \\ 0 \\ 0 \\ 0 \\ 0 \\ 0 \\ 0 \\ \tfrac{1}{5} \end{bmatrix}
$$

$$
v_a^3 = \begin{bmatrix} \tfrac{1}{6} \\ -\tfrac{5}{6} \\ 0 \\ -\tfrac{5}{6} \\ 0 \\ -\tfrac{5}{6} \\ -\tfrac{1}{6} \\ -\tfrac{2}{3} \\ \tfrac{1}{6} \\ -\tfrac{1}{6} \\ -\tfrac{1}{2} \\ -\tfrac{1}{2} \\ -\tfrac{1}{6} \\ \tfrac{1}{6} \\ -\tfrac{1}{3} \\ -\tfrac{1}{6} \\ -\tfrac{2}{3} \\ -\tfrac{1}{2} \\ -\tfrac{1}{6} \\ -\tfrac{1}{2} \\ -\tfrac{5}{6} \\ -\tfrac{1}{2} \\ -\tfrac{1}{6} \\ -\tfrac{5}{6} \\ -\tfrac{1}{2} \\ 0 \\ -\tfrac{1}{2} \\ -\tfrac{1}{6} \\ -\tfrac{5}{6} \\ -\tfrac{1}{3} \\ -\tfrac{1}{2} \\ -\tfrac{1}{2} \\ \tfrac{1}{3} \\ -\tfrac{1}{6} \\ -\tfrac{1}{3} \\ -\tfrac{1}{2} \\ \tfrac{1}{6} \\ -\tfrac{1}{6} \\ 0 \\ 0 \\ 0 \\ 0 \\ \tfrac{1}{6} \\ -\tfrac{1}{2} \end{bmatrix}
\qquad
v_b^3 = \begin{bmatrix} -\tfrac{1}{3} \\ 0 \\ \tfrac{1}{3} \\ \tfrac{1}{3} \\ 0 \\ -\tfrac{1}{3} \\ 0 \\ 0 \\ \tfrac{1}{6} \\ 0 \\ 0 \\ 0 \\ -\tfrac{1}{3} \\ 0 \\ 0 \\ 0 \\ 0 \\ 0 \\ 0 \\ 0 \\ 0 \\ 0 \\ 0 \\ 0 \\ 0 \\ 0 \\ 0 \\ 0 \\ -\tfrac{1}{3} \\ 0 \\ 0 \\ 0 \\ 0 \\ 0 \\ 0 \\ 0 \\ 0 \\ 0 \\ 0 \\ 0 \\ 0 \\ 0 \\ 0 \\ -\tfrac{1}{6} \end{bmatrix}
$$

$$
v_a^4 = \begin{bmatrix} \tfrac{7}{11} \\ -\tfrac{7}{11} \\ \tfrac{2}{11} \\ -\tfrac{9}{11} \\ -\tfrac{8}{11} \\ \tfrac{4}{11} \\ \tfrac{9}{11} \\ \tfrac{7}{11} \\ 0 \\ \tfrac{5}{11} \\ \tfrac{3}{11} \\ -\tfrac{1}{11} \\ 0 \\ \tfrac{5}{11} \\ \tfrac{9}{11} \\ \tfrac{8}{11} \\ \tfrac{2}{11} \\ \tfrac{1}{11} \\ -\tfrac{4}{11} \\ \tfrac{2}{11} \\ -\tfrac{4}{11} \\ -\tfrac{7}{11} \\ -\tfrac{1}{11} \\ \tfrac{5}{11} \\ -\tfrac{4}{11} \\ -\tfrac{8}{11} \\ \tfrac{2}{11} \\ \tfrac{6}{11} \\ \tfrac{8}{11} \\ \tfrac{7}{11} \\ \tfrac{4}{11} \\ \tfrac{3}{11} \\ \tfrac{5}{11} \\ \tfrac{8}{11} \\ \tfrac{1}{11} \\ \tfrac{2}{11} \\ \tfrac{9}{11} \\ \tfrac{9}{11} \\ 0 \\ 0 \\ 0 \\ 0 \\ -\tfrac{1}{11} \\ -\tfrac{4}{11} \end{bmatrix}
\qquad
v_b^4 = \begin{bmatrix} \tfrac{4}{11} \\ -\tfrac{6}{11} \\ 0 \\ 0 \\ 0 \\ -\tfrac{3}{11} \\ 0 \\ -\tfrac{6}{11} \\ 0 \\ -\tfrac{6}{11} \\ -\tfrac{6}{11} \\ -\tfrac{6}{11} \\ 0 \\ 0 \\ 0 \\ 0 \\ 0 \\ 0 \\ 0 \\ 0 \\ 0 \\ -\tfrac{6}{11} \\ -\tfrac{6}{11} \\ \tfrac{7}{11} \\ 0 \\ -\tfrac{6}{11} \\ 0 \\ 0 \\ 0 \\ 0 \\ 0 \\ 0 \\ 0 \\ 0 \\ 0 \\ 0 \\ 0 \\ 0 \\ \tfrac{3}{11} \end{bmatrix}
$$

$$\delta_1 \qquad\qquad \delta_2 \qquad\qquad \delta_3 \qquad\qquad \delta_4$$

Cocycles.

Bibliography

[1] J. F. Adams, Lecture on Lie groups, Benjamin, New York, 1969

[2] D. Anosov, The Nielsen numbers of maps of nil-manifolds. Uspekhi Mat. Nauk., **40** (1985), 133-134. English transl.: Russian Math. Surveys, **40** (1985), 149-150

[3] C. Allday, V. Puppe, Cohomological methods in transformation groups. Cambridge Studies in Advanced Mathematics, 32. Cambridge University Press, Cambridge, 1993

[4] M. F. Atiyah, V. K. Patodi, I. M. Singer, Spectral asymmetry and Riemannian geometry I, Math. Proc. Cambridge Philos. Soc. **77** (1975), 43-69

[5] L. Auslander, An account of the theory of crystallographic groups, Proc. AMS, **16** (1965), 1230-1236

[6] L. Auslander, Bieberbach's theorems on space groups and discrete uniform subgroup of Lie groups I, Ann. of Math. **71** (1960), 579-590

[7] L. Auslander, Bieberbach's theorems on space groups and discrete uniform subgroup of Lie groups II, Amer. J. Math. **83** (1961), 276-280

[8] M. Belolipetsky, A. Lubotzky, Finite groups and hyperbolic manifolds, Invent. Math. **162** (2005), 459-472

[9] N. Berline, E. Getzler, M. Vergne, Heat kernels and Dirac operators, Springer, Berlin, 1992

[10] A. L. Besse, Einstein Manifolds, Springer, Berlin, 1987

[11] L. Bieberbach, Über die Bewegungsgruppen des n-dimensionalen euclidischen Raumes mit einem endlichen Fundamentalbereich, Gött. Nachr. (1910), 75-84

[12] L. Bieberbach, Über die Bewegungsgruppen der Euklidischen Räume I, Math. Ann. **70** (1911), 297-336

[13] L. Bieberbach, Über die Bewegungsgruppen der Euklidischen Räume II, Math. Ann. **72** (1912), 400-412

[14] L. Bieberbach, Über die Minkowskische Reduktion der positiven quadratischen Formen und die endlichen Gruppen linearer ganzzahliger Substitutionen, Gött. Nachr. (1912), 207-216

[15] W. Browder, Surgery on simply-connected manifolds, Springer-Verlag, New York, 1972

[16] H. Brown, R. Bülow, J. Neubüser, W. Wondratschek and H. Zassenhaus, Crystallographic groups of four-dimensional space, New York, 1978

[17] H. Brown, J. Neubüser, H. Zassenhaus, On integral groups III: Normalizers, Math. Comp. **27** (1973), 167-182

[18] K. S. Brown, Cohomology of groups, Springer, New York, 1982

[19] P. Buser, H. Karcher, The Bieberbach case in Gromov's almost flat manifold theorem, Global differential geometry and global analysis proceedings (Berlin 1979), Lecture Notes in Math. 838, Springer, Berlin, (1981), 82-93

[20] P. Buser, H. Karcher, A geometric proof of Bieberbach's theorems on crystallographic groups, L'enseignement Math. **31** (1985), 137-145

[21] P. Buser, H. Karcher, Almost flat manifolds, Asterisque **81**, Paris, 1981

[22] E. Calabi, Closed, locally Euclidean, 4-dimensional manifold, Bull. American Math. Soc. **63** (1957), 135

[23] L. S. Charlap, Bieberbach groups and flat manifolds, Springer, New York, 1986

[24] C. Cid, T. Schulz, Computation of five and six dimensional Bieberbach groups, Exp. Math. **10**(1) (2001), 109-115

[25] Ch. H. Yiu, N. Petrosyan, Fibering flat manifolds of diagonal type and their fundamental groups, Internat. J. Algebra Comput. **32**(7) (2022), 1411-1445

[26] G. Cliff, A. Weiss, Torsion free space groups and permutation lattices for finite groups, Contemp. Math. **93** (1989), 123-132

[27] S. Console, R. Miatello, J. P. Rossetti, $\mathbb{Z}_2$-cohomology and spectral properties of flat manifolds of diagonal type, J. Geom. and Physics, **60** (2010), 760-781

[28] D. E. Cohen, Combinatorial group theory: a topological approach, volume 14 of London Mathematical Society Student Texts. Cambridge University Press, Cambridge, 1989

[29] J. Conway, J. P. Rossetti, Describing the platycosms, preprint, Princeton University, arXive:math.DG/0311476

[30] W. Craig, P. A. Linnell, Unique product groups and congruence subgroups, J. Algebra Appl. **21**(2) (2022), Paper No. 2250025, 9 pp.

[31] W. C. Curtis, I. Reiner, Representation theory of finite groups and associative algebras, Wiley, New York, 1962

[32] W. C. Curtis, I. Reiner, Methods of representation theory, vol. II, Wiley, New York, 1987

[33] K. Dekimpe, Almost-Bieberbach Groups: Affine and Polynomial Structures, Lecture Notes in Math. 1639, Springer-Verlag, Berlin Heidelberg, 1996

[34] K. Dekimpe, B. De Rock, W. Malfait, The Anosov theorem for infranilmanifolds with cyclic holonomy group, Pacific J. Math. **229**(1) (2007), 137-160

[35] K. Dekimpe, B. De Rock, H. Pouseele, The Anosov theorem for infranilmanifolds with an odd-order abelian holonomy group, Fixed Point Theory Appl. 2006, Special Issue, Art. ID 63939, 12 pp.

[36] K. Dekimpe, M. Hałenda, A. Szczepański, Kähler flat manifolds, J. Math. Soc. of Japan **61**(2) (2009), 363-377

[37] K. Dekimpe, M. Sadowski, A. Szczepański, Spin structures on flat manifolds, Monatsh. Math. **148** (2006), 283-296

[38] L. Dornhoff, Group representation theory, (Part A), Marcel Dekker, New York, 1971

[39] B. Eckmann, G. Mislin, Rational representations of finite groups and their Euler class, Math. Ann. **245** (1979), 45-54

[40] L. Evens, Cohomology of Groups. Oxford University Press, 1992

[41] F. T. Farrell, S. Zdravkovska, Do almost flat manifolds bound?, Michigan J. Math. **30** (1983), 199-208

[42] F. T. Farrell, W. C. Hsiang, The topological Euclidean space form problem, Invent. Math. **48** (1978), 181-192

[43] D. R. Farkas, Crystallographic groups and their mathematics, Rocky Mountains J. Math. **11**(8) (1981), 511-551

[44] T. Filar, Flat Riemannian manifolds as torus bundles, Comm. Algebra **42**(6) (2014), 2380-2387

[45] T. Friedrich, Die Abhängigkeit des Dirac-Operators von der Spin-Struktur, Coll. Math. **47** (1984), 57-62

[46] T. Friedrich, Dirac operators in Riemannian geometry, AMS, Graduate Studies in Mathematics, vol. 25, Providence, Rhode Island, 2000

[47] G. Frobenius, Über die unzerlegbaren diskreten Bewegungsgruppen, Sitz. ber. der Preuss. Acad. Wissen., Berl. Ber. Berlin (1911), 654-665

[48] S. M. Gagola, Jr., S. C. Garrison, III, Real characters, double covers, and the multiplier, J. Algebra **74** (1982), 20-51; II, J. Algebra **98** (1986), 38-75

[49] A GAP package, http://www.gap-system.org

[50] G. Gardam, A countrexample to the unit conjecture for group rings, Ann. of Math. (2) **194**(3) (2021), 967-979

[51] W. Gaschütz, Zum Hauptsatz von C.Jordan über ganzzahlige Darstellungen endlicher Gruppen, J. reine angew. Math. **596** (2006), 153-154

[52] O. N. Golowin, On factors without centers in direct decompositions of groups, Mat. Sbornik **6** (1939), 423-426

[53] R. L. Griess, Jr., A sufficient condition for a finite group to have a non-trivial Schur multiplier, Not. Amer. Math. Soc. **17** (1970), 644

[54] M. Gromow, Almost flat manifolds, J. Differential Geometry **13** (1978), 231-241

[55] N. Gupta, S. Sidki, On torsion-free metabelian groups with commutator quotients of prime exponent. Internat. J. Algebra Comput. **9** (1999), 493-520

[56] W. Hantzsche, H. Wendt, Dreidimensional Euklidische Raumformen, Math. Ann. **110** (1934-35), 593-611

[57] J. Hempel, Branched covers over strongly amphicheiral links, Topology, Vol **29**, No. 2 (1990), 247-255

[58] P. De La Harpe, Topics in geometric group theory, Chicago, IL, 2000

[59] A. Hatcher, Algebraic Topology (Cambridge Univ. Press, 2002)

[60] A. Hatcher, Spectral Sequences in Algebraic Topology. Chapter 1: The Serre Spectral Sequences. Unpublished, 2004

[61] H. Hiller, Bieberbach groups with cyclic holonomy groups, preprint 1986

[62] H. Hiller, Flat manifolds with $\mathbb{Z}_{p^2}$ holonomy, Enseign. Math. **31** (1985), 283-297

[63] H. Hiller, C. H. Sah, Holonomy of flat manifolds with $b_1 = 0$, Quarterly J. Math. **37** (1986), 177-187

[64] H. Hiller, Z. Marciniak, C. H. Sah, A. Szczepański, Holonomy of flat manifolds with $b_1 = 0$ II, Q. J. Math. **38** (1987), 213-200

[65] J. Hillman, Flat 4-manifold groups, New Zealand, J. Math. **24** (1995), 29-40

[66] J. A. Hillman, Four-manifolds, geometries and knots, Geometry & Topology Monographs, 5. Geometry & Topology Publications, Coventry, 2002

[67] G. Hiss, A. Szczepański, Flat manifolds with only finitely many affinities, Bull. Sci. Acad. Pol. Math. **45** (1997), 349-357

[68] G. Hiss, A. Szczepański, Holonomy groups of Bieberbach groups with finite outer automorphism groups, Arch. Math. **65** (1995), 8-14

[69] G. Hiss, A. Szczepański Spin structures on flat manifolds with cyclic holonomy, Comm. Algebra **36**(1) (2008), 11-22

[70] J. Hołdys, Bieberbach's Theorem, Seminar, Gdańsk University, 1999/2000

[71] B. Huppert, Endliche Gruppen I. Berlin-Heidelberg-New York, 1967

[72] D. Husemoller, J. Milnor, Symmetric bilinear forms, Springer-Verlag, Berlin-Heidelberg-New York, 1973

[73] B. Iversen, Lectures on Crystallographic groups, Matematisk Institute Aarhus Universitet, Lectures Notes Series 60, 1990/1991 (Fall 1990)

[74] D. L. Johnson, Presentation of groups, L.M.S. Lecture Note Series **22**, Cambridge Univ. Press, 1976

[75] D. L. Johnson, J. W. Wamsley, D. Wright, The Fibonacci groups, Proc. London Math. Soc. **29** (1974), 577-592

[76] F. E. A. Johnson, E. G. Rees, Kähler groups and rigidity phenomena, Math. Proc. Camb. Phil. Soc. **109** (1991), 31-44

[77] F. E. A. Johnson, Flat algebraic manifolds, Geometry of low-dimensional manifolds 1, (Durham 1989), London Math. Soc. Lecture Notes Ser. 150, Cambridge University Press, Cambridge, 1990, 73-91

[78] D. D. Joyce, Compact manifolds with special holonomy, Oxford Univ. Press, New York, 2000

[79] E. Kleinert, Units of classical orders: a survey, L'enseignement Math. **40** (1994), 205-248

[80] M. Kneser, Quadratische Formen, Springer, Berlin, 2002

[81] S. Kobayashi, K. Nomizmu, Foundations of Differential Geometry, vol. 2, Wiley-Interscience, 1969

[82] A. Kolpakov, B. Martelli, Hyperbolic four-manifolds with one cusp, Geom. Funct. Anal. **23**(6) (2013), 1903-1933

[83] S. Lang, Algebra, Revised Third Edition, Springer-Verlag, 2002

[84] H. Lange, Hyperelliptic varieties, Tohoku Math. J. **53** (2001), 491-510

[85] R. Lee, R. H. Szczarba, On the integral Pontrjagin classes of a Riemannian flat manifold. Geom. Dedicata, **3** (1974), 1-9

[86] D. D. Long, A. W. Reid, On the geometric boundaries of hyperbolic 4-manifolds. Geom. Topol. **4** (2000), 171-178

[87] D. D. Long, A. W. Reid, All flat manifolds are cusps of hyperbolic orbifolds, Algebraic and Geometric Topology, Vol. 2 (2002), 285-296

[88] R. Lutowski, Z. Marciniak, Affine representations of Fibonacci groups and flat manifolds, Comm. Algebra **46**(6) (2018), 2738-2741

[89] R. Lutowski, J. Popko, A. Szczepanski, Spin$^{\mathbb{C}}$ structures on Hantzsche-Wendt manifolds, J. Geom. Phys. 171 (2022), Paper No. 104394, 17 pp.

[90] R. Lutowski, A. Szczepański, Crystallographic groups with trivial center and outer automorphism group, Math. Proc. Camb. Phil. Soc. **164** (2018), 363-368

[91] R. Lutowski, On symmetry of flat manifolds, Exp. Math. **18** (2009), 201-204

[92] S. Mac Lane, Homology, Springer-Verlag, 1963

[93] W. Malfait, The Nielsen numbers of virtually unipotent maps on infranilmanifolds, Forum Math. **13** (2001), 227-237

[94] W. Malfait, A. Szczepański, The structure of the (outer) automorphism group of a Bieberbach group, Compositio Math. **136** (2003), 89-101

[95] A. Manning, There are no new Anosov diffeomorphisms on tori, Amer. J. Math. **96**(3) (1974), 422-429

[96] Z. Marciniak, A. Szczepański, Extensions of Crystallographic Groups, preprint Gdańsk-Warszawa, 1987

[97] G. Maxwell, Compact Euclidean Space Forms, J. Algebra **44** (1977), 191-195

[98] D. B. McReynolds, Controlling manifold covers of orbifolds, Math. Res. Lett. **16** (2009), 651-662

[99] R. J. Miatello, R. A. Podestá, Spin structures and spectra of $\mathbb{Z}_2^k$-manifolds, Math. Zeitschrift. **247** (2004), 319-335

[100] R. J. Miatello, R. A. Podestá, The spectrum of twisted Dirac operators on compact flat manifolds, Trans. Amer. Math. Society **358**(10) (2006), 4569-4603

[101] R. J. Miatello, J. P. Rossetti, Isospectral Hantzsche-Wendt manifolds, J. Reine Angew. Math. **515** (1999), 1-23

[102] R. J. Miatello, J. P. Rossetti, Hantzsche-Wendt manifolds of dimension 7. Differential geometry and applications (Brno, 1998), 379-390, Masaryk Univ., Brno, 1999

[103] R. J. Miatello, J. P. Rossetti, Comparison of Twisted P-Form Spectra for Flat Manifolds with Diagonal Holonomy, Ann. Global Anal. Geom. **21** (2002), 341-376

[104] J. Milnor, J. D. Stasheff, Characteristic classes, Annals of Math. Studies 76, Princeton Univ. Press, Princeton, 1974

[105] H. Minkowski, Diskontinuitätsbereich für arithmetische Äquivalenz, J. reine angew. Math. **129** (1905)

[106] B. E. Nimershiem, All flat three-manifolds appear as cusps of hyperbolic four manifolds, Topology and its Appl. **90** (1998), 109-133

[107] W. Nowacki, Die Euklidischen, dreidimensionalen, geschlossenen und offenen Raumformen, Comment. Math. Helv. **7** (1934), 81-93

[108] R. K. Oliver, On Bieberbach's analysis of discrete euclidean groups, Proc. AMS, **80**(1) (1980), 15-21

[109] J. Opgenorth, W. Plesken, T. Schulz, CARAT - Crystallographic Algorithms and Tables, `https://lbfm-rwth.github.io/carat/`

[110] J. Opgenorth, W. Plesken, T. Schulz, Crystallographic algorithms and tables, Acta Cryst. Sect A **54**(5) (1998), 517-531

[111] D. Passman, The algebraic structure of group rings, Interscience 1977

[112] F. Pfaffle, The Dirac spectrum of Bieberbach manifolds, J. Geom. Phys. **35**(4) (2000), 367-385

[113] W. Plesken, Kristallographische Gruppen, group theory, algebra, and number theory, Colloquium in memory of Hans Zassenhaus, Saarbrucken, June 4-5, 1993, de Gruyter, Berlin, 1996, 75-96

[114] W. Plesken, Minimal dimensions for flat manifolds with prescribed holonomy, Math. Ann. **284** (1989), 477-486

[115] J. Popko, A. Szczepański, Cohomological rigidity of oriented Hantzsche-Wendt manifolds, Adv. in Mathematics **302** (2016), 1044-1068

[116] J. Popko, A. Szczepański, Properties of the combinatorial Hantzsche-Wendt groups, Topology and its Appl. **310** (2022), Paper No. 108037, 15 pp.

[117] H. L. Porteous, Anosov diffeomorphisms of flat manifolds, Topology **11** (1972), 307-315

[118] S. David Promislow, A simply example of a torsion-free, nonunique product group, Bull. London Math. Soc. **20**(4) (1988), 302-304

[119] B. Putrycz, Commutator subgroups of Hantzsche-Wendt groups, J. Group Theory **10** (2007), 401-409

[120] B. Putrycz, A. Szczepański, Existence of spin structures on flat four manifolds, Adv. in Geometry **10**(2) (2010), 323-332

[121] D. Quillen, The Mod 2 Cohomology Rings of Extra-special 2-groups and the Spinor Groups, Math. Ann. **194** (1971), 197-212

[122] J. G. Ratcliffe, Foundations of hyperbolic manifolds, Graduate Text in Math. **149**, Springer, New York, 1994

[123] J. G. Ratcliffe, S. T. Tschantz, The volume spectrum of hyperbolic 4-manifolds, Exp. Math. **9** (2000), 101-125

[124] J. P. Rossetti, A. Szczepański, Generalized Hantzsche-Wendt flat manifolds, Revista Iberoam. Mat. **21**(3) (2005), 1053-1079

[125] M. Sadowski, A. Szczepański, Flat manifolds, harmonic spinors, and eta invariants, Adv. in Geometry **6**(2) (2006), 287-300

[126] J. P. Serre, Representations lineaires des groupes finis, Hermann-Paris, 1967

[127] C. L. Siegel, Discontinuous groups, Ann. of Math. **44** (1943), 674-689

[128] E. H. Spanier, Algebraic Topology, McGraw-Hill Inc., 1966

[129] I. Stewart, D. Tall, Algebraic number theory, Chapman and Hall, London, 1987

[130] P. Symonds, Flat manifolds. Ph.D. thesis, Cambridge University, 1987

[131] A. Szczepański, Aspherical manifolds with the ℚ-homology of a sphere, Mathematika **30** (1983), 291-294

[132] A. Szczepański, Decomposition of flat manifolds, Mathematika **44** (1997), 113-119

[133] A. Szczepański, Crystallographic groups, Algebraic Topology Seminar, Warsaw University, October 1980

[134] A. Szczepański, Five dimensional Bieberbach groups with trivial centre, Manuscripta Math. **68** (1990), 191-208

[135] A. Szczepański, Holonomy groups of crystallographic groups with finite outer automorphism groups, Group Theory and Low Dimensional Topology, German-Korea Workshop, Pusan 2000, (J. Mennicke, Jung

Rae Cho (eds.)) in Research and Exposition in Mathematics, Vol. 27 (2003), 163-166

[136] A. Szczepański, Kähler flat manifolds of low dimension, Preprint, IHES, M 43, 2005

[137] A. Szczepański, Outer automorphism groups of Bieberbach groups, Bull. Belgian Mathem. Soc. Simon Stevin, **3** (1996), 585-593

[138] A. Szczepański, Problems on Bieberbach groups and flat manifolds, Geometriae Dedicata **120** (2006), 111-118

[139] A. Szczepański, Properties of generalized Hantzsche-Wendt groups, J. Group Theory **12** (2009), 761-769

[140] A. Szczepański, Flat manifolds with the first Betti number equal to zero, Thesis, Warsaw University, 1987 (in Polish)

[141] A. Szczepański, The Euclidean representations of the Fibonacci groups, Q. J. Math. **52** (2001), 385-389

[142] W. P. Thurston, Three dimensional manifolds, Kleinian groups and hyperbolic geometry, Bull. Amer. Math. Soc. **6** (1982), 357-381

[143] P. Tirao, Primitive compact flat manifolds with holonomy group $Z_2 \oplus Z_2$, Pacific J. Math. **198**(1) (2001), 207-233

[144] J. Tits, Free subgroups of linear groups, J. Algebra **20** (1972), 250-270

[145] A. T. Vasquez, Flat Riemannian manifolds, J. Diff. Geom. **4** (1970), 367-382

[146] J. H. C. Whitehead, On the asphericity of regions in a 3-sphere, Fund. Mathematicae 32 (1939), 146-166

[147] R. Waldmüller, A flat manifold with no symmetries, Exp. Math. **12** (2003), 71-77

[148] J. Wolf, Spaces of constant curvature, MacGrow Hill, New York, 1967

[149] H. Zassenhaus, Beweis eines Satzes über diskrete Gruppen, Abh. Math. Sem., Univ. Hamburg **12** (1938), 289-312

[150] H. Zassenhaus, When is the unit group of a Dedekind order solvable?, Comm. Algebra **6** (1978), 1621-1627

[151] H. Zassenhaus, Über einen Algorithmus zur Bestimmung der Raumgruppen, Comment. Math. Helvet. **21** (1948), 117-141

[152] H. Zieschang, Cystallographic groups, Lecture Notes in Mathematics 875, Springer, Berlin, 1981, 101-133

Index